Maya and Catholic Cultures in Crisis

UNIVERSITY PRESS OF FLORIDA

Florida A&M University, Tallahassee
Florida Atlantic University, Boca Raton
Florida Gulf Coast University, Ft. Myers
Florida International University, Miami
Florida State University, Tallahassee
New College of Florida, Sarasota
University of Central Florida, Orlando
University of Florida, Gainesville
University of North Florida, Jacksonville
University of South Florida, Tampa
University of West Florida, Pensacola

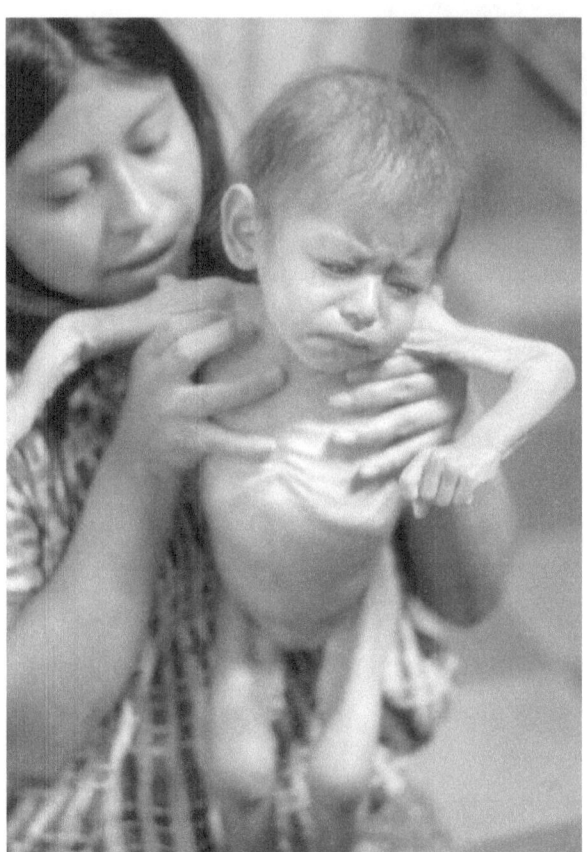

The Maya Crisis

Maya and Catholic Cultures in Crisis

JOHN D. EARLY

University Press of Florida
Gainesville · Tallahassee · Tampa · Boca Raton
Pensacola · Orlando · Miami · Jacksonville · Ft. Myers · Sarasota

Copyright 2012 by John D. Early
All rights reserved
Printed in the United States of America on acid-free paper

The publication of this book was funded in part by a grant from the Department of Anthropology at Florida Atlantic University.

This book may be available in an electronic edition.

21 20 19 18 17 16   6 5 4 3 2 1

First cloth printing, 2012
First paperback priting, 2016

Library of Congress Cataloging-in-Publication Data
Early, John D.
Maya and Catholic cultures in crisis / John D. Early.
p. cm.
Includes bibliographical references and index.
ISBN 978-0-8130-4013-4 (cloth: alk. paper)
ISBN 978-0-8130-5404-9 (pbk.)
1. Mayas—Religion. 2. Mayas—Rites and ceremonies. 3. Maya cosmology. 4. Indian Catholics—Latin America—History. 5. Catholic Church—Missions—Latin America—History. 6. Christianity and culture—Latin America. 7. Christianity and other religions—Latin America. I. Title.
F1435.3.R3E36 2012
299.7'842—dc23
2012001057

The University Press of Florida is the scholarly publishing agency for the State University System of Florida, comprising Florida A&M University, Florida Atlantic University, Florida Gulf Coast University, Florida International University, Florida State University, New College of Florida, University of Central Florida, University of Florida, University of North Florida, University of South Florida, and University of West Florida.

University Press of Florida
15 Northwest 15th Street
Gainesville, FL 32611-2079
http://www.upf.com

Contents

List of Illustrations   vii
List of Tables   ix
Preface   xi
List of Abbreviations   xv

PART I. INTRODUCTION   1

1. The Research   3

PART II. THE BACKGROUND OF THE CRISIS IN MAYA COMMUNITIES AT MID-TWENTIETH CENTURY   9

2. The Traditional Maya Worldview as Influenced by Later Evangelization   11
3. Retention of Maya Culture through Periods of Domination   35
4. Growing Inability of Maya Communities to Provide Subsistence   49
5. Long-Standing Strains within Maya Communities   62
6. The Maya Crisis and the Search for Answers   72

PART III. RENEWED EFFORTS OF THE CATHOLIC CHURCH IN MAYA COMMUNITIES   85

7. Worldview of Tridentine Catholicism   87
8. Presentation and Maya Reception of the Tridentine Worldview   99

PART IV. CRISIS WITHIN THE CATHOLIC WORLDVIEW   121

9. Beyond Tridentine Belief and Ritual: Worldview of Vatican Council II   123
10. Crisis and Reaction in Latin America: The Liberation Movement   133
11. Maya Dioceses Reorganize for Action Catholicism   145
12. Maya Communities Organize for Social Action   160

PART V. LIBERATION CONSCIOUSNESS ASSISTED BY BIBLICAL REFLECTION   175

13. The Bible and Its Worldview as a Cultural Document   177
14. Methods of Reflecting on the Bible   189
15. Biblical Reflections in Maya Communities   199
16. Biblical Reflections in Lacandón Migrant Communities   216

PART VI. THE WORLDVIEWS OF INSURGENCY AND COUNTERINSURGENCY 233

17. Guatemala: The Role of the Maya in the Worldview of Marxist Insurgency 235
18. Guatemala: The Maya in the Military's Worldview of Counterinsurgency 249
19. Militarization in Guatemala 262
20. Chiapas: The Role of the Maya in the Worldview of the Zapatista Insurgency 273
21. Militarization in Chiapas 293

PART VII. THE IMPACT OF THE MAYA CRISIS ON THE WORLDVIEWS OF PASTORAL WORKERS 313

22. Two Pastoral Workers Evolve 315
23. Social Justice by Sacramental Observance 331
24. Social Justice by Maya Empowerment 347
25. Social Justice by Armed Rebellion 361
26. Liberation Catholicism: Its Relation to the Morality of Armed Rebellion 375

PART VIII. THE SEARCH FOR A REVITALIZED MAYA WORLDVIEW 393

27. Choices Faced by Catholic Maya in a Turbulent Society 395
28. A Bishop's Evolving Worldview 408
29. The Movement for a Maya Christianity 430

PART IX. CONCLUSION 449

30. A Look Backward and Forward 451

Notes 461
Bibliography 465
Index 487

Illustrations

Figures

3.1. Early socialization in Maya worldview   44
4.1. View of Santiago Atitlán and the San Pedro volcano   55
4.2. Milpas on side of San Pedro volcano   56
4.3. Erosion on slope of San Pedro volcano   57
4.4. Milpa on a mountainside in highland Chiapas   57
4.5. Departure for coastal plantations   60
5.1. Cofrade drinking cane liquor   67
6.1. Severe case of malnutrition, front view   76
6.2. Severe case of malnutrition, back view   76
6.3. Malnutrition patients   77
6.4. Malnutrition patients   77
6.5. Funeral of small child   78
6.6. Funeral of an adult   78
12.1. Father Stafford in Atiteco dress   163
12.2. Father Carlin washes the feet of cofrades   163
12.3. Poster for two literacy lessons (cartela)   165
12.4. Dr. Elizabeth Nick, Director of Radio Schools   167
12.5. Graduation Day   167
12.6. A Montessori student   168
13.1. The biblical pattern   180
22.1. The dynamo nun   316
22.2. Cooperative truck   326
23.1. Shrine to Father Rother   337
23.2. Sculpted image of Father Rother   338
23.3. Inscription at the foot of the shrine   338
24.1. Father Ricardo Falla   354
25.1. Fernando Hoyos   367

26.1. Guadalupe procession and protest march 388
26.2. A community group with their placard 389
26.3. Guadalupe image on the Mexican flag 389
26.4. A placard 389
26.5. A placard 390
26.6. A placard 390
28.1. Bishop Ruíz García 422

Maps

1. Chiapas, Mexico, with Its Three Catholic Dioceses; and Guatemala, with Its Twenty-Two Departments Comprising Fourteen Dioceses, Two Archdioceses, and Two Vicariates 5
2. Some Maya Communities in the Chiapas Highlands 100
3. Some Maya Communities in the Guatemalan Western Highlands 101
4. The Lacandón Region of Chiapas 217

Tables

3.1. Approximate Size and Ratio of Spanish and Maya Populations, Guatemala, 1550–1680   40

4.1. Rough Estimates of the Guatemalan Maya Population and Percent Average Annual Rates of Increase, 1520–2002   53

4.2. Household Agricultural Production Units by Size of Holdings and Percentage of National Agricultural Lands, Rural Guatemala, 1950, 1973   58

6.1. Life Table Functions, Santiago Atitlán, 1940–1969   75

10.1. Structure of the Liberation Movement   137

Preface

In 2006 I published *The Maya and Catholicism: An Encounter of Worldviews*. That work began with the Spanish evangelization of the Maya in the sixteenth century and ended with a description of the results of the Catholic efforts to convert the Maya during the colonial and early national periods. This book is a sequel to that work. It adds some material about the colonial and national periods as background for the interaction of the worldviews from about the middle to the end of the twentieth century. I have attempted to make this work stand by itself, but I strongly urge the reader to consult the previous volume for the background of this volume. Chapter 2 about the Maya worldview, is a revision of chapter 5 of the previous volume.

The writer wishes to acknowledge the invaluable assistance of the following experts on the various topics covered in the chapters. Without their help, this book could not have been written. The following people read earlier versions of the entire manuscript and made many helpful suggestions: Allen Christenson; Christine Eber; Christine Gudorf; Christopher Lutz; Brent Metz; Judith Maxwell; Norman Schwartz; and James Walsh, SJ. Others commented on parts of the manuscript pertinent to their specializations and/or furnished research materials: Garrett Cook; Ricardo Falla, SJ; Christine Kovic; Thomas and Marjorie Melville; Heidi Moksnes; Elizabeth Nick; Thomas Stafford; David Stoll; Kay Warren; John Womack; and Alexander Zatyrka, SJ. I am grateful for their help, but responsibility for the final product is mine alone.

The maps are the work of Clifford Brown. Duncan Earle furnished the introductory photographs to chapters 17, 18, and 19; Jeanne Simonelli those for chapters 20 and 21. Thomas and Marjorie Melville furnished the introductory photographs to chapters 22, 26, and figures 22.1, 22.2. Allen Christenson furnished those for chapter 23 and figures 23.1, 23.2, and 23.3; Ricardo Falla for figure 24.1; and María Pilar de Hoyos de Asig for chapter 25 and figure 25.1. I took all the remaining photographs and did all translations from the Spanish. Biblical translations are from the Standard Edition of the New Jerusalem Bible. At times the text makes temporal references with the words "currently," "contemporary," or similar designations. For any future readers, they usually refer to the years 2000 to 2010.

The older spellings of Maya languages based on Spanish orthography were changed in 1987 by the Guatemalan government's acceptance of the alphabet proposed the Academia de Las Lenguas Mayas de Guatemala. The writer has used these newer spellings when referring to the languages. But the older spellings have been retained when the references are to geographical places, rather than the predominant language of a place. This is more in accord with the spellings in the older literature and geographical usage.

Important documents from the Vatican, bishops' conferences, and some United Nations' publications make use of a citation system that eases the problem of finding a cited source. Vatican documents are written in Latin, then translated into many of the world's languages. In addition, there may be more than one translation for the same language, or multiple editions of a document or translation by different publishers. The end result is many possible editions with differing paginations for the same document. To avoid the problem of finding a citation only in the edition used by the writer, each section or paragraph of the original document is numbered and these numbers are repeated in all subsequent editions and translations. Here, after I cite the page number(s) of the edition I used, I've added the paragraph numbers(s), preceded by the number (#) sign.

Over the years, some have misunderstood my involvement with the Maya. This is reflected in statements such as, "I was the parish priest in Zinacantán or Santiago Atitlán," or "I witnessed many of the events described in the chapters." I am a former Jesuit: ordained a priest in 1957, and resigned in 1969. I have had little personal contact with the Catholic church or the Jesuits since that time.

In 1960-62, I was a graduate student in the Social Relations Department at Harvard. Joining the Harvard Chiapas Project directed by Professor Evon Vogt, I arrived in Zinacantán in December 1962. In the preface to the previous volume (Early 2006: xiv), I explained my situation in Zinacantán during the years 1963-64: "I did not function as the town priest, although I would sometimes help the non-resident priest with masses and baptisms. Most of the time I was just another Harvard graduate student spending time with the *fiscales* and members of the cargo system, learning what they did." This was field work to gather material for my dissertation. The non-resident priest was the pastor of eight other parishes and swamped with ritual duties. Hence the occasional request for help. Since I was living in Zinacantán during that period and the pastor only made occasional visits, I had to be very careful to avoid any other priestly functions so that people would not see me as taking his place. As a member of the Harvard Chiapas Project, I was not involved in any type of empowerment activities.

During this period I occasionally conversed with Bishop Samuel Ruíz García between sessions of Vatican Council II. He was very kind and gave me logistical help whenever I needed it. On the few occasions when we both were in San Cristóbal at the same time, he would invite me for Sunday supper. The bishop, his father, mother, and myself would sit around the kitchen table in a family atmosphere and chat. His parents were wonderful, "salt of the earth" people. I am sure that his later evolution had its basis in the values he took from them. I was probably the first cultural anthropologist the bishop knew, but we did not talk much about anthropology. Before going to Chiapas, I had been warned by friends that San Cristóbal was an extremely conservative diocese, and given what were considered my "far out" theological ideas, it was best to be quiet. I thought the diocese's effort among the Maya at that time made little sense, either culturally or theologically, but I did not express my thoughts. Given Don Samuel's Tridentine theological thinking at that time, I think I received good advice. Our evolutions were not synchronized.

I received the PhD in 1965, spent a year at Georgetown University, and in 1966 was assigned to teach moral and pastoral theology at the Jesuit house of studies then located in Woodstock, Maryland. My formal duties there required my presence for only one semester a year. Father Carlin, the pastor of the Oklahoma mission in Santiago Atitlán, invited me to join their group. He wanted me to help develop a Maya Catholic liturgy that incorporated Mayan themes and symbols. From 1967 to 1969, I spent a semester each year in Santiago. I worked on the liturgy project, but quickly realized it was a long-term project, much longer than Father Carlin realized.

I also became involved in the other programs being developed by members of the group. For the medical program I did demographic research and field studies to help understand and plan for a medical situation that was unknown in the developed world. I helped the credit cooperative by spending long, boring hours checking its books. The Atiteco in charge of the books was careless at times with his addition and subtraction. Nothing can ruin a cooperative quicker than incorrect books. Finally, the Peace Corps assigned an accounting student to visit all the cooperatives in the area to moniter their books. I also worked with two members of the group to develop a literacy program and train personnel for the radio station, TGDS. The program was based on Paulo Freire's work. I held meetings with the teachers for the purpose of their designating the community problems to be discussed in the various lessons of the curriculum. These problems would also furnish the key words for the Freire method. (The details of the curriculum are described in Early 1973.) In the summers of 1969 and 1970, I returned to assist the development of the program, which was under the capable direction of Elizabeth Nick. In the decline of the Oklahoma program described in chapter 23, I was identified with the "culturally sensitive" group. After 1970, I was not personally involved in any of the Catholic activities in Guatemala. During the late 1960s and early 1970s, I helped train Peace Corps volunteers for work in Guatemala. Also during the 1970s, I made several trips to Guatemala City to gather demographic data for my 1982 book and as a demographic consultant to several development organizations. My work for Maya empowerment in Santiago led to suspicions that I was involved with subversive groups. This resulted in surveillance and phone tapings by U.S. intelligence services. Conversely, some thought that I was a government informant about leftist activities. Neither suspicions were correct.

Why have I written the two volumes about the Maya and Catholicism? That religion plays a vital role in the lives of many peoples around the globe is an irrefutable fact regardless of one's personal evaluation of this fact. From an anthropological viewpoint, I think the two volumes provide an important case history showing the longitudinal dynamics of two religious worldviews interacting with each other. The case is part of my wider interest in comparative religions and their role in fomenting war or peace. I have seen religion do great good and great harm. In the eighth decade of my life, I can say that about most human phenomena. To ignore the serious study of religion leads to superficial knowledge of many worldviews and the cultures they help to define.

Abbreviations

(C = Chiapas; G = Guatemala)

AC	Acción Catolica (Catholic Action), G
ANCIEZ	Alianza Nacional Campesina Independiente Emiliano Zapata (Emiliano Zapata Independent Peasant Alliance), C
ARIC	Associación Rural de Interés Colectivo (Rural Association of Collective Interest), C
CCRI	Comité Clandestino Revolucionario Indígena (Clandestine Committee of Indigenous Revolution), C
CEDECAS	El Centro de Capacitación Social (Center for Social Empowerment), G
CIAS	Centro de Información y Acción Social (Center of Information and Social Action), G
CIOAC	Central Independiente de Obreros Agrícolas y Campesinos (Independent Confederation of Agricultural Workers and Peasants), C
CNC	Congreso Nacional Campesina (National Peasant Confederation), C
CONASUPO	Compañía Nacional de Subsistencias Populares (National Company of Popular Subsistence), C
CPR	Comunidades de Población en Resistencia (Communities of Population in Resistance), G
CUC	Comité de Unidad Campesina (Committee for Peasant Unity), G
DCG	Democracia Cristiana Guatemalteca (Guatemalan Christian Democracy), G
EGP	Ejército Guerrillero de los Pobres (Guerrilla Army of the Poor), G
EZLN	Ejército Zapatista de Liberación Nacional (Zapatista Army of National Liberation), C
FAR	Fuerzas Armadas Rebeldes (Armed Forces in Rebellion), G

FLN	Frente de Liberación Nacional (National Liberation Front), C
FZLN	Frente Zapatista de Liberación Nacional (Zapatista National Liberation Front), C
INI	Instituto Nacional Indigenista (National Indigenous Institute), C
LP	Línea Proletaria (Proletarian Line), C
OCEZ	Organización Campesina Emiliano Zapata (Emiliano Zapata Proletarian Association), C
OID	Organización Ideológico Dirigente (Directing Ideological Organization), C
ORPA	Organización del Pueblo en Armas (Organization of the Armed Community), G
PGT	Partido Guatemateco de los Trbajadores (Guatemalan Party of the Workers), G
PP	Política Popular (Politics of the People), C
PRI	Partido Revolucionario Institucional (Institutional Revolutionary Party), C
URND	Unidad Revolucionaria Nacional Guatemalteca (Guatemalan Nation Revolutionary Unit), G
UP	Unión del Pueblo (Union of the People), C
UU	Unión de Uniones Ejidales y Grupos Campesinos Solidarios de Chiapas (Union of Unions), C

I

Introduction

1

The Research

The Catholic church has had an impact on Maya culture since the sixteenth century. This can be seen in the churches it constructed in many Maya communities, the many saint images in them, the invocation of Jesus, Mary, and the saints in traditional Maya ceremonies, and the Maya insistence on the Catholic mass and the baptismal ceremony performed by a Catholic priest. The Maya incorporated these elements into their own worldview, with its distinctive concepts and logic, which changed their meanings from those of Catholic orthodoxy. At the same time, the Maya perceived that these imported Catholic elements still required the services of the Catholic priest, for reasons seen in the previous volume (Early 2006). This accounts for the continual Maya desire for his presence in their communities into the twentieth century.

The suppression of the church by liberal governments beginning in the nineteenth century resulted in the almost disappearance of priests from

Maya communities. With a change in the political climate, the church began to recover around the middle of the twentieth century and to reestablish its presence. But soon thereafter the Catholic worldview underwent changes that resulted in the church taking a more active role in Maya communities. To adequately understand Maya communities in this period, it is necessary to understand the presence and impact of Catholicism in them, regardless of one's personal evaluation of that presence. This involves understanding the changing worldview of Catholicism, its presentation to the Maya, the Maya perceptions of it in terms of their own worldview, and the consequent communal actions resulting from these perceptions. This is the intent of this book.

The Questions of the Research

Both Maya culture and Catholicism are defined by their theologies as the foundations of their worldviews. Catholicism's missionary activity has sought either to attract or to impose its worldview on those with different worldviews. Among the Maya, the consequent interaction has taken place for over four hundred years. The previous volume examined the topic from the early Spanish efforts to evangelize the Maya in the sixteenth century to the relationship in the early twentieth century. This volume continues the research for the Maya of Guatemala and Chiapas during the latter part of the twentieth century. The four main questions are: What were the conditions in Maya communities around the mid-twentieth century when the church returned to them? How did the Catholic Church present itself to the Maya then and in the following decades? How did the Maya perceive these presentations? What activities took place in Maya communities as a result of their perceptions?

To answer these questions, this research presents a synthetic view of the principal trends of the interaction during this period. It does not pretend to be a comprehensive historical work on the relationship between the Catholic Church and the Maya in these areas. It takes an anthropological look at the dynamics of the interactions between two religious systems when incarnated in reality. This distinguishes an anthropological approach from that of systematic (dogmatic) theology that examines the formal characteristics of religious systems, abstracting from their insertion into reality. The research draws on and synthesizes the writer's research as well as ethnographic and historical accounts of specific topics by others. The significance of individual events can be better understood when seen in a relational context and longitudinal perspective. As much as possible, the

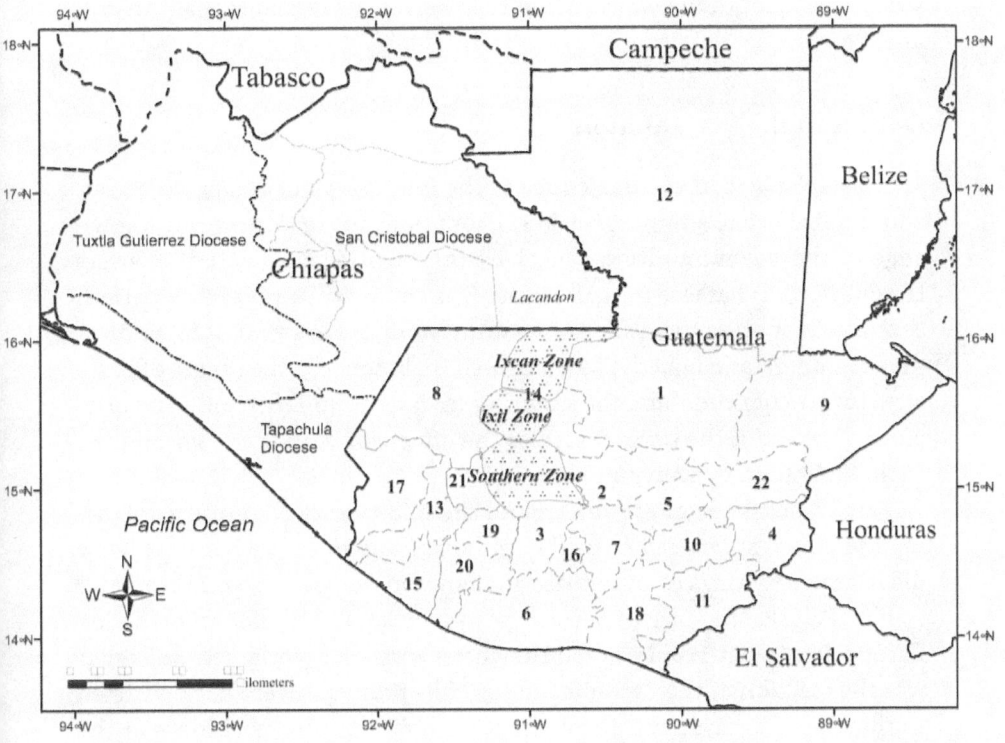

1. Alta Verapaz and 2. Baja Verapaz
5. El Progreso and 11. Jutiapa
6. Escuintla
7. Guatemala and 16. Sacatepequez (archdiocese)
8. Huehuetenango
9. Izabal (vicariate)
10. Jalapa
12. Peten (vicariate)

13. Quezaltenango and 21. Totonicapán (archdiocese)
14. Quiché with its three geographical zones
15. Retalhuleu and 20 Suchitepequez
17. San Marcos
18. Santa Rosa
19. Sololá and 3. Chimaltenango
22. Zacapa and 4. Chiquimula

Map 1. Ecclesiastical Structures of Chiapas and Guatemala—the three dioceses of Chiapas and the twenty-two departments of Guatemala that form the boundaries of two archdioceses, ten dioceses, and two apostolic vicariates. Beginning in the western part of the country, they are: 8 Huehuetenango; 17 San Marcos; 14 Quiché, with its three distinct geographical zones; 13 Quezaltenango and 21 Totonicapán (archdiocese); 15 Retalhuleu and 20 Suchitepequez (single diocese); 19 Sololá and 3 Chimaltenango (single diocese); 1 Alta Verapaz and 2 Baja Verapaz (single diocese); 7 Guatemala and 16 Sacatepéquez (archdiocese); 6 Escuintla; 18 Santa Rosa; 10 Jalapa, and 5 El Progreso, and 11 Jutiapa (single diocese); 22 Zacapa and 4 Chiquimula (single diocese); 9 Izabal (a vicariate possibly prior to becoming a diocese); 12 Peten (a vicariate).

study uses quotations where Maya themselves speak about their actions and reactions.

Outline of the Presentation

The various parts of the book answer the four main questions. To understand Catholicism's impact on Maya communities requires prior knowledge of these communities, especially their worldviews. Part II answers the question: What was the culture of Maya communities at mid-twentieth century? It is often described as a culture sunk in poverty. But it is necessary to understand how the Maya viewed their poverty, the causes of it, its structure, magnitude, and the resulting crisis as exemplified in the frontispiece. The rest of the book describes the efforts of both the Maya and the Catholic Church to overcome this crisis.

Part III poses the question: Around the mid-twentieth century, following the liberal suppressions of the Catholic Church, with what worldview did it return to Maya communities and how did the Maya first interpret it?

Part IV asks: In spite of the acceptance by some Maya of an orthodox Catholicism, what problems did they have with that worldview and what was the reaction of the Catholic Church? The problems were not restricted to the Maya, but systemic throughout Catholicism. This part looks at the theological evolution of Catholicism that took place at Vatican Council II, at the Second General Conference of Latin American Bishops in Medellín, and in local Maya dioceses that initiated Catholic Action and the liberation movement as an answer to the Maya crisis.

Since these movements were based on biblical theology, part V examines the Bible as a cultural document, and Maya reflections on it. It asks: How did biblical reflections help to raise Maya social consciousness and empower social action to confront their crisis?

Part VI inquires how this aroused social consciousness was a factor in some Maya Catholic communities engaging in armed insurrection against the Mexican and Guatemalan governments, even as other Catholic communities rejected this course of action.

Part VII examines the other direction of the Maya-Catholic interaction. What impact did the experience of working with the Maya during their crisis have on the worldviews of non-Maya Catholic priests and nuns?

Part VIII returns to another aspect of the Maya crisis: the doubts about the Maya traditional worldview raised by the subsistence aspect of the crisis. Given the persistence of the Maya worldview in the background of many Maya Catholics, this part looks at attempts to formulate a Maya

Christian theology based on the traditional Maya worldview. It would become the basis of an autochthonous Maya Catholic Church.

There is a rough chronology underlying these questions. The Maya crisis had been building for many years. Its results are examined as of 1950. The renewed presence of the Catholic Church began in the 1940s in Guatemala, and in the 1960s in Chiapas. Prior to and during this same period, a crisis had been building within institutional Catholicism itself that paralleled the Maya crisis. The institutional reaction to that crisis reached Guatemala in the late 1960s and 1970s, Chiapas in the 1970s and 1980s. The armed rebellion in Guatemala was a long, drawn-out conflict that reached its greatest intensity and Maya involvement in the late 1970s and first part of the 1980s. The short-lived insurgency in Chiapas took place in 1994. While the Teología India movement had its beginnings in the 1970s, the attention of the Maya and the church was focused on the pressing issues of social justice, so that the movement began to formalize only in the 1990s.

Catholicism was not the only alternative to which the Maya turned during their crisis. But it had a long history of contact with the Maya, and at the time of its renewed presence, it was the predominant alternative. While some Protestant groups had a presence among the Maya going back to the nineteenth century, their numbers were small. Later, especially after the 1976 earthquake, various evangelical groups entered the Maya area in force (Bogenschild 1992; Garrard-Burnett 1998). The same is true for other religious groups. In reaction against all types of missionary activity, some Maya intellectuals initiated the pan-Maya movement, a revitalization effort seeking to reestablish the traditional Maya worldview (Fischer and Brown 1996; Montejo 2002; Warren 1998, 2002). Except for an occasional mention, this research abstracts from these other alternatives in an effort to keep the task within manageable limits.

The Importance of Worldview

Like the previous volume, this research emphasizes worldviews and their evolutions. When dealing with problems of change of an indigenous group, there is a tendency to dwell on political and economic factors. These are seen as forcing change on a minority group, with little agency on their part. Analytically, this tendency looks at religious worldviews with Marxian glasses, as a passive superstructure determined by political and economic factors. This problem arose in the nineteenth-century studies of European culture stemming from the rise of capitalism. Marxian analyses explained its origins using an analytical model of economic and political factors. It was presented as a comprehensive metaphysic, thereby excluding all other explanations. Max Weber saw this shortcoming and insisted on the importance of worldview as an additional important explanatory factor. To confirm this position, he produced his famous work, *The Protestant Ethic and the Spirit of Capitalism*. There is a dialectical relationship between worldview and socioeconomic structure demanding that both receive attention.

A somewhat similar situation has arisen regarding studies of the Maya experience during the last half of the twentieth century. While the importance of political and economic factors can never be discounted, an analysis restricted to these factors alone has serious shortcomings (Comaroff and Comaroff 1991, 1997). It downplays the agency of the Maya themselves and their worldview in determining how they react to the pressures confronting them. This study emphasizes worldview, while political and economic factors have been described at length by other writers. They will be referred to here only in summary form.

II

The Background of the Crisis in Maya Communities at Mid-Twentieth Century

While life as part of a Spanish colony and later as the lowest segment of newly independent countries was always difficult, the Maya managed to retain many aspects of their traditional culture. Part II briefly examines the traditional culture, its retention, and the forces that began to undermine it, leading to its crisis as seen at mid-twentieth century.

2

The Traditional Maya Worldview as Influenced by Later Evangelization

This chapter presents a logical outline of a fundamental part of the Maya worldview abstracted from its many concrete expressions and descriptions in myths, stories, rituals, artistic expressions, early Spanish chroniclers, and ethnographic descriptions. It is a Weberian construct, a model that has many variations when the abstract logic is manifested in specific actions and objects (Weber 1949).

The Maya Metaphysical Paradigm

The Maya view ultimate reality as a power, a force, an impersonal god existing in itself and responsible for the coming into existence of all other beings. It is the divine principle of unity within all existing things, an infinite power-god that emanates directly or indirectly into all other cosmic entities and continually, but conditionally, sustains them for an allotted period of time. Otherwise, they would immediately cease to exist. Several phrases have been used to designate this infinite entity, such as the One, the cosmic force, and the spiritual force.

Emanation means that the power-god extends part of its own being as the sustaining power, the inner force of the outward covering of the emanated being.[1] This is a pantheistic conception that is sometimes inadequately defined as "god in all things" or labeled "animism." There are degrees of emanation giving rise to different kinds of beings. Emanation is seen as an evolutionary process. More-differentiated beings emerge from less-differentiated ones. Emanation is different from the concept of creation, the coming into existence of beings from nothing by a god that is transcendent to all created beings. This difference is systemic, giving rise to theological systems structured by differing concepts and logics.

The Two Worlds of the Cosmos

The One, the power or force god, emanates out to form a cosmos that has been through various epochs—prolonged cycles of time. In the present epoch, the cosmos is composed of two worlds, the visible world of humans' everyday experience and the invisible realm of emanated gods, both sustaining power-gods and threatening power-beings. To distinguish the two worlds, Arias, a Tzotzil Maya and a trained anthropologist, translates the Tzotzil terms as the "surface sky-earth," the realm of everyday experience, as distinguished from "internal sky-earth," the realm of the gods and other invisible emanated beings (Arias 1991: 39). There is a difference in the use of directional symbols between Maya thought and some world religions. For the Maya, the realm of the invisible world is usually within the surfaces of the experienced sky-earth, whereas other religions symbolize this realm as "up there," "in heaven above," etc. The two worlds interact with each other forming a single cosmos.

The Popol Vuh outlines the emanation of the visible cosmos (Tedlock 1996: 64–71, 145–49). It describes a primal sea with a vault of an empty sky above it. Emanating and differentiating from the primal sea is the earth, that in turn is differentiated by its mountains, valleys, and rivers. "Let it be this way, think about it: this water should be removed, emptied out for the formation of the earth's plate and platform, then should come the sowing, the dawning of the sky-earth."

Metaphysical Characteristics

All emanated beings on surface sky-earth and in internal sky-earth have a common metaphysical characteristic. Traditionally, the Maya, as agriculturists, lived close to nature. To them, all existents manifested themselves as a series of cycles. There was the daily cycle of day and night. Especially important was the yearly cycle, as it gave rise to lesser cycles, the seasons of the year, the latter in turn defining the corn cycle, the Maya staff of life. All plant life and all fauna had their cycles. Looking to the skies, there were astrocycles—the stars, the planets, and the cycle of the moon evolving through its various phases. There was the human life cycle, and within it, the female ovarian cycle and the embryonic gestational cycle. There was the intergenerational cycle of grandchildren replacing their grandparents. All existence showed itself as a cycle of some kind.

Therefore, the Maya concluded that the existence of all emanated beings on surface and in internal sky-earth have a cyclical existence with a common structure: birth, then death, leading to rebirth/regeneration, which begins a new cycle. Without death there can be no regeneration. Corn grows, is consumed, but its seed regenerates a new crop and another cycle begins. The sun lives during the day, dies at night, and regenerates the next day. Human beings live, then die, but their sustaining powers may be regenerated in other beings. Since a cycle goes through this pattern, it has a duration, time. "Time permeates all and is limitless. Time is . . . an attribute of the gods: they carry it on their backs. In a word, *kinh* [sun-day-time] appears, as the heart of all change, filled with lucky and unlucky destinies within the cyclic reality of the universe and most probably inherent to the essence of divinity itself" (León-Portilla 1973: 33). Since the existence of all emanated beings is cyclical, the existence of each has an allotted length of time, often determined by the conjunction of cosmic forces on the day of its inception. However, this allotted time can be cut short for reasons to be seen.

Another characteristic of the cosmos is its ideal of harmonious balance. The cosmos is conceived as a teleological entity in which all its constituent parts in both the visible and invisible worlds should be in a harmonious relation with each other. This results in the well-being of each part, and of the cosmos itself. But this harmonious balance can be upset by the actions of some emanated beings. Disharmony brings sicknesses, sufferings of various kinds, and premature deaths.

The Emanation of Humans

After the emanation of the cosmos, the Popol Vuh describes four epochs in which the emanated gods attempted to fashion creatures that would provide them with the praise and nourishment they needed to sustain their existence for their allotted time. In the first epoch, the gods made the animals, including deer and birds. They commanded them: "Talk, speak out. Don't moan, don't cry out. . . . Name now our names, praise us. We are your mother, we are your father." When the animals failed to do this, they were condemned to live in the forests, where they were to be hunted as food for humans.

The emanated gods continued their efforts for providers of food and praise. "It must simply be tried again. The time for planting and dawning is nearing. For this we must make a provider and nurturer. How else can we be invoked and remembered on the face of the earth? We have already made our first try at our work and design, but it turned out that they didn't keep our days, nor did they glorify us." The gods began experimenting with human figures made out of several materials. They tried mud, which quickly crumbled. The divine makers were disappointed and asked themselves: "What is there for us to make that would turn out well, that would succeed in keeping our days and praying to us?"

In the next epoch, they made human figures of wood, who multiplied and covered the earth. Again the gods were disappointed, because "there was nothing in their hearts and nothing in their minds, no memory of their mason and builder. They just went and walked wherever they wanted. Now they did not remember the Heart of the Sky." Since the wooden figures could not praise their makers, they were killed in a great flood and turned into monkeys. These epochs were cycles of birth, death, and rebirth. The death of each cycle was due to the failure of its creatures to fulfill the purpose for which they came into existence, the praise and nourishment of the emanated gods.

Before narrating the next epoch, the Popol Vuh tells the story of some emanated gods who disturbed the harmony of the cosmos by pride, murder, envy, family discord. For these actions, they were punished by other gods. While the Popol Vuh emphasizes the ritual offerings needed by the emanated gods, this section shows that moral actions in addition to ritual offerings are necessary for the harmonious balance and continued existence of the emanated gods.

According to the Popol Vuh: "But there will be no high days and no bright praise for our work, our design, until the rise of the human work, the human design.... Let's try to make a giver of praise, giver of respect, provider and nurturer." In this epoch the gods made the skin of humans out of corn, and blood from water. They were brought to life by emanated sustaining powers, extensions of the being of the power-god itself. This time, the gods were successful. "They talked and they made words." The words indicated their ability to think and choose. "Truly we thank you doubly, triply that we were created [emanated], that we were given our mouths and faces. We are able to speak and listen. We are able to ponder and move about. We know much, for we have learned that which is far and near. We have seen the great and the small, all that exists in the sky and on the earth. We thank you, therefore, that we were created, that we were given frame and shape" (Christenson 2003: 199). The humans had complete knowledge of all things in the cosmos. The gods started to worry about this. Therefore, they dimmed the humans' vision with imperfect knowledge. These were the ancestors of the Maya.

The Human Condition

Humans are composed of a corporeal body and the emanating invisible sustaining power. "The body primarily belongs to the visible world, but the abode of the sustaining power is mainly in the invisible world.... While the body lives, the sustaining power remains in it as a transient who will emigrate to the other world when it leaves the body. The sustaining power comes from the other world and should return to it. In addition, the sustaining power can easily escape from the body and frequently gets lost ... [due to] anything that lessens self-consciousness; fright, abstractedness, or concentration on something foreign to the ego, even distractions, are described in terms of the sustaining power abandoning the body" (Arias 1991: 40–41).

At birth, the emanating power is barely in the body. Over time, it needs to be securely implanted. "Education is a long process that begins with the birth of the child and lasts until the last moments of life. Education is the slow but constant, step by step acquisition of the sustaining power (*ch'ulel*).... Tzotzil and Tzeltal communities consider the formation of a personality as the constant and diligent building of what is culturally considered to be the ideal man and the ideal woman.... Since the *ch'ulel* does not immediately possess it, it is the duty of each person to orient oneself toward the ideal world formed by the ancestors and living elders because only they are in full possession of their sustaining powers.... The childish and unproductive activities of six or seven year old children are attributed to the fact that their sustaining powers have not yet fully developed, it is the age of games and trivia" (Arias 1991: 28–29).

The individual person goes through their life cycle from birth to death with a destiny, an allotted number of days to live as determined by the conjunction of cosmic forces on the day of birth. Whether this destiny unfolds in a favorable or unfavorable manner can be influenced by a person's dealing with the gods and other humans. Human regeneration after death is vague in Maya thought, and there is no unanimity about it. The sustaining power withdraws and can be recycled in various ways. Some see the powers later becoming the sustaining powers of future infants. A Maya shaman expressed his view in this manner: "The Maya religion is not like all the other religions where you return to God when you die. The Maya religion is to hope that, after death, one returns so as not to appear dead. Simply put, death is a journey, or a change of clothes, because the spirit [sustaining power] never dies. We pass to a star of the firmament. The spirit retains its existence and power and immortality" (Pieper 2002: 56).

Maya thought does not abstract individuals from their community. Education consists in teaching and living up to the norms of the ideal man and woman as defined by the community, who received the norms from the ancestors. At the beginning of the present epoch, the founders of the different Maya groups were great ancestral shamans whose sustaining powers conversed with the gods in internal sky-earth. From them, they learned the divine purpose and design of the present epoch, including the templates of harmonious balance with which to mold their communities physically, culturally, and morally. When they followed these templates, surface sky-earth was in harmonious balance, an equilibrium that should be continued without change by every generation until the end of its allotted cycle. In this sense, the "is" of the template becomes the "ought to be" for all generations. The Popol Vuh describes an ordered linkage comprising the harmonious balance. Plants are seen as the food of higher creatures; animals are to be food for humans; and human beings exist to praise and feed the gods by rituals and moral lives that contribute to the desired harmony within the community. In philosophical language, human beings, their community, and the surface of sky-earth are teleological, a hierarchy of finality (Arias 1991: 36).

Maya communities chronicled their history in terms of the cycles of their respective gods. This can be seen on various stone monuments and in the Book of Chilam Balam of Tizimin, a compilation of Yukatek Maya historical writings from both pre-Conquest and post-Conquest periods (Edmunson 1982). The Maya realized that if they could quantitatively describe the cycles, they could understand the cosmic forces of the future, since they would be analogous repetitions of the past. This led to their development of mathematics and sophisticated calendars used by shamanic diviners and ritual specialists. When the Maya encounter a new situation, they attempt to interpret it in terms of a logic drawn from an appropriate past cycle.

The surface of sky-earth is a dangerous place. There are many threats against the allotted life cycle of both the individual and the community. These threats can be unleashed by the gods for behavior that upsets the harmony of the community. The forms of punishment often center on the sources of food. They can be damaged or destroyed by disruption of the normal cycles of nature—droughts, floods, winds, insects. Deadly disease may strike the normal balance of bodily functions at any time. Man can cause disharmony in his community by kin disputes, stealing, and murder. People and their communities need protection to survive their allotted time as they traverse through their dangerous cycles of existence.

Internal Sky-Earth

There is a plurality of emanated power-gods living in internal sky-earth. Many are identified with the forces of nature, such as rain, wind, fertility, etc. Others are ancestral gods who designed the templates of harmonious balance that are the moral criteria to be implemented by Maya communities. All the gods are emanations of the One, not simply a polytheistic pantheon. However, some Maya fundamentalists may literally interpret the separate names or functions as independent gods in a polytheistic sense.

Also in internal sky-earth live animal powers, extensions of the sustaining powers of humans. Each human has such a counterpart. This is possible since all entities are emanations of the One. There is a reciprocal relationship between an individual's sustaining power and its animal counterpart. Whatever happens to one also happens to the other. These animals can be attacked, or suffer illnesses that will be transferred to the person who shares its sustaining power. Shamans in Cancuc (Pitarch 2010 [1996]: 40–57, 214, passim) see additional powers (*lab*) within the human body with corresponding counter-powers in internal sky-earth that can take on various outward appearances. Each individual has one to thirteen of these powers that play a role in forming the personality characteristics of individuals. Their many possible combinations and interrelationships in both sky-earths impact the body, often as sickness or misfortune. They are another source of the many dangers that individuals meet during their passage through their allotted time.

The dynamics and cultural logics of internal sky-earth are different from those of surface sky-earth. Arias suggests the dynamics of dream activity as a helpful analogy to understand these logics. When humans dream, their sustaining powers (ch'ulel) leave their bodies and journey to the internal sky-world. There, they may see the emanated gods and converse with them. Upon the return of their sustaining power to their bodies, their experiences or visions in internal sky-earth can influence judgments and actions of everyday life.

The Divine Condition

Since the emanated gods are cyclical, and therefore finite, their cycles must be sustained for them to live their allotted time. If the emanated gods are not nourished, they will decay and die before the allotted time of their cycle. Their nourishment consists of candles, alcoholic drinks, incense, music, dances, animal blood, and human blood, and in some historical periods this logic was taken to the extreme of offering them decapitated human heads and thumping human hearts. "In the Maya scheme of things, man's existence depended on a host of sacred beings who controlled the universe . . . [who] in turn were dependent on human attention for their welfare, if not their very existence. Mesoamerican gods were like extremely powerful infants . . . likely to go into terrible tantrums and eventually expire if neglected. They had to be housed and cared for, diverted with music, dance, and colorful paintings and—most especially—fed" (Farriss 1984: 286). The rituals were not simply petitions or commemorations. They repeat the emanation of the cycles that brought the gods into existence, thereby sustaining them.

But the gods are mortal, and eventually their cycles enter the death phase. "In Atiteco theology, Martín is the deity who embodies the positive, life generating aspects of the world. Although Martín is a powerful deity, he also has his limits. Like other Maya gods, he ages, falters and dies on an annual basis. No single [emanated] god . . . has the advantage of omnipotence. In the end everything, including gods in all their manifestations, must periodically give way to darkness and chaos before they can be reborn to new life" (Christenson 2001: 157). In other words, every emanated god has a cycle with a pattern of birth, death, and rebirth as another cycle takes its place as part of the endless cycles of existence.

The Maya Covenant of Reciprocity

Given the human and divine conditions, human beings and the gods mutually need each other to survive their allotted cycles. The gods need to be praised and nurtured by humans. Praise and nurture mean not only ritual food offerings, but also the offerings of lives conducted in conformity with the divine templates of harmony given by the ancestors: the moral code. Human beings, in turn, need the protection of the gods to survive their allotted time in the face of all the dangers that beset the human life cycle, especially sickness and lack of food. Because of these mutual needs, the gods and Maya communities enter into a sacred covenant of mutual support. By the covenant the gods will protect and sustain humans in return for humans praising, nurturing, and sustaining the gods.

The covenant is modeled on reciprocity agreements, a common mechanism for the exchange of goods and services in many non-market societies. It is a personal agreement based on kinship or friendship, in which a good or service is given to another when in need with the expectation of some kind of future return when the giver is in need. Usually, there is no strict accounting. If a receiver does not eventually make a return, especially when the previous giver is in need, the pact is broken. Reciprocity is a way of spreading the many risks of the Maya life cycle, a type of insurance. It is an important part of daily Maya life, and they use it to express their view of the god-human relationship.

The founding shamans of Maya communities continue to live in mountain-caves as assistants to the gods, or sometimes fused with them. They are the sacred ancestors who monitor the observance of the divine templates, frequently expressed by the oft-repeated phrase, "tradition of the ancestors." Arias constantly emphasizes the teleology of the Maya community, its covenant and the role of the ancestors by the use of such phrases as "given order," "destiny of the world," "established order," "norms of tradition," or "transgressions of the rules linked to tradition" (Arias 1991: 38, 42, passim). Observance of covenant obligations results in harmonious balance within a community. There is an important difference between Maya thought and theological worldviews that have a preoccupation with "the saving of one's soul" or "going to heaven" or "one's spiritual perfection," etc. In such views, the soul is the center of attention and the body is depreciated. Daily life in a community is a means to an afterlife. In Maya thought, "the interaction . . . with the world and its surroundings is not considered a means but an end in itself; interaction with the invisible world is oriented toward the maintenance of harmony in the tangible *habitat* of humanity" (Arias 1991: 52–53). Arias is echoing what Landa (1941: 129) found in Yucatán in the sixteenth century: "This people had a great and excessive fear of death, and they showed this by the fact that all the services, which they performed for their gods, were for no other end nor for any other purpose than that they should give them health and life and sustenance."

Covenant Rituals

The offerings made during rituals nurture the gods, while the accompanying prayers and songs praise them. Consequently, rituals are an essential ingredient of the covenant. Their structures, the kinds of food offerings and the exact words of the prayers and songs, were originally learned from the gods themselves by the first ancestors in internal sky-earth (Arias 1991: 35, 38, 43). "A spiritual contract between the people of the village and the Gods said that they would keep life coming to us if we promised to send them remembrance. The fruit of our remembrance was this earth and our lives, and we had to send them some of its deliciousness by means of ritual" (Prechtel 1998: 106).

As previously mentioned, for the Maya, ritual is not just praise for the gods, not simply a request for a desired result. By the logic of the covenant, the ritual performance itself brings about the original emanating act of the world, thereby sustaining it. "When a Maya priest-shaman performs a ritual at the proper time and in the proper manner, he is able to recreate the world just as it was at the first dawn of time [an epoch]. . . . Traditionalist Atitecos believe that the regenerative nature of the earth is controlled through ritual first established by the ancestors of the community" (Christenson 2001: 24, 78). The introductory photograph shows Zinacanteco shamans and cargo holders praying, singing, and making offerings atop a sacred mountain, at an altar with crosses, the entrance to the cave-home of an ancestral god.

Special Ritual Sites

The Maya perceived many mountains to be the outer shells of large interior caves that are both intrusions into internal sky-earth and extrusions into surface sky-earth. "For the Maya, the cave acted as a passage of transition between the real and invisible world.... The archeological record suggests that caves as sites of ritual activity were important as far back as the Preclassic" (Bassie-Sweet 1991: 240). The Maya also erected pyramids conceived in much the same fashion as mountain-caves. Several symbols were used to depict these points of linkage between the two worlds of the cosmos: the serpent's throat, the cracked turtle shell, the jaw of a jaguar. From these portals flowed the sustaining powers of the emanated gods. Corn from which Maya bodies are made was found in a cave. Its propagation depends on rain, wind, and fire, all of which are forms of emanated gods who live in their cave-homes. The physical characteristics of caves in the Maya area lend themselves to this interpretation. "The cave was naturally thought to be the source of surface water, for most of the caves in the Maya region are wet and many contain springs, streams, rivers, waterfalls, pools of water and small lakes. The Maya also believed that rain, mist clouds, thunder, lightning and wind were produced in caves. There is often a cool or even cold wind blowing from the opening as the warm air of the outside draws the cool air from the cave. This natural observation surely led to the conclusion that wind was created in caves. Moist environment and temperature changes often create mist. It is not unusual to find clouds of mist at the mouth of the cave and at vertical openings. In addition, clouds form on the slopes and tops of mountains. These phenomena have been used to explain why clouds and rain were thought to be formed in caves.... During storms, lightning bolts flash around the mountain tops, giving the impression that lightning, a natural source of fire, originates there" (Bassie-Sweet 1996: 10).

Types of Rituals

The Indians of Guatemala "have two types of sacrifices.... One is general in as much as the whole town together offers sacrifices during the festivals that they celebrate. The other type consists of the private sacrifices which each ... private person offers according to his own devotion and for whatever necessity he has" (Casas 1958: 148; also in Ximénez 1929: 81). For community festivals, a pre-Columbian K'iche' town had a central plaza where there was a four-sided pyramid of stone—an artificial mountain. On top of the pyramid the K'iche' erected a small chapel where they placed an image of their god, Tojil.

> The community sacrifices are usually offered during the festivals that are held in some provinces five or six times a year, and in others more

or less according to the devotion and custom of each one. Additional sacrifices are ordered when there is a drought or an epidemic or a war or other common misfortunes. When a sacrifice is to be performed either for a coming festival or when some necessity requires it, the person in charge of the sacrifice consults with all the distinguished elders. They in turn consult with the high priest and others about the festival until they have conferred with the diviners who tell them the opportune time. . . .

With the day decided, the priests begin a vigil in which young and old live apart from women even if they are married. The vigil lasts seventy eight to a hundred days depending on the solemnity or necessity of what is asked [of the gods]. Each day everyone must perform a sacrifice by drawing blood from the soft tissue of their arms, tongues and other parts of their bodies at certain times during the day and night while burning incense and other things. The men do not bathe. They paint themselves with black soot so that they look like the Devil. This was as a sign of penance. No one can sleep in their houses, but in those constructed near a designated temple during the penitential period. Next to the temple fires burned in braziers. All observed these ceremonial prescriptions because if it was known that anyone omitted them, they would be severely punished. They have an overwhelming fear of quickly dying as they consider it a grievous sin if they omit any part of these ceremonies. . . .

For these festivals and sacrifices, they construct and adorn their idols with much gold, precious stones and the most valuable things they have. They put them on platforms and carry them in processions with unspeakable devotion to the accompaniment of drums, trumpets and other musical instruments. They place them in a spacious town square which every community has where ball games are held. There, in front of the idols as part of the festival, the prominent elders play the ball game. In some places they place the idols in temples at the start of the vigil fasting. There they offer them small sacrifices such as birds, incense, cacao molido and other like offerings. In other places, they leave them in their caves where they are usually kept and send them gifts and offer the same kind of sacrifices. (Casas 1958: 148–49; also in Ximénez 1929: 81-82).

A table by Sharer (1994: 541) shows the rituals and their specific purposes for seventeen of the nineteen Maya months comprising the cycle of the solar year. Ritual elements included: recitation of prayers, burning of incense; drinking fermented beverages; offering food, ornaments, valuables, and blood; and during the major festivals performing dances ending with feasting usually culminating in drunkenness. Following the Spanish evangelization and the adoption of the saints as additional emanated gods, many of these festivals were held in conjunction with the feast days of the saints as determined by the Catholic liturgical calendar (Early 2006: 11–58). They retained a number of the ritual elements mentioned by Las Casas and Sharer.

There were a number of private rituals. The Maya had statues of their gods and were attentive in their care of them. With great reverence, they placed them on altars and in sacred places, such as caves, springs, hilltops, certain large trees, and crossroads. There they performed rituals in which they burned incense, killed birds and other animals, sprinkled blood drawn from various parts of their own bodies, and made offerings of cotton, cacao, salt, chili pepper, or some other valuable gift. By these rituals, they had recourse to the gods for protection or provision of their necessities: a new house, not encountering danger while traveling or while searching for ornamental feathers, protection for any kind of planting, for guarding fruit orchards, when weeding their grain fields, at harvest time, at times of sickness, the ability to have children, protection at the birth and weaning of a child, when the child begins to crawl on all fours, when it begins to talk, when it has its first haircut, on birth days, when young girls around age eight leave to live in the temple until married (Casas 1958: 154–57; Ximenez 1929: 85, 364). A number of these rituals continue to be performed by local shamans.

The Words of Ritual Prayers

Prayers composed of fixed formulas were an integral part of Maya rituals. The efficacy of the rituals depended on the recitation of the exact words of the prayer formula as learned by the first generation of shamans from the gods in internal sky-earth (Arias 1991: 38). The importance of specific words is found in the Popol Vuh. Emanations took place by saying the proper words. "The earth arose because of them, it was simply their word that brought it forth. For the forming of the earth, they said 'Earth.'" In describing the making of the wooden figures, the Popol Vuh says, "The moment they spoke, it was done." When the gods changed from making animal to human forms, the Popol Vuh's description is "We have changed our word" (Tedlock 1996: 65, 67, 70). Chamulan tradition also sees words bringing forth the first world (Gossen 2002: 21, 221). These passages imply that the utterance of the word brings forth the specific thing that the word signifies. This conception of the power of words is restated in Chilam Balam books from prior to and after the Conquest (Edmonson 1986: 31). It is also echoed in the Santiago Atitlán version of the Popol Vuh account. "Two Deities, an Old Man and an Old Woman, sat in each of these [five] creation layers, and they were the ones who assembled you.... As each pair of Old People did this... they uttered special magical words and phrases. These words became the very things they described. The Gods spoke the world into life by continuously repeating their names. When they reach the fifth creation [surface sky-earth], the Deities' names take on physical form and function.... We are made of words, and those words are places in our bodies and on the earth, too. The fifth world gave all these Old People's words a place to have form and run around, happy to be alive and eating together" (Prechtel 1998: 105–6). Words have this characteristic because they have a sustaining power of their own. In Santiago Chimaltenango, *naab'l* has "sustaining power" as one of its meanings. It is used only for human beings. But there is one exception. "Significantly, words are the only non-human beings possessing *naab'l, tnaab'l yoal* meaning a word's 'meaning'" (Watanabe 1992: 82).

Comparing pre-Columbian glyphs on a stela with the Popol Vuh, Freidel has noted: "As the opening quote from the Popol Vuh explains, Creation began with the utterance of a word and the appearance of the thing embodied by the word. The ancient Maya apparently thought of the process in the same way.... Words are a fundamental medium for Maya communion between this world and the Otherworld. They are not merely preamble to some magical action or a way of describing things that are manifested through tangible supernatural and natural forces like rain, lightning, and thunder. They are, rather, an essential conduit of those forces" (Freidel et al. 1993: 65, 179–80). The cosmic power emanates itself into words as their sustaining power, so that, when spoken, it extends existence to the entity they signify. This cultural logic explains the necessity of the recitation of the exact words of a prayer formula for the effectiveness of ritual and the consequent fulfillment of the covenant.

Rituals and Physical Representations of the Gods

In Yucatán, Landa found god images to be made of wood, clay, and stone. The making of them was a ritual.

One of the things, which these miserable people regarded as most difficult and arduous, was to make idols of wood, which they called making gods. And so they had fixed a particular time for this ... consulted the priest first, having taken his advice, they went to the workmen who engaged in this work. And they say the workmen always made excuses, since they feared that they or someone of their family would die on account of the work, or that fainting sickness would come upon them. When they had accepted, the Chacs [ritual specialists] whom they had also chosen for this purpose, as well as the priest and the workman, began their fastings. While they were fasting the man to whom the idols belonged went in person or else sent someone to the forest for the wood for them, and this was always cedar. When the wood had arrived, they built a hut of straw, fenced in, where they put the wood and a great urn in which to place the idols and to keep them under cover, while they were making them. They put incense to burn to four gods called Acantuns, which they located and placed at the four cardinal points. They put what they needed for scarifying themselves or for drawing blood from their ears, and the instruments for sculpturing the black gods, and with these preparations, the priest and the Chacs and the workmen shut themselves up in the hut, and began their work on the gods, often cutting their own ears, and anointing these idols with the blood and burning their incense, and thus they continued until the work was ended, the one to whom (the idols) belonged giving them food and what they needed, and they could not have relations with their wives, even in thought, nor could anyone come to the place where they were....

When the idols were finished and perfected, the owner of them made the best present he could of birds, game and their money in payment of the work of those who had made them; and they took them from the little house and placed them in another arbour, built for this purpose in the yard, where the priest blessed them with great solemnity and plenty of fervent prayers, he and the workmen having first cleansed themselves of the soot with which they had anointed themselves, since they said that they fasted while they were making, and having driven off the evil spirit as usual, and having burned the blessed incense, they placed the new images in a little hamper, wrapped up in a cloth, and handed them over to their owner, and he received them with great devotion. The good priest then preached a little on the excellence of the profession of making new gods, and on the danger that those who made them ran, if by chance they did not keep their abstinence and fasting. After this they ate very well and got drunk still more freely. (Landa 1941: 159–61)

In this account, the image belonged to a private individual, not the community. While some idols were kept in the temples, more appear to have been kept in numerous private residences. "They had a very great number of idols and of temples, which were magnificent in their own fashion. And besides the community temples, the lords, priests and the leading men had also oratories and idols in their houses, where they made their prayers and offerings in private" (Landa 1941: 108).

Violating the Covenant

If the Maya constantly fail in their covenant obligations, either deliberately or accidentally, the covenant is broken. Because of their mutual dependence on each other, both the gods and humans will prematurely end their cycle of existence. In the Atiteco version: "A forgotten God was an angry God, or a dead God. In either case, the life sap [sustaining power] would stop flowing, and all this life would be as if it had never existed. We would cease. All of our rituals in the village, whether personal or public, were memory feasts for the spirits. Being remembered was their food" (Prechtel 1998: 107). For occasional instances of neglect, earth tremors from the angry gods would warn the Maya of their failures and the danger to the covenant.

The Maya fully realize their shortcomings with regard to the covenant, that their community is not always in harmonious balance, that daily life frequently does not mirror the "ought to be" of the template of the original emanation. Studying the themes of numerous Zinacanteco tales, Laughlin (1977: 2) finds: "Despite the Zinacantecs' lack of compulsion to recall an origin for every aspect of the present world, to an extraordinary degree their tales reveal the Zinacantec musing over his cosmic journey.... In tale after tale the Zincantec wrestles with the problem of his fall from grace. 'Where does the responsibility lie?' he asks. His answer is contradictory, as many-angled, as the historical facts warrant. 'It is the Ladino, the wenching priest, who brought divine punishment upon our town. But, too, the negligence of the elders—or was it the shamans, or even the entire town which shares the guilt?" The "wrestling with the problem of his fall from grace," that is, from harmonious balance, creates tensions in Maya lives.

The unfolding of one's destiny can take place in various ways, some good and some bad. Ritual performances not only praise and feed the gods, but also are a means to cope with these tensions. Shamanic diviners and curers are asked for assistance in obtaining a favorable unfolding of one's destiny. "The time universe of the Maya is the ever-changing stage on which are felt the aggregate of presences and actions of the various divine forces which coincide in a given period. The Maya strove by means of computations, to foresee the nature of these presences and the result of their various influences at specified moments. Since *kinh* is essentially cyclic, it is most important to know the past in order to understand the present and predict the future" (León-Portilla 1973: 54).

The Maya observed and measured the cycles of their epoch, especially the Sun, Moon, Venus. They formulated cyclical, interlocking calendars of days, months of twenty days, years, and groupings of twenty years. The days, with their corresponding gods, were also assigned numbers from one to thirteen, and each had an astrological connotation. Therefore, each specific day could be defined by the multiple divine influences comprising it. The diviners used the calendars to understand the forces impinging on an individual or community. Then they decided the resultant of the forces for various days and chose one that was favorable for corrective action, usually a ritual in which the shaman intercedes for his worried client. Once the original order or balance of the cosmos has been broken, it is the function of the shamans to restore it by their ceremonies. Their shamanic power enables them to enter the internal sky-earth and plead directly with the gods to remove the punishment and to restore order and harmony. Some ceremonies may include confession of wrongdoing either to the gods or to the person offended (Arias 1991: 43–44).

Plurality of Covenants

The Popol Vuh narrates the epic of the K'iche' Maya, although the underlying thought patterns were common to many Maya and other indigenous groups. The term "Maya" can be misleading, as it implies a unity that never existed among the groups designated as "Maya." Consequently, there is no single covenant between the gods and the Maya people, but many covenants. Elements of the worldview described above appear to be common to most groups, but there are differences in specification arising from differences between and within various Maya groups.

Independence of Maya Communities

Most Maya groups saw themselves as emanating from internal sky-earth at their portion of surface sky-earth, which they called the navel of the world or some other designation for a center. In Maya thought there was a close unity between this physical place, the ancestral gods who lived in the nearby mountains, and saint gods who resided in the local church. Each Maya group looked upon their gods as tribal, belonging only to their community. There was little sense of identity or belonging beyond this. Other Maya communities had their covenants with the gods, but their gods were different. "In some areas they have the custom of guarding the images in very hidden places. This creates greater reverence because if the images are viewed many times, the people lose their reverence and respect for them. This is also done so that neighboring groups will not steal them. These people become very envious when they think there are stronger images in some areas than others. People would risk their life to steal them. The priests have for assistants the sons of nobility and relatives of the town leaders who are young and unmarried. These alone know where the images are concealed and they have the responsibility of guarding them and bringing them the offerings of the faithful. When it is decided to bring out the images from the ravines and caves and to have a procession through town, the young nobles carry them. Stopping from time to time, they sacrifice the offerings made during the procession. The boxes in which the images are carried are lined with branches and flowers so as to make them as beautiful as possible" (Ximénez 1929: 82).

There was frequent boasting by the residents of Maya communities that their gods were better than the gods of another community, who were not worth anything, meaning they were unable to deliver what was asked for (Ximénez 1965: 714). The different Maya groups with their tribal gods waged war against each other, sometimes to obtain a vassal group that would supply tribute and slaves to the conquering group. Some of the slaves would be used for ritual sacrifice. Such warfare was a clash of gods as well as men because of the tribal nature of the gods. This was how the battles with the Spaniards were conceptualized, leading to the adoption of the Spanish gods (Early 2006). Maya communities are careful to lock and secure their churches at night for fear that neighboring communities might attempt to steal their saint images.

Social Strata within Maya Groups

Historically, there was a Great and a Little Tradition, a lengthy and a condensed version of the worldview held by different stratified groups. An elite would sometimes be formed by one kin group occupying the ruling offices and arranging for members of their group to succeed them. Among the elite offices was that of the high priest, the carrier of the Great Tradition. In Yucatán,

> they had a high priest who was called Ah Kin Mai, and by another name Ahau Can Mai, which means the Priest Mai, or the High-Priest Mai. He was very much respected by the lords and had no *repartimiento* of Indians [workers], but besides the offerings, the lords made him presents and all the priests of the towns brought contributions to him, and his sons or his nearest relatives succeeded him in his office. In him was the key of their learning and it was to these matters that they dedicated themselves mostly; and they gave advice to the lords and replies to their questions. He seldom dealt with matters pertaining to the sacrifices except at the time of the principal feasts or in very important matters of business. They provided priests for the towns when they were needed, examining them in the sciences and ceremonies, and committed to them the duties of their office, and the good example to people and provided them with books and sent them forth. And they employed themselves in the duties of the temples and in teaching their sciences as well as writing books about them. They taught the sons of other priests and the second sons of the lords who brought them for this purpose from their infancy, if they saw that they had an inclination for this profession. (Landa 1941: 27)

The content of the priestly worldview was a more developed and sophisticated version of the worldview than that of the commoner. "The sciences which they taught were the computation of the years, months and days, the festivals and ceremonies, the administration of the life cycle rituals, the fateful days and seasons, their methods of divination and their prophecies, events and cures for diseases, and their antiquities and how to read and write with the letters and characters, with which they wrote, and drawings which illustrate the meaning of the writings" (Landa 1941: 27–28). In later Maya communities, respected elderly shamans performed some of these functions.

The Little Tradition was the commoners version of the worldview. It was focused on their role as producers of corn that was "el cuerpo y la Emanación de nuestro Señor" (the body and emanation of our Lord) (Arias 1991: 23). Corn was originally found in the recess of a cave-home of a god and is an emanation of the god. Arias's explanation of the Maya cognitive world shows the central role of furnishing corn as the basic necessity for fulfilling the covenant. In San Pedro Chenalhó, corn comprised 80 percent of all consumed food, and without it there would be starvation. Prestige for community service was based on the amount of corn cultivated, since it supplied the resources to obtain the ritual offerings. The intricacies of corn farming comprised a large part of a boy's education. A boy was not considered a man until he had shown that he could produce a crop on his own. This meant that he was ready for marriage and family responsibilities. Corn, the local community, and performing service for it, were at the heart of the commoner's worldview, along with the two main threats to their successful accomplishment: crop failure from any source and disease. In learning and fulfilling these duties, a person's sustaining power entered more fully into their bodies.

During the colonial and into the national periods, everyone in Maya communities underwent hardships as their economic base was greatly reduced by loss of land and the requirements of the forced labor laws and other taxes. This meant that the stratification system was significantly compressed. This was especially true of the elite priests. Their books were burned by the Spaniards. Their economic base was eroded since the commoners and subordinate priests were no longer able to support them. They became milpa farmers themselves, like the commoners. The tradition was now carried on by individuals in a series of oral myths and stories. For this reason, the contents of the tradition became diffuse, losing some of its uniformity. It became a matter of individual interpretation by anyone who claimed to have knowledge of the tradition. It was passed down through their family or received in dream-visions from the ancestors. This led to many variations, but most traditions retained the basic concepts and logics outlined here. As a contemporary ethnographer observed: "I have found that a cycle of mythic tales concerning revered ancestors and divinities is widely known in Santiago Atitlán and repeated often, although not codified in any known written text. Such tales are never told the same way twice, even by the same person; however, the basic core concepts embedded in each vary little" (Christenson 2001: 15).

The Maya Worldview Interprets the Spanish Evangelization

The logics embedded in the concepts of emanations and cycles give the Maya worldview great flexibility to absorb new events, while retaining its essential characteristics. For analytical purposes, two types of cycles can be cited, although the distinction is not always clear-cut. Epochs are longer cycles of time, periods of actual or mythical history in which the main outlines of the cosmological or cultural system remain stable, while cycles, of shorter duration, take place within them. Both types have the structure of birth, death, rebirth. Rebirth is not always an exact return to the original conditions of birth, but can be the beginning of a new cycle related to the former one.

As explained in chapters 12 and 13 of the previous volume, the Maya saw their battles against the Spanish invaders not only as a clash of armed forces, but also as a battle between Spanish and Maya gods. Since the Spanish were victorious, this meant that the Spanish gods were stronger than the Maya gods. Consequently, the Maya wished to obtain the protection of these stronger gods by adopting them into their covenant that had failed them. But they needed to know what the Spanish gods required from the Maya to seal the covenant. To find out, the Maya carefully observed the Spaniards. This visual communication was how they formed their perceptions of the Spanish gods and their needs. The Spaniards demanded the erection of cave-like structures, churches, as the homes of their gods. Inside these structures they placed their gods, a few wooden statues for whom they performed rituals, especially the mass. Therefore, the statues were animated by sustaining power that allowed them to see and hear what was happening in the community. Each was a distinct god, even if some carried the same name or were different representations of the same person in Catholic theology, especially those of Christ and Mary. The Spaniards usually placed images of Christ, Mary, and the community patron in more prominent places in the churches, indicating some hierarchy among the Spanish gods.

Each of these emanated gods also had their cycles of existence whose continuation needed nourishment. The Spaniards did this in a ritual, the mass, in which bread and wine were offered to the various gods in the churches on certain days of the year. These gods carried the cycles of time, just as the Maya gods. The Spaniards were also very insistent on the ritual of baptism for newborn babies, a ritual the Maya interpreted as a plea to the Spanish gods to protect the infants from sickness. The Spaniards had specialists who conducted their rituals, much the same as Maya shamans perform Maya rituals. In the Maya view, the Catholic priests derive their power from their knowledge of the exact prayer formulas needed for the validity of the new rituals. None of the Maya knew these prayer formulas, so the Catholic priest was incorporated into their system along with the Spanish gods and rituals.

The Maya petitioned the Catholic priests to perform the rituals that nourished the images so that they would help maintain harmonious balance in the community. In this way, the images joined the ancestral gods in protecting the community. Sometimes the saints were fused with the emanated gods or the ancestors. There were numerous variations on the ways in which the basic logic was employed. The Maya grew very attached to these new gods, and were responsible for placing an ever increasing number of images in the churches (Early 2006: 201–5).

In summary, the Maya took the logical structure of their pre-Columbian worldview and applied it to the Catholic evangelization and its perceived pantheon, which they adopted. Although the military defeat had shown a weakness of the traditional Maya covenant, it did not destroy it. There was no disavowal of the ancestors. Rather, there was a revitalization of the covenant by the addition of more emanated gods who were perceived as strengthening its protective power. "Mayas placed the Spanish invasion, and the violence and epidemics it brought, within the larger context of history's cycles of calamity and recovering, relegating the Conquest to a mere blip in their long term local experience" (Restall 2003: 122). Seeing covenant protection as weakened and then revitalized by adopting the Spanish gods is another type of cyclical thinking characteristic of the Maya worldview. Its categories can be labeled covenant protection, partial failure, revitalization. More examples of this type of cyclical thinking will be seen in upcoming chapters.

Summary

This chapter has presented a synthetic construct, an ideal type, a model of what the author considers a predominant logic of the traditional Maya worldview and how it was used to incorporate Catholic elements after the defeat by the Spaniards. This worldview provides a metaphysical viewpoint from which to understand other aspects of Maya thought and behavior. Anthropologists have used various semantics to describe such metaphysical descriptions: "deep, generative roots" (Gossen 1986: 4–5); "underlying foundation," "cognitively deep" (Fischer 2001: 17–20); "durable transposable dispositions" (Bourdieu 1977: 22, quoted in Fischer 2001). The remainder of Part II will show how the Maya retained the main outlines of this worldview throughout the colonial and national periods.

Maya Religion: A Comparative Perspective

The Maya were not unique in their metaphysical views. Analogous systems exist among a number of groups, especially those associated with versions of Hinduism and Buddhism. This is not surprising since American indigenous groups came from similar Asiatic roots. Eliade's book, *The Myth of the Eternal Return* (also published as *Cosmos and History*), contains many examples. "This book undertakes to study certain aspects of archaic ontology—more precisely, the conception of being and reality that can be read from the behavior of the man of the premodern societies.... The metaphysical concepts of the archaic world were not always formulated in theoretical language; but the symbol, the myth, the rite, express, on different planes and through the means proper to them, a complex system of coherent affirmations about the ultimate reality of things, a system that can be regarded as constituting a metaphysic" (Eliade 1959: 3).

With regard to the emanated sustaining powers, the cosmic force: "If we observe the general behavior of archaic man, we are struck by the following fact: neither the objects of the external world nor human acts, properly speaking, have any autonomous intrinsic value. Objects or acts acquire a value, and in so doing become real, because they participate, after one fashion or another, in a reality that transcends them." (Eliade 1959: 3–4).

Eliade describes the cyclical nature of the cosmic force. "Belief in a time that is cyclic, in an eternal returning, in the periodic destruction of the world and mankind to be followed by a new world and a new, regenerated mankind—all these beliefs bear witness primarily to the desire and hope for periodic regeneration of the time gone by, of history. A cosmic cycle includes a 'creation,' an 'existence' (or 'history,' wearing-out, degeneration) and a 'return to chaos' ... [then] hope of a total regeneration of time that is evident in all the myths and doctrines involving cosmic cycles" (Eliade 1958: 407).

Eliade has also shown how widespread is the view that the structure of ritual, including the words, must follow a template associated with the emanation of surface sky-earth: "rituals and significant profane gestures which acquire the meaning attributed to them, and materialize that meaning, only because they deliberately repeat such and such acts posited *ab origine* by gods, heroes or ancestors" (Eliade 1959: 5–6). Also, "*in illo tempore*, in the mythical period ... the creation and arranging of the cosmos took place, as well as the revelation of all the archetypical activities by gods, ancestors, or cultural heroes.... From the point of view of primitive spirituality, every beginning is *illud tempus*, and therefore an opening into the Great Time, into eternity. . . . every one of these 'religious things' indefinitely repeats the archetype; in other words, repeats what took place at the 'beginning,' at the moment when a rite or religious gesture was being revealed, at the same time expressed in history" (Eliade 1958: 395–96). These passages place the Maya worldview within the panorama of world religions.

3

Retention of Maya Culture through Periods of Domination

This chapter describes the traditional social organization of Maya communities and how it enabled them to retain many elements of their worldview during periods of colonial and national rule. Physically, many communities were clusters of houses surrounded by communal lands, with trees, which provided fuel for cooking and house supports, a water source, and cleared land for crops. The everyday life of the Maya was centered on their homes and fields.

> The traditional thatch-roofed, wattle-and-daub house . . . provides shelter. With a fire constantly burning on the hearth, it is also the most important focus of family life. Here tortillas and beans are cooked for the family's meal; babies are conceived and born; children learn Zinacanteco customs; marriages are planned and wedding ceremonies occur; old people die. In short, almost all family interaction takes place here. Zinacanteco women spend most of their lives in and near their houses, working in the patios immediately outside, or

fetching wood and water nearby.... Human life in Zinacantán would be impossible without the maize field. Nearly all Zinacantecos utilize some other food sources (wild plants ... or commercial products purchased with money earned from the sale of surplus maize or from wage labor), but the overwhelming proportion of calories consumed comes from the fields of cultivated maize, with the supplementary crops of beans and squash. Some of this produce comes from small plots located around the houses, but most is grown in large fields located either on ejido [communal] land on the flanks of the Highlands or on rented land in the lowlands of the Grijalva River valley. Men and their older sons spend much of their lives in the fields—cutting and burning brush in the ancient method of swidden agriculture, planting with digging sticks, harvesting the crops. (Vogt 1976: 51)

This recent account describes centuries of the day-to-day life of many Maya except the more recent wage labor due to the shortage of land.

Using land either to grow a crop or to obtain wood and adobe for a house was a sacred act, since the fertility of the land was a godly emanation. Both tasks involved rituals that marked the four corners and the center of the cultivated plot or house site. The ritual offerings were reciprocity returns for the emanated elements used (Guiteras-Holmes 1961: 26, 42–47; Vogt 1976: 51–59). The community defined itself as a group who should conduct themselves according to the moral dictates of the covenant handed down from the ancestors who entrusted the house sites and communal lands to them. By so doing, they would bring about a state of harmonious balance that would ensure the survival of its inhabitants and the community itself to the end of their designated cycles.

Social Organization

Maya life was organized by several hierarchical segments that continued the pre-Conquest structure, with modifications to accommodate colonial masters and their national successors.

Within a community, the basic units of social organization were extended family groups that formed a clan or patrilineage or a similar type of kinship organization. A community may have been comprised of a single kin unit, or several of them. They were headed by one or more elderly men, *principales*, who had previously served in civil-religious posts, charged with the welfare of the community.

Kin units belonged to *parcialidades* (also *chinamit*) (Hill and Monaghan 1987; Hill 1992: 38–47; Cook 2000). "It was a key principle of organization. The chinamit was basically a territorial unit, ruled by an aristocratic core family. . . . The head of the chinamit was aided by council of elders and a small staff of messengers or criers. . . . Apart from the ruling family, the rest of the chinamit's population was essentially peasant-like in status and was burdened with a high degree of corporate control and numerous obligations" (Hill 1992: 39). The parcialidad/chinamit was responsible for ritual observance in honor of its patron god. It held land in common. A family group had a right to use, but not own, a portion of the community's land. At death the land either reverted back to the parcialidad to be redistributed, or the right to use could be passed on as inheritance, but in no case was this ownership. With the sharp decline of the Maya population in the sixteenth century, the communal lands initially were sufficient to provide the resources for the community needs. When the Spanish created towns as centers of their congregation policy, they were composed of several parcialidades, giving rise to conflicts between them.

Later, national governments revised the Spanish political structure. The country was divided into twenty-two departments. Within each department, there were a number of *municipios*, with a town designated as its administrative center. Outlying communities were classified according to the size of their populations. A council of officials took care of the day-to-day administrative functions of the municipio (Carmack 1995: 89–96, 185–90, 309–18, e.g.). These men were selected for a year or more of community service by the principales. Their duties involved dealing with outside Spanish or national authorities, but most of their time was spent arbitrating disputes within the community. "Zinacantecos appear to commit the standard range of crimes that are found around the world: murder, assault, theft, adultery, witchcraft, rape, insult, abuse of public office, etc. . . . But most common by far are marital disputes, which are almost always discussed as instances of wife beating. There are also many quarrels between kinsmen over rights to inherited property and disputes between families of an engaged couple over the delicate bride-price transaction" (J. Collier 1973: 11). A number of crimes involved the abuse of alcohol.

Another member of the council was in charge of public works and collecting taxes. He was an advisor to the head of the council and acted in his place when necessary. Other council members had specific duties, such as investigating crimes or representing the community in dealing with other communities or colonial or national authorities. The council had assistants to help them carry out their various functions. There was a scribe who could read and write documents in the Spanish language, necessary for dealings with outside authorities. There were representatives to the council from the outlying communities and landholding units. Others were policemen, to make arrests, take charge of prisoners, keep order at hearings if necessary, and to run errands.

In keeping with the dictates of the Maya covenant, the council attempted to restore harmony in the community when it was disrupted by disputes and criminal deviance. The Maya use arbitration as their method of conflict resolution (J. Collier 1973; González 2002: 107–38). "Zinacantecos were not concerned with crime and punishment. They cared about ending conflicts to forestall supernatural vengeance. I finally understood why civil officials looked puzzled and gave inconsistent answers when I asked 'What do you do when someone commits a specific type of crime?' They didn't *do* anything. Their job was to wait for a plaintiff to state his wishes and then try to suggest a compromise solution acceptable to both sides" (J. Collier 1973: viii).

In Tzotzil, an act that breaches community harmony is called *mulil*. It "covers the concepts expressed in the English words 'guilt,' 'crime,' 'sin,' 'blame' . . . , but a mulil is best understood as any act that displeases the gods and can provoke supernatural retaliation. . . . When someone commits a mulil, an explosive situation exists. Not only is the guilty person vulnerable to supernatural punishments, but there is a victim of the wrong who holds anger in his heart. . . . It still cries out for vengeance" (J. Collier 1973: 92, 93). The aim of conflict resolution is to come to an agreement by both parties about what needs to be done to right the wrong. This calms the hearts of all concerned and restores harmony. It seeks to avoid revenge that only leads to more mulil and more disharmony. A judicially assigned punishment would not necessarily calm all hearts. Two cases that may appear identical in the eyes of the Western legal tradition could be solved in radically different ways by the Maya arbitration method. This method is primarily concerned with the psychological dimension, the heart, and the restoration of harmony that is the goal of the covenant.

Similar to the Spanish medieval worldview, the Maya worldview made no distinction between religious and secular worldviews, laws, types of authority, but fused them together as elements of a unitary system. The council members usually had minor ritual obligations as a necessity for properly carrying out their civil functions. There was another set of officials whose sole duty was to perform covenant rituals for the maintenance and protection of the community. The *cargo/cofradía* systems were the

Maya response to the request of the friars for an organization to financially support the churches and the centers for teaching the catechism. The Maya already had similar organizations for their traditional gods from pre-Columbian times (Early 2006: 72). They added to them a new patron, a Catholic saint, and celebrated his feast day with a mass in addition to festivities used for their traditional gods (Cook 2000: 21–103). The parcialidades did much the same. The ceremonies also involved the shamans of the community (González 2002: 139–178). Called to their role by the ancestors in a dream or some other psychic experience, they conducted covenant rituals asking the emanated gods, including the saints, for protection of the community, especially by forgiving their transgressions. The shamans did not form an organized group, but acted on an individual basis, and were called upon when needed by individuals or by a unit of the social organization. Shamans could also occupy any role of the social organization. The introductory photograph shows a group of Zinacanteco shamans talking with the senior *alférez*, the head man of the cargo system who, in this case, was also a shaman.

Retention of the Traditional Culture

How were Maya communities able to retain so much of their traditional culture during the colonial and national periods? It was the result of both externally imposed and internally maintained isolation from Spanish and ladino society.

Spanish Avoidance of Maya Communities

From the Spanish point of view, there were practical as well as philosophical reasons for keeping their distance from Maya communities. The colonies existed to enrich the Spanish monarch and the Spaniards involved in the Conquest. The king needed resources to fight European wars. For the conquistadors, wealth was the necessary means for them to become upper-class Spanish noblemen. Most of them were not interested in becoming permanent residents of the colony; they sought merely a temporary presence to extract wealth. Later Spanish settlers came for the same reason. Some remained and sought to attain upper-class status within the colony itself. The implementation of these goals required control of the conquered people as a labor force to work the land and produce the wealth.

Table 3.1 shows the immediate problem faced by the Spaniards. In the early years of the colony there was one Spaniard for every 856 Maya. In

Table 3.1. Approximate Size and Ratio of Spanish and Maya Populations, Guatemala, 1550–1680

Year	Populations Spanish	Maya	Ratios Spaniards to Maya
1550	500	427,850	1:856
1575	2,500	236,540	1:95
1620	5,000	128,000	1:26
1680	5,000	242,020	1:48

Sources: Adapted from Lovell and Lutz 1994: 136; MacLeod 1973: 218.
Note: Spanish population totals were calculated by using the number of heads of households multiplied by five.

spite of the decrease in this ratio, as Maya populations were decimated by infectious diseases and as more Spaniards arrived, the Spaniards were greatly outnumbered. How could the few Spaniards control and extract wealth from so many Maya? It was done by indirect rule, an often-used mechanism of colonial systems. Forceful conquest, or the constant threat of it, brought about the submission to the king by Maya authorities, who continued in office from pre-Conquest times. "Insofar as colonial rule impinged on the Maya world, it did so through the Maya elite, who mediated almost every form of contact between Spaniards and the Maya masses. Their attitudes and motives were therefore crucial in determining the effect on the Maya of the many pressures, direct and indirect, exerted by the colonial regime.... Once the military conquest was complete, all the Maya leaders cooperated with the new rulers. They had no other choice if they wished to remain in power" (Farriss 1984: 96). This was the usual case, but there were occasional rebellions against Spanish rule.

These elites worked as consensus builders among the principales of the various kin groups in their area. "Each town had its own 'cacique' who was its ruler, but he did not have absolute power. By himself he could not execute any matter except that commanded by the district leaders. First he had to summon the heads of the descent groups. Only after discussing the matter with them and reaching an agreement, could it be executed. These heads of descent groups collected from each household its share of the tribute.... If someone was involved in a dispute or accused of a crime, the head of his descent group was summoned to act as his defender so that no injustice was done to him" (Ximénez 1929: 104).

Indirect rule allowed the Spaniards to live in the towns they built apart from the Maya. This isolation sprung from the Spanish view that equated town living with civilization and rural life with an undisciplined way of life similar to animals. This was the philosophical principle behind the policy of congregation, in addition to the practical benefits of tax collection and religious instruction (Early 2006: 135). The Spaniards looked down upon the Maya as a barbaric, uncivilized, lower-caste people. No Spaniard would wish to live among such people. The exceptions were the friars who lived in the Maya areas.

The Implicit Colonial Compact

The Spanish colonists and later Creole nationalists needed Maya labor to attain their goals—taxes for the king or the national government, production of agricultural products for export and the local food supply, and construction and maintenance of public works. As long as the Maya could obtain minimal subsistence from lands belonging to their communities, they would not voluntarily furnish this labor at any price. To solve this problem, the colony and national governments enacted a series of labor laws that forced the Maya to periodically work on non-Maya plantations and public works (McCreery 1994: 186–94). In the colonial and national periods, the Maya performed this labor under an implicit compact—a tolerable level of exploitation in return for being left alone in their communities the rest of the time. "By the early eighteenth century rural Guatemala was characterized by a laboriously worked out, if largely unarticulated, agreement between the mass of the population and the state and elites, a 'colonial compact' codified in custom. This custom, the product of centuries of struggle, defined a level of exploitation with which the communities and the rural population could live. After the collapse of cacao in the seventeenth century, the state and the landowning and office holding elites left the rural population largely to its own devices, subject only to the exploitative forms such as taxes and repartimientos [forced labor] mediated by the local community, and to the demand of formal adherence to ideologies of Crown and church" (McCreery 1994: 326). This outward adherence did not involve conversion to orthodox Catholicism, as most evangelization efforts failed (Early 2006: 115–89). For the colonial and national elites, this arrangement had the economic advantage of turning Maya communities into labor reserves. The demand for Maya labor varied with the type of crop and its agricultural cycle, but it was most acute at harvest time. During the remainder of the year, Maya returned to the isolation of their communities, where they maintained themselves. In this way, the Spaniards were not required to support a permanent labor force—a much more expensive factor of production.

The mechanisms of isolation described above continued well into the nineteenth century. "The Hapsburgs [1524–1700] had left the villages largely to their own devices so long as they paid their taxes and gave at least outward obedience to the dictates of the church. . . . The Bourbons [1700–1821] and the post independence Liberal [1821–37] and Conservative [1837–70] states largely failed in their efforts to promote commercial and export agriculture. . . . Left largely to their own devices, the Indians and castas . . . enjoyed an autonomy unrivaled since the sixteenth century" (McCreery 1994: 18, 48; see also 13, 19–24, 40).

Self-Imposed Isolation

The Maya themselves reinforced the externally imposed isolation from the Spaniards and ladinos. Maya left their communities under forced-labor drafts or to trade or visit other communities. These experiences could involve encountering gods and attitudes at variance with the local worldview, thereby raising possible doubts about it. The Maya worldviews took account of this possibility and erected defensive measures against it. If a Zinacanteco should be absent from his community for more than two or three months, the ancestral gods would appear to him in his dreams and tell him to return. If he refused after several such dreams, the ancestral gods would cease to take care of his animal counterpart, resulting in sickness and possible death of the absent person. If it was necessary for him to be absent for a longer period of time, he must ask permission of the ancestral gods. His family must perform rituals during his absence to ward off sickness and death (Early 1965: 192).

Another mechanism that mitigated the danger from temporary out-migration was the portability of the worldview. When Maya left their communities for outside work, they frequently went with and remained in a group from the same community, often kin. Together, they mentally remained within their customary worldview and continued to offer the traditional rituals that helped to ward off any threats from the non-Maya world. To permanently leave a community was a rejection of its gods. "When a person abandons once and for all his community, all the divine beings of that place feel offended because the parting implies a renunciation of the obligations to them, and they feel rejected because that person can now offer what had been theirs to some other group of powerful beings in another place. . . . [The gods] reason in this way: if you disregard me, I will disregard you; if you abandon me, I will abandon you and we will see who wins" (Arias 1991: 64). Sickness was sometimes seen as vengeance by the local gods. The covenant had been broken.

In later years, a threat to the traditional worldview arose from ladinos moving into Maya communities. They were usually middle-men, such as minor government officials, traders, and labor contractors for the plantations. To protect traditional community customs from their presence, these ladino immigrants could not receive an allotment of communal land. Later, when land began to be privatized, initially, alienation to ladinos was not allowed by community authorities. However these measures proved futile in the long run, as ladinos illegally seized Maya land.

When ladinos took possession of Maya lands and formed plantations from them, the plantations included the Maya communities that lived on them (Montagu 1970; Bobrow-Strain 2007). Their inhabitants now comprised the permanent labor force of the plantation and became the responsibility of the plantation owner. A system of feudal relationships was established. In exchange for their labor, the plantation owner allotted the Maya land for their milpas, medicines when needed, and protection against outside authorities and interests. All outside visitors needed the owner's permission to enter the communities. For the few required negotiations with government authorities, such as taxes, reporting of serious crimes, etc., all was taken care of by the plantation owner. Foremen for the owner assigned the labor tasks. Otherwise, the Maya communities were left to function in their traditional way guided by their own principales. Given the complete control of the owner and his foremen over the Maya, conditions varied. In the plantations studied by Montagu, the villagers were required to work one out of every three weeks for the plantation owner in place of a wage. Women worked in maintaining the Casa Grande (the Big House). Some owners were benevolent in their care of their subjects, but many were not. Among the Maya, stories abound of economic and sexual exploitation by ladino owners and overseers, often accompanied by physical abuse.

Generational Transmission of the Worldview

Given the social structures that protected the local culture, how was the worldview, with its moral order, transmitted within the community itself? It was primarily done within the family unit. Young children attended the various community and private rituals in which their families participated, as seen in figure 3.1. They would hear the prayers with the names of the

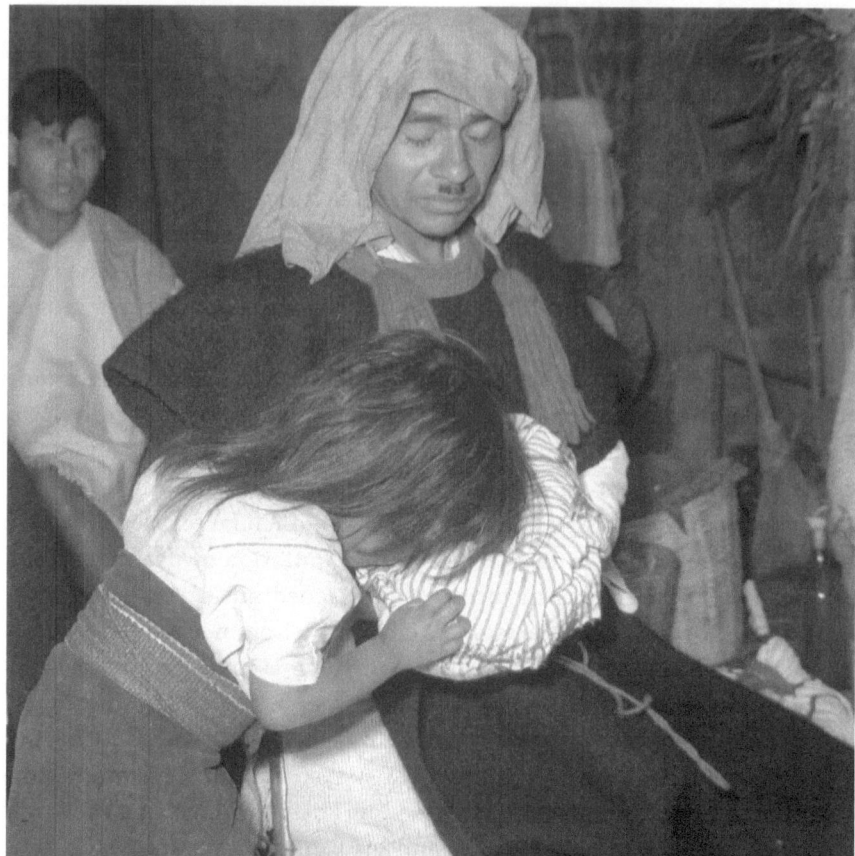

Figure 3.1. In Zinacantán, the Mayordomo Rey's young daughter kisses the bundle with the saints' necklaces. Maya are socialized at an early age in the community worldview.

emanated gods, of the ancestors, and learn the proper behavior the gods expected of them. The myths state the contents of the moral order in a concrete fashion. There is no abstract summary, such as in the Ten Commandments. From this writer's observation, the Maya moral order is similar to the Hebraic Ten Commandments applied to the conditions of Maya life. It included an etiquette of respect for authority. The myths frequently take the contents of the moral order for granted, as something everyone knows, and are more concerned with explaining the foundation of the moral order, or what happens when it is broken.

Grandchildren were often seen as the replacements (*k'e?s*) of their grandparents and usually named after them (Mondloch 1980). If the grandparents were dead, some saw their souls returning in the bodies of the new born. Grandparents were considered the most knowledgeable ones about the myths and stories comprising the worldview, with its moral guidance. Living grandparents were important in a child's learning the moral order, especially when left in their care while the parents were working. The grandparents symbolized all the ancestors, the way of life learned from years of experience, lessons they were now passing on to the younger generation. In this way, the children were socialized in the traditional way of life from a source with whom they usually had a bond of affection. This helps to explain the psychological power of the phrase "the wisdom of the ancestors." From it, the innermost psyches of Maya individuals drew their sense of self-identity at an early age. If later challenged, this identity was spontaneously defended, and any weakening of it caused psychological trauma.

Elaboration of the Traditional Worldview by Biblical Stories

The Maya worldview during the colonial and early national periods was not static. Over time, Maya communities reinterpreted various biblical stories using concepts of their own worldview, especially "emanations" and "cycles." These interpretations became part of the oral corpus of Maya myths. They were related as individual stories whose details were sometimes not clearly connected, or if so, contained logical disconnects that were of concern to no one. The main themes remained consistent. In chapter 2, an ethnographer noted these characteristics in the way the worldview was conveyed in Maya villages. Biblical names and accounts became additional cycles in the traditional post-Conquest worldview. This can be seen in a version from San Andrés Semetabaj, whose inhabitants are called Trixanos (Warren 1978: 33–41). Warren sees it as a creation model. This writer prefers to interpret it as an emanation model. The themes of the stories are guided by the purpose in relating them, here the "recognition of the individual as a moral being with an independent life cycle." The stories describe the unfolding of an undifferentiated cosmos by the emanations of more-differentiated beings, until finally arriving at individual human beings with the power of moral choice. This unfolding takes place during three epochs: God the Father, Jesus Christ, and the Spaniards. The following is an interpretation and highly condensed version of Warren's description.

In the first epoch, the abiding pantheistic One, called God the Father, emanates as a primeval, undifferentiated world, "the interpenetrating levels of the universe—the seas, earth and sky." "In the epoch of the Eternal Father all was united. There was very little land. . . . The world was an undifferentiated unity because the categories of existence interpenetrated each other. . . . The sky and earth were also united because God lived in both places, coming to the earth to direct creation and to oversee all that happened and returning to the sky when he was tired from his labors." Despite picturing the epoch as an undifferentiated cosmos, the stories do find a creature within this epoch who lacked intelligence, "had little knowledge of the world and no volition because good and evil were not clearly distinguished as separate forms of behavior." The Eternal Father was not satisfied with the cosmos in this condition. "In a world of incomplete order, his 'humble and obedient' creations became increasingly disobedient. . . . [This] was a natural tendency of the world and Trixanos offer no further explanation of it. There were periodic destructions of it. The Eternal father gave humans a long life of one hundred to five hundred years, but, as evil increased with age, God . . . destroyed the entire world by flood." Here, the biblical story of the flood is adapted to this context. Finally, God the father became bored with his labors and withdrew back into himself as the eternal One.

In the next epoch, God the Father emanates as Jesus Christ to continue the work of differentiating the cosmos. He begins to define the natural environment by separating the earth from the primeval waters and differentiating humans from animals. This epoch also saw the emergence of a moral order. A conflated biblical story is used as its explanation. Satan, an apostle of Jesus, tries to assume his powers of emanation. Jesus calls on the angel Gabriel to punish Satan by transforming him into a less-differentiated being with both animal and human characteristics. Christ would dominate in the heavens, mountaintops, and towns, where his followers would live, while Satan oversaw the wilds and a part of the underworld. The moral order is established by individuals now endowed with the ability to choose which path they will follow. Those who suffer but "adore God, behave according to His precepts and love their fellow men, are rewarded on earth with a long life, many children, and successful plantings and harvests." Those who "make a pact with Satan to gain riches without suffering through hard work [a ladino characteristic] . . . are transformed into human-animal beasts of burden, which must work eternally without rest."

Along with separating humans from animals, Jesus differentiated various "races"—the "rich," who were town-dwelling "foreigners"; and the "poor," who were "natives" living in the wilds wearing animal skins and feathers. Foreigners had knowledge of God and of the world through exploration. Natives did not know God and worshiped stones and roots as deities. This epoch was ended with the death of Jesus Christ.

In the third epoch, Spaniards were knowledgeable foreigners about the work of Jesus Christ. "They civilized the native peoples by introducing a school to impart knowledge and a religion centering on the worship of God and the saints so that humans would have a choice between good and evil. . . . They gave the Indians of each town a distinctive kind of clothing to take the place of the traditional dress of animal skins and feathers . . . introduced a new religion which centered on the celebration of God and the saints . . . renamed each [town] after the saint that became its guardian. Images of the saints were given to the Indians, who were organized into brotherhoods to celebrate the new religion. . . . . Yet Trixanos point out that in the process of civilizing the tribes, the Spaniards made people suffer and sacrifice themselves 'like beasts of burden' for the new religious and social orders. . . . It is said that only the priests treated the poor well by defending, curing and teaching them religion" (Warren 1978: 40). This version of the post-Conquest worldview was influenced to a much greater degree by the colonial catechesis than versions in some other Maya communities.

Therefore, the San Andrés mythology gives two accounts of the origin of the moral order—the traditional one that emphasizes the role of the ancestors, and the one just described that emphasizes the role of Jesus and the Spaniards. Warren (1978: 50) remarks that these accounts "may appear contradictory; but holding these contradictory ideas has allowed Indians a sense of continuity and separatism in the face of the domination that followed the conquest."

Summary

The previous chapter examined the Maya worldview as it first absorbed Spanish religious elements. This chapter has described how during the colonial and national periods the Maya were able to retain much of their traditional colonial worldview. Spanish social philosophy isolated Spaniards from contact with Maya communities. The implicit labor compact allowed the Maya to control their own villages. These external pressures for isolation were reinforced by the Maya worldview itself, which provided defense mechanisms against any danger to the culture as a result of outside contact. Communities elaborated more-detailed versions of the cyclical emanations of the cosmos and its moral order using biblical stories from the Catholic evangelization. This was evolution within the traditional worldview whose main characteristics the Maya retained for at least four hundred years after the Spanish Conquest. Later chapters will examine some aspects of its further evolution.

4

Growing Inability of Maya Communities to Provide Subsistence

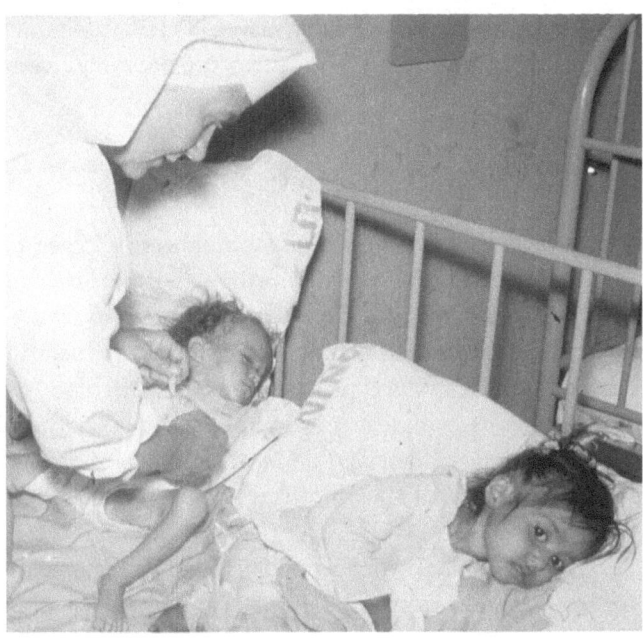

Beginning in the colonial period and continuing into the first part of the twentieth century, an accumulation of external and internal pressures on Maya communities finally resulted in their inability to obtain subsistence from what remained of their lands. The introductory photograph shows children suffering from malnutrition being cared for by nuns in the government hospital in Huehuetenango, an example of the result of this inability.

In the highlands, Maya communities occupied lands perceived as passed down from their ancestors. Many communities used adjacent piedmont and lowlands to supplement the lesser yields of the colder highlands. The method was slash-and-burn agriculture, primarily used for planting maize, beans, and squash. Land would be used for several years; then, as the yields began to fall, it was left in fallow, as the family moved to another piece of land not already in use or fallow. The fertility of land was an indwelling divine force that belonged to the whole community. This routine was broken by the time needed to perform compulsory labor for non-Maya outside the community. This was arduous work, but under the colonial compact, tolerable for the price of being left alone in their own communities. A family group, as members of a parcialidad, had the right to use any land not under cultivation at that time by another family. If any disputes should arise, the principales and the parcialidades would mediate them. Since land was identified with the community, no land could be transferred to anyone outside the community.

Acquisition of Maya Land by Ladinos

Under Spain's policy of indirect rule, no ladinos were allowed to reside in Maya areas. In spite of this prohibition, ladinos began moving into Maya territory. "Over the course of the eighteenth century, however, the calculus of Crown rule was altered by the demographic expansion of ladinos within traditionally indigenous zones—the Western Highlands. This trend received added impetus from a succession of earthquakes that destroyed Guatemala's colonial seat, Santiago de los Caballeros, during the 1770s, and prompted a mass exodus of ladinos toward the indigenous hinterlands to the west" (Reeves 2006: 141). It reached the Cuchumatán communities after 1871 (Davis 1997: 38). The purpose of this ladino migration was to obtain land, Maya land, since they composed almost all the population of these regions.

There is a long and convoluted history of land loss to ladinos (McCreery 1994: 236–64; González 2002: 35–106, 185, 199). But there was an underlying trend to its many twists and turns. Initially, a Maya community was required by law to obtain titles for their communal lands. They could be titled as "communal," but guided by the Enlightenment philosophy of economic progress, particular sections of the communal lands had to be assigned to specific family groups to ensure the land's productivity and to minimize disputes resulting from population pressures. Previously, a group simply used a specific area for a couple of years before abandoning it to fallow and moved on to another area not under cultivation.

Under the government's legal climate of titled private property, the assigned plots gradually came to be considered restricted private property. They could be inherited or sold to other members of the community, but under no circumstances would community authorities allow them to be sold to an outsider, either a Maya from another community or a ladino. Still later, as population growth, poverty, and ladino political pressures mounted on the shrinking Maya land base, full privatization, with the right of alienation to anyone, including ladinos, was accepted.

To circumvent Maya rule of their own communities, ladinos brought with them national laws and administrative directives, many of which were used to seize Maya land in an illegal manner. Later, they introduced national politics and political parties that were used for the same purpose. Land was acquired by wealthy nationals through fraud, intimidation, or purchase from a Maya seller in extreme circumstances. For the most part, the Maya were illiterate, had no understanding of national law, and were easily tricked into accepting documents whose legal implications were completely unknown to them. Although there was legal protection for indigenous lands, it was never enforced. Extreme circumstances often arose because of sickness made worse by increased labor demands and alcohol abuse. As a result of this history, many Maya either sold their now privatized holdings to ladinos, or were illegally dispossessed of them.

This pattern of land loss was accelerated by the Bourbons, who changed the payment of tribute from goods to cash in 1747. This amounted to an increase of the tribute tax. It required the Maya to spend more time doing forced labor to the neglect of their own fields, for the minimal subsistence they provided. The result was increased poverty, requiring some to give up what little land that had been the basis of their survival. Land loss was further accelerated in the latter part of the nineteenth century when coffee became the primary export crop for the nation's economy. Coffee was a more labor-intensive enterprise than the previous export crops and created an unprecedented demand for Maya labor. In Guatemala, the liberal government after 1871 responded to this need. "The governors of the departments were now to grant planter requests for agricultural labor drafts from the Indian communities. The law provided that they could dispatch groups of up to sixty workers at a time, for fifteen days if the property was in the same department or thirty if outside it. Orders could be renewed if the employer requested" (McCreery 1994: 188). Workers received small wage advances stipulated by the law. Sometimes this was immediately spent on alcohol. In addition, from these advances were subtracted the worker's living expenses while working on the plantations, so that many became indentured workers. This debt was often inherited by the family upon the death of the worker. These added burdens accelerated forced sales of land and eroded the colonial compact, along with the isolation it had afforded Maya communities.

Internal Population Growth

Table 4.1 shows the size and rate of change of the Guatemalan Maya population over a 482-year period. The table indicates three different trends: rapid population decrease to about 1625; a very slow increase until about 1950; then a more rapid increase into the early years of the twenty-first century. The rapid decrease during the early years of the colony was primarily due to infectious diseases unknowingly brought by the Spaniards. The conditions of forced labor imposed by the Spaniards added to the toll. From the arrival of the Spaniards in 1524, until 1550, the population decreased at an average annual rate of 5 percent per year, with at least four pandemics (MacLeod 1973: 98). At this rate of decrease, the population was being cut in half every fourteen years. After 1550, the population continued to decrease until around 1625, but at a slower rate. In this period, there were at least three pandemics and five local epidemics (98, 99). The average annual rate of decrease for all of the first phase, 1524–1625, was 2.58 percent per year, which translates into the population diminishing by 50 percent every twenty-seven years. In one community, Totonicapán, there were at least six epidemics during this period (Veblen 1982: 97). In addition to infectious disease, health conditions were poor, resulting in a population with weakened powers of resistance. The *encomienda* system frequently degenerated into serf-like conditions. Its attempted reform by the New Laws, in 1542, lessened some of its effects, but it remained a coercive labor system that increased morbidity and deaths of workers. Also, migrations occasioned by numerous failed efforts at congregation contributed to poor health conditions.

The second phase, from 1625 to 1950, was a 325-year period of very slow growth at an average annual rate of 0.78 percent. There were at least four pandemics and many local epidemics during this period (MacLeod 1973: 99, 100; Veblen 1982: 97; McCreery 1994: 149). Yet the population was doubling in size on an average of every eighty-nine years.

How were the Maya able to sustain the slow growth rate of the second phase, 1625 to 1950, in the face of continued epidemics and high mortality? In San Pedro Laguna, from 1810 to 1816, the crude death rate (per 1,000 population) was a very high 44.3. Yet during this same period, there was a small, 15.5 rate of population increase (1.6 percent). The reason was the very high crude fertility rate of 59.8 (Early 1982: 129–30). Santiago Atitlán in the 1940s had a high crude mortality rate (35 per 1,000 population), with an infant mortality rate of 130 per 1,000 births. Of all deaths in the decade, 67.7 percent were infants and children ages 0 to 4. But this was offset by a high crude fertility rate of 57. Married women who survived to

Table 4.1. Rough Estimates of the Guatemalan Maya Population and Percent Average Annual Rates of Increase, 1520–2002

Year	Population	% Average Annual Rate of Change	
		In Interval since Previous Listed Year	During a Phase
PHASE I			
ca. 1520	2,000,000	—	
1550	427,850	-5.01	
1575	236,540	-2.34	
1595	133,280	-2.83	
1625	128,000	-0.13	-2.58 since 1520
PHASE II			
1684	242,020	+1.09	
1710	236,208	-0.09	
1770	220,500	-0.11	
1778	248,500	+1.51	
1804	292,000	+0.62	
1820	350,000	+1.41	
1830	428,744[a]	+2.05	
1840	525,700	+2.96	
1850	592,900	+1.21	
1860	665,700	+1.16	
1870	756,000	+1.28	
1880	844,384	+1.11	
1893	1,005,767	+1.35	
1914	1,462,721[a]	+1.80	
1921	1,343,283	-1.21	
1940	1,560,000	+0.79	
1950	1,611,928	+0.33	+0.78 since 1625
PHASE III			
1964	2,185,679	+2.20	
1973	2,680,178	+2.29	
1980	3,320,393	+2.70	
1988	4,000,000	+2.71	
1994	3,566,042	-1.90	
2002	4,592,000	+3.21	+2.03 since 1950

Sources: Population figures are from Lovell and Lutz 1994: 136; and Lovell 2000: 170; 2005: 22.
Note: a. Author-derived by interpolation for missing (1914) or improbable (1830) data.

menopause had an average of 11.6 live births (98–102, 132). Therefore, it was the very high fertility barely offsetting the high mortality that resulted in the slight population increase.

But what accounted for such a high rate of fertility? In non-contracepting populations with extended periods of lactation as long as three years, and where there is high mortality of infants and children, the high mortality can paradoxically contribute to high fertility. The death of an infant (defined as a child in the first year of life) or of a very young child means that the mother stops nursing. As a result, the frequent but not infallible contraceptive effect of nursing ceases because prolactin is no longer secreted. Consequently, the mother will probably conceive again sooner than if she had continued to nurse a surviving child, given the probable continuation of the contraceptive effect. In other words, in these circumstances, the high mortality of infants and young children can lead to shorter birth intervals and the consequent possibility of a larger number of births within the female reproductive span than in a situation of low infant and child mortality. The contrast can be seen among the Hutterites, another non-contracepting, high fertility population. Their crude death rate was only 4.4 per 1,000 population, with a low infant death rate 45 per 1,000 births (Eaton and Mayer 1954). This low infant death rate meant longer birth intervals. Even though the crude birth rate was 45.9, it is considerably lower than the crude birth rates of almost 60 among the Maya populations.

Atiteco women try to control their high fertility (Nick 1975; Early 1982: 138–39). They attempt late-term abortions by consuming certain liquids, tightening a band around their stomachs, or deliberately falling down. Another method is "thinning of the flock." In Santiago Atitlán health workers have noted how parents would assiduously take advantage of medical and nutrition programs to save some children and completely neglect others. Families would deliberately withhold food and medical attention from some sick children in the first few years of life and let them die. This usually occurs when a child, more often a female than a male, is the closely spaced middle child between an older and a younger sibling. It is a way of attempting to cope with the lack of resources to raise so many children. Some must die so that others may live.

The third phase, from 1950 onward, was one of significant growth. If sustained, the 2.03 percent rate of increase means the Maya population will double in size every thirty-four years, rather than the every eighty-nine years of the previous phase. This is the result of Maya communities beginning to receive health services and to become involved in community development programs, to be described in later chapters.

Signs of Pressure on the Land

In addition to ladino encroachment, the internal growth of Maya populations created an additional pressure on the size of communal and private landholdings. With each generation, less and less land, or in some cases no land, was inherited by family groups. These smaller holdings also became fragmented due to different sources of inheritance or the way the inheritance was structured. Land cannot be distributed equitably simply in terms of area. Quality of soil, altitude above sea level, availability of timber and water, access to markets—all these and other factors must be taken into consideration. Eventually, a family's usage area can be fragmented into a series of small, distantly scattered plots requiring an inefficient use of labor. As population pressure builds, fallow periods are shortened or omitted, resulting in soil exhaustion possibly followed by erosion. The use of fertilizer, natural or chemical, will delay exhaustion, but its use on scattered small plots is usually not economical.

Figures 4.1 to 4.4 are a pictorial essay about the land shortage in Guatemala and Chiapas. These pictures were taken in the mid-1960s and show how the land shortage forced families to plant their milpas in such unfavorable locations as steep mountain slopes. Once a patient was brought

Figure 4.1. Santiago Atitlán is in the foreground on a small bay off of Lake Atitlán. In the background is the San Pedro volcano with its forested top section and relatively treeless lower section.

56 · Part II. The Background of the Crisis in Maya Communities

Figure 4.2. A closer look at the side of San Pedro shows the treeless lower section is configured by numerous Atiteco milpas.

into the clinic in Santiago Atitlán with a bad head wound. The writer inquired about what had happened. The reply: "He fell out of his corn field." When working under such conditions, men usually drive a heavy stake into ground, tying one end of a rope to the stake and the other end around their waists. If they should slip and fall, the rope will break their fall and prevent them from tumbling to the bottom of the mountain. In this case, something had happened to the patient's rope.

The Result: Land Crisis

The land still held by the Maya had been pressed beyond the limits of its ability to sustain the Maya population. Maya communities gradually lost the self-sufficiency that had allowed them to retain their culture for three centuries. Beginning in the 1930s, the land shortage was recorded in many Guatemalan Maya communities: in 1932 in Santa Eulalia and

Figure 4.3. A still closer look shows how being forced to farm in such unfavorable locations leads to erosion and deterioration of the soil.

Figure 4.4. In Chiapas, beyond Chiapa de Corzo in the Grijalva Valley, rise the highlands, with a Maya milpa on a steep slope of the mountain.

Table 4.2. Household Agricultural Production Units by Size of Holdings and Percentage of National Agricultural Lands, Rural Guatemala, 1950, 1973

Acres Classification	1950		1973	
	% Units	% Agric. Land	% Units	% Agric. Land
0: Landless	32.3	0	41.3	0
<17.3: Insufficient	59.7	14.3	52.6	16.1
17.4–111: Sufficient	6.4	13.3	4.8	18.6
112–1,120: Excess	1.2	21.0	1.2	29.8
>1,120: Superabundant	0.2	51.4	0.1	35.5
Total	100%	100%	100%	100%

Source: Condensed from Early 1982: 67, 69. For data sources, definitions, and assumptions, see Early 1982: 67–71.

Sololá (LaFarge 1947: 5; McBryde 1933: 33); in 1935–36 in Santa Catarina Palopó (Tax 1946: 124); in 1940–41 in Chichicastenango and San Antonio Palopó (Bunzel 1959: 88–90; Redfield 1946: 50); and in 1945 in the Quetzaltenango Basin and Todos Santos (Horst 1956: 163; Oakes 1951: 40). Continuing into the 1960s, shortages were noted in San Marcos, San Juan La Laguna, and San Lucas Tolimán (Rojas Lima 1968: 291, 310; Woods 1968: 207); in 1966 in Santiago Atitlán (Douglas 1968: 247; Early 1970b); and in 1967–69 in Sumpango and Patzún (Elbow 1972: 46, 183, 185). The total impact by 1950 can be seen in table 4.2. It estimates the size of land holdings for the rural population of Guatemala, including both Maya and ladino. The Maya profile would probably show a slightly more precarious position if the data had distinguished Maya from non-Maya. In 1950, 32 percent of rural households were landless and another 60 percent had holdings insufficient to absorb the normal productive effort of a rural family to provide for their needs—two adults working the greater part of the year using the normal agricultural practices of the region. Therefore, at least 92 percent of Maya families were unable to sustain themselves by agriculture alone in their traditional communities. This was the combined result of losing land to ladinos and internal population pressures.

The social injustice of the system can be seen in the extreme imbalance of land holdings. Table 4.2 shows that in 1950 and 1973, the Guatemalan rural population with insufficient land for self-sufficiency held only 15 percent of the national agricultural land, while those with excess and superabundant holdings held between 65 percent and 72.4 percent. Within this latter group, plantations of more than 1,120 acres were owned by only 15 percent of the population but occupied 51.4 percent of the national agricultural land.

The same picture emerges for Chiapas, which was part of Guatemala until 1821, when it became independent for a few years and then joined Mexico. Soon after 1821, ladino plantation owners took over lands that had been property of the Spanish Crown but traditionally farmed by Maya communities. In the 1890s, the state government, with the backing of the national government, decreed that indigenous communal lands should be sold, although the decree was not universally enforced. Where it did occur, usually ladino buyers created plantations, forcing the Maya to work on them (Rus and Collier 2003: 58, note 4). By 1900 the Zinacantecos had lost 40 percent of their ancestral lands. They recovered some of it as *ejidos* under the Agrarian Reform Act of 1940. Combined with their communal lands, which they had been able to retain, they enjoyed a brief period of self-sufficiency. But due to population growth they were again land-short by 1960. Chamula was less fortunate. It received little land under the 1940 Agrarian Reform Act. By the 1940s and 1950s, the vast majority of Chamulans held less than two-and-a-half acres. By 1980, 40 percent were without land (37–38). Situations similar to Zinacantán and Chamula existed in all the highland municipios of Chiapas. A mere 2.4 percent of landowners held 25 percent of the available land, while 50 percent of landowners, mostly Maya, held less than 1 percent of the available land. Each of forty-four plantations had 57,000 or more acres (Benjamin 1996: 226). Extreme latifundia and minifundia exist in symbiotic relationship to each other. The latifundia exist by the use of agricultural labor supplied by Maya forced to do plantation labor at unjust wages due to minifundia conditions of their home communities. With cheap labor available, by the 1980s there was no incentive for plantation owners to invest in agricultural technology and increase production. It became a stagnant system supported by the poverty of the Maya.

Alternative Sources of Subsistence

Faced with very little land from which to draw their traditional subsistence, the Maya turned to outside sources for work. In 1934, the labor laws in Guatemala were overhauled and all debt servitude abolished. A new set of laws were passed, the Vagrancy Laws. Unless an agriculturist could prove that he worked three or four manzanas of his own land (one manzana can vary between 1.75 and 2.33 acres), he was required to do wage labor on the plantations for a fixed number of days each year. Many Maya could not meet these requirements. When Arévalo became president in 1945 and initiated his social revolution, the Guatemalan Congress did away

Figure 4.5. Atiteco workers depart to work on the coastal plantations.

with all forced-labor laws. "Guatemala was the last country in the Western Hemisphere to end state-sanctioned coerced labor" (McCreery 1994: 322). With insufficient land, the Maya needed cash for their purchases of maize, clothes, animals, house materials, medical expenses. For many, cash could be obtained only by working as agricultural laborers on the commercial plantations in the piedmonts and lowlands (figure 4.5). The Maya became part of the market economy on the lowest rung of the national society.

Others sought wage labor in many occupations previously closed to them, but always at the lowest entrance positions. A merchant tradition had long existed among the Maya. Now, more became merchants, and those who were already established, expanded their operations. Some became quite wealthy (Cancian 1992). These experiences in the national marketplace took the Maya into a new cultural world, one that was not structured by the Maya covenant and reciprocity agreements. In many respects, it was a contradiction of them.

While for most Maya poverty was the result of the land loss, some profited by it. They bought land from their impoverished neighbors and used it to solidify their position of greater wealth. In Aguacatán, irrigation machinery and trucks were acquired to produce garlic and onions for the national and international markets (Brintnall 1979). In San Antonio Ilotenango, merchants expanded their inventory and purchased trucks to greatly expand their network of markets (Falla 2001 [1978]). In Panajachel, Maya helped build an infrastructure for a tourist trade and lakeside chalets for the wealthy national elite (Hinshaw 1975). "The economic transformation, in short, created a multitude of potential sources of tension within the traditional order. It eroded the economic underpinnings of Ladino dominance and Indian subordination; it created major socio-economic differences among a people that had once shared a common oppression and a unitary system of political and ritual life, and it subverted the social realities implicit in the custom of land inheritance by installments [described in the next chapter], a practice which was the economic basis for the age-grading of hierarchies, the respect for age and the cult of the dead" (Brintnall 1979: 116).

Summary

This chapter has shown how many Maya agriculturists no longer had sufficient land to fulfill their traditional role in the Maya covenant as the provider of maize to feed their families and the gods in ritual ceremonies. They were forced to seek any type of wage labor, often outside the home community. Usually, it was as laborers on commercial plantations or as small-scale merchants in the national economy, both areas whose cultural logics were quite different than those of the traditional community. "Ya los santos no dan" (The gods no longer provide). This complaint was often heard by the writer in the 1960s. The covenant was no longer functioning. The traditional worldview based on communities of small-scale subsistence agriculturists was entering a period of crisis as it provided no orientation in a culture becoming more and more market defined.

5

Long-Standing Strains within Maya Communities

Maya culture was never a social utopia, as it is sometimes depicted in ethnographic and advocacy accounts. As the Maya writer Montejo (2002: 125) observed, "I am willing to criticize my own culture because like any other civilization, it has its own problems and little hidden monsters and demons." While the Maya remained relatively isolated from the national society and the traditional worldview remained intact, these internal problems were tolerated. Once the subsistence crisis had precipitated questioning many aspects of the culture, they were no longer taken for granted.

Dominance of Elders: Conflict of Generations

The traditional Maya community was a male gerontocracy with the principales as the ultimate authority of the community. To be a principal, a man had proven his ability by having successfully occupied a series of age-graded roles in the cargo/cofradía system. There were no young men catapulted to the top of the system in a short period of time. But that did not stop some of the younger men, especially those with ability, from wishing they could be. They would chafe under the dominance of the elders, but in the traditional community there was little they could do about it.

Perhaps the most frustrating situation for some young men was the dominance of the family father. Under the communal system, the father controlled the land allotted to the family and the work on it. With privatization came personal inheritance (Brintnall 1979: 83–85; Guiteras-Holmes 1961: 38–40). In most Maya communities, land was inherited only by sons. The delicate question became, When would the inheritance take place? Some fathers would dominate to the end by not giving land to his sons until near death. This meant that they worked for him until that time. At the other extreme, some fathers would loosen their hold by giving a son his inheritance a year or so after marriage, although the family often continued to work together as a unit. Still others doled out the inheritance in installments as resources for the sons rising in the age-graded hierarchy. Another aspect of the inheritance problem was concern about the fairness of the distribution. With the land shortage and fragmentation of holdings, frequently it became impossible to make an equitable distribution because of the small size of the plots, differing qualities of soil, location, rainfall, etc. In addition, favoritism toward a particular son was another source of tension.

Difficulties also arose over the dominance of elders in the selection of a marriage partner for their children. Traditionally, the parents of the male would select a girl to be his wife. With the help of a marriage petitioner, negotiations would begin with her parents until an agreement was reached and sealed by extensive gifts to the future bride's family. In some families, a young man or woman was allowed to express his or her wishes, which might or might not be seriously considered by the parents. To escape the authority of their parents, some couples would elope by leaving the village for a short time. Later, they would return as a married couple and attempt a reconciliation.

Expenditures for Covenant Rituals

As the powerful gods of the covenant were mortal, they required ritual offerings in order to survive their allotted cycle. These were to be given and used with a grateful heart, which was to be shown by an ostentatious abundance of candles, incense, food, music, dances, and alcoholic drinks (Early 2006: 22–41). Prestige in a community was gained by the amount expended for its festival calendar. It was a costly undertaking primarily supported by the *cargueros/cofrades*. In Panajachel the most expensive cargo costs considerably more than a man could earn in a full year of wage work. In Zinacantán the most expensive cargo required the equivalent of ten years' wage work. In San Pedro Chenalhó, one of the lesser religious offices implies an expenditure equivalent to three to four years of a man's daily wage. Higher positions demand that the Pedrano expend four or five times this amount (Cancian 1967: 288). There are accounts of men selling their land to raise the necessary funds.

A concomitant problem was the amount of time required to fulfill the obligations of cargo service. In Zinacantán, a cargo holder spent about 40 percent of the year performing ritual duties (Early 2006: 19). This was time taken away from cultivating his maize fields, and involved additional expenditures to hire laborers in his place. All this resulted in poverty, hardship, and suffering for a number of families. It raised questions about the validity of a worldview that imposed such a burden.

Additional expenses for rituals were paid to shamans for their ceremonies, especially curing ceremonies. In Maya communities, shamans were regarded with a certain amount of ambivalence. As a result of their shamanic power, all shamans were looked upon as potential performers of sorcery rituals that would result in sickness or death. Also, there were questions about their ability to cure. A resident of Santiago Chimaltenango observed, "If the *chmaan* [shaman] had been true, they would have gladly been given their tortillas (payment), but many times they just drank their tragos (cane liquor) and didn't do a thing" (Watanabe 1992: 207). Another observed "When we went to *chmaan* . . . sometimes they would say that the sick person would live, other times that they would die. Sometimes the *chmaan* were right, sometimes they were wrong. They didn't know" (207). Another skeptic remarked that a *chmaan*, hidden in the enclosure surrounding his *mees* [divining] altar and covered with a blanket in a darkened house, could easily fake an encounter with an emanated god (Early 2006: 246–48). "Once when I was young, I went to a *chmaan* with my mother because one of her cousins was sick. He got under his blanket and began calling the *witz* [mountain gods]. I don't know how he knew so many names! Then he changed his voice and began speaking in Spanish. The *chmaan* called me by my older brother's name since he had accompanied my mother on her previous visit. When I told the spirit I wasn't my brother, he said that I should have more respect!" (Watanabe 1992: 208). Some shamans were leeches. In the crisis of a drawn-out illness, some would continually prescribe expensive curing ceremonies that the distraught family thought necessary for recovery of the patient following the logic of the covenant. The abuse would cease only with the death of the family member or the family's inability to continue paying (Falla 2001 [1978]: 118, 169). Instances such as these raised doubts about the validity of the shamans' claims and consequently about the worldview on which they were based.

Witchcraft

Most traditional Maya communities believe in witchcraft as part of their worldview. As seen in chapter 2, powerful curers can perform rituals in which they talk to the emanated gods and seek favors from them in return for ritual offerings. The emanated gods control the powers of nature, keeping them in balance so that a community prospers with an adequate food supply and protection against diseases. But this process can be reversed. Powerful curers can make offerings that ask the gods to disrupt the harmony of the bodily processes of an individual person, leading to sickness and possibly death (witchcraft). It is a common belief that many sicknesses are caused by witchcraft.

If a person believes that their sickness is the result of witchcraft, then that person may kill the curer who performed the witchcraft ceremony or the one who requested the ceremony. If the person can convince the community that they have been a victim of witchcraft, there will be no punishment for the homicide. It is justifiable self-defense. Witches and those who resort to witchcraft are seen as threats to the community. There appears to be great variation in the practice of witchcraft in Maya communities. Nash (1970: 245–47) studied a Maya community in Chiapas where witchcraft was a common practice. She examined thirty-six homicide cases from 1938 to 1965 and found that 53 percent of them involved witchcraft accusations, and that 25 percent of the persons killed were curers.

As seen in chapter 3, Maya communities have deviant behaviors typical of communities worldwide. These can lead to hatreds and the need to avenge. Also, traditional Maya communities, like many peasant communities, perceive that the amount of the desired things in life is fixed, limited, and cannot be expanded (Foster 1965). Therefore, if someone in the community is perceived as substantially increasing any of the desired things, especially wealth in any form, it means that they must be taking it from someone else in the community. It is a zero-sum situation. This leads to dislikes, hatreds, and envy of those persons perceived to be bettering themselves. They are seen as threats to the rest of the community and may become objects of witchcraft.

Alcohol Abuse

During the performance of the extensive rituals by both cargueros/cofrades and shamans, abundant amounts of cane liquor were consumed. In Zinacantán, at the beginning, middle, and end of each segment within a ritual, a shot glass of cane liquor was poured for each ritual performer and those assisting by an attendant as his sole function during the ceremony (Early 2006: 25). As the festivals lasted three or four days, at the end, many were marked by drunken cargueros/cofrades sprawled on the church steps or in the town square or in front of the cofradía house. Others, barely able to move, were helped home, with their wives holding them up on one side and their children on the other. In some communities, where the women drank in the same fashion, the scene was chaotic. The introductory photograph depicts a painting of a typical scene in the cofradía housing the Atiteco god, Maximon, who will be described in chapter 8. A close-up in figure 5.1 shows one of the attendants with a bottle of cane liquor in hand and two empty at his feet.

In Guatemala, the cofradía houses were as important as the church for festival rituals. "In order to defray the expense of fiestas the cofradías put on *zarabandas*. For these they hire a *marimbero* to play, and they sell drinks and dance in the *sala* where the saint is kept. The principal drink served on this occasion is *chicha* [corn liquor] which is made by the women in the cofradía. They also buy or make *aguardiente* [sugar cane liquor] for sale on these occasions. The *sala* is cleared of all furniture except the altar and the benches along the wall. In one corner is the marimba. . . . Near it is the counter where the drinks are sold. . . . Both men and women drink a great deal, and become very drunk. There is a great deal of laughing and shouting and erotic behavior. . . . For the drinking in the cofradías is an act of worship, like dancing or singing in other places. The gods want to see people happy, the joy of drink is, therefore, pleasing to them" (Bunzel 1959: 255–56). In San Pedro Chenalhó, "when we talked about rum [cane liquor] in traditional fiestas, like Carnival, Domingo and Antonia told me

Figure 5.1. A close-up of the cofrade in the introductory photograph consuming three bottles of cane liquor in the Maximon cofradía house.

that in this context, rum is not the Devil's but God's. 'It's always been our tradition,' they said, 'When leaders of the fiestas drink, they do so because God wants them to. God wants the fiestas to be joyful. Merriment and rum come from God.' they said" (Eber 2000: 56).

While festivals were the occasions for several days of extended drinking, alcohol played a pervasive role in daily Maya life as well. All family celebrations and rituals involved drinking. Every time a person asked a favor of another, it was preceded by a gift of a bottle of cane liquor, and the agreement was sealed by drinking some or all of it. Friends meeting on the trail would often exchange drinks from a bottle carried at the hip in a small net bag.

It is difficult to know how many Maya men were compulsive alcoholics, as distinguished from drunkards. It is also difficult to know if abusive drinking was more prevalent in Maya society than in others. But there can be no doubt that it was a serious problem. Many days of subsistence work were lost to the effects of alcohol. "Since the wage scale is low and liquor is relatively high, one prolonged spree can wipe out an ambitious man's savings for a whole year and leave him with debts to work off during the next. During the fiesta in July, 1937, several men spent six to ten dollars on *aguardiente* [cane liquor], as much as they could earn at one coffee harvest, sixty to a hundred days of common labor in the village, or from six to ten *quintales* of maize sold at the moderate price of one dollar per hundred" (Wagley 1969 [1941]: 77). A quintal equals 100 pounds. This describes a spree at one fiesta. These figures need to be multiplied by the number of festivals in a year.

As in any society, chronic abuse of alcohol created a web of destruction beyond the individual drinker. When the abuser was male, its most immediate impact was on his wife. She had to become the family provider, going out and obtaining maize by working the fields herself or imposing on kin reciprocity. Further destruction involved repeated cases of wife beatings and other abuses by drunken husbands. "Marital disputes typically flare up when the huband comes home drunk and beats his wife . . . almost any type of marital discord will be attributed to wife beating before deeper causes are sought" (J. Collier 1973: 11). Sometimes the beatings included small children. Marital infidelities also took place in these circumstances, putting further strains on family relationships.

The web of destruction went still further. Alcohol abuse was one of the means by which ladinos kept the Maya subservient and provided cheap labor for plantations. At fiestas, labor contractors set up small stands near ladino liquor stores. Maya in need of cash for liquor borrowed money from the labor contractor in return for signing a work contract to go to the coastal plantations for thirty days or more. During that time he would work off the debt. Consequently, he would return home to his family with nothing to show for his work. Sometimes he would incur additional debts on the plantation for his meals or more alcohol. At that point, stealing, as a means of coping with debt, might be added to the destructive web.

The legitimation for this excessive use of alcohol was the Maya covenant. In return for the protection of the gods, the Maya must feed them during the festivals. Cane liquor, chicha, or some other kind of fermented drink was one kind of required food. The carguero/cofrade must celebrate the festival with a joyous heart, otherwise the offerings would not be accepted. For many, a joyous heart was shown by the provision and consumption of excessive amounts of alcohol. As the Chenalhó woman earlier quoted said: "God wants the fiestas to be joyful. Merriment and rum [cane liquor] come from God." In the isolated traditional community, the abuse of alcohol and its web of destruction was tolerated, often explained away with the words "That is the way things are and should be." This is not to say that all in the community accepted the custom. But as long as the community maintained its relative isolation and protection of traditional ways, there was little that could done about it.

The Position of Women

The life of the traditional Maya woman was one of constant work. Before generators and electric milling machines were introduced, preparing tortillas by hand from corn kernels took at least five hours a day. "Their whole day is taken up with work. The wife of a Maya farmer rises at three in the morning to make corn posol and breakfast for the men. . . . She spends the whole day this way until night comes, seven days a week. . . . 'We begin as small children to carry our younger brothers and to help with grinding corn, making tortillas, sweeping the house or helping with laundry. There is no opportunity to go to school even if there is one in the community'" (Rovira 1994: 220–21).

Women were seen as subordinate to their husbands. In the village or on the trail, a woman walked behind her husband, had little to say in mixed company, and avoided eye contact with others while covering her mouth with a shawl. At meal time, men were served first. Few women spoke Spanish. But there was not complete subservience. Some Maya women had a strong sense of self-identity and independence.

> The majority said that at least some times in their marriages they and their husbands made all household decisions jointly. Moreover, women have well-honed market skills as traders and customers. In their homes or in private groups, women's lively talk and opinions are noteworthy, and their sense of self is demonstrated in their humor, aesthetic sensibilities, and their bodily presences. Their once important public roles either in the community alongside their husbands in the cofradía or as active agents in their own right as Maya priests, healers, midwives, and mediators in traditional marriage patterns reveal their social independence and worth as members of their households, families, and communities. (Green 1999: 98)

However, for a number of Maya women, this independence and worth meant nothing on the occasions when they were beaten and abused by a drunken husband. In some households, the beatings took place whenever the husband was displeased about some household matter (Freyermuth Enciso 2003: 37–40). The situation was made even worse when the women themselves abused alcohol to celebrate festivals or to escape family problems (Eber 2000: 115, 121–22, 127).

Strains from Increased Exposure to the Market Economy

Once the Maya community could no longer provide independent subsistence, cash became a necessity. Men were forced to go outside the community to seek work for wages.

> The balance between subsistence and wage labor has shifted, leading to women's increasing dependence on men as wage earners. As cash has become integral to the viability of the household economy—essential for shoring up the entire bundle of economic tactics—the opportunities to earn income available to men and women differ vastly . . . has negatively affected rural women's well-being . . . resulted in increasing social vulnerability. . . . its [women's domestic work] contribution to household survival has been devalued and marginalized as cash has taken on increasing importance in the domestic economy . . . [leading to] their husband's irresponsible behavior toward their families. . . . The women say that they had little choice but to accept the situation. (Green 1999: 96–97)

The Ubico regime introduced two changes of great significance for the Maya. The stranglehold of the plantation owners, ladino labor contractors, and money lenders on Maya communities was broken by the cancellation of all debts of agricultural workers, thereby eliminating debt servitude. This freed the Maya to seek whatever type of work they could find.

The other important change was the construction of a national road system. This allowed more Maya to become merchants, and for those already engaged, to expand their market reach. Merchants bought trucks of their own and became sophisticated businessmen (Falla 2001 [1978]: 49–74, 100). They began to understand the universal logic of the marketplace—that any good or service was simply a commodity governed by the blind laws of supply and demand. This contradicted the traditional logic of reciprocity. In their travels to other towns, especially Guatemala City and Quetzaltenango, they came to view their own communities as backward in their ways. This set them on a path to modernize their communities that would clash with the dictates of the traditional worldview. Other Maya seeking wage labor traveled over a wider area. They became familiar with more acculturated Maya communities and their adaptations to ladino culture. They also began to think of introducing acculturating ways in their own communities.

Conclusion: The Desire to Modernize

Every culture has internal strains that are tolerated as long as the stronger cohesive forces of the culture remain in place. Among the cohesive forces, a long-accepted worldview is one of the most important. This chapter has described some of the internal strains in the traditional Maya communities. But given the land-subsistence crisis seen in the previous chapter, and the weakening of the traditional worldview as a result of it, these strains began to break through the customary restraints and cause some to seek alternatives. To escape dishonest shamans or those who were ineffective, some sought other ritual specialists who called on stronger gods. Others turned to the services of health clinics where they were available. Some young men sought to escape the gerontocracy of the elders or their fathers by actively seeking a place in the national economy, and the financial independence it would furnish. The economic pressures of cargo service caused many Maya to seriously questioned its validity and that of the Maya covenant for guaranteeing the food supply. Numerous Maya began to seek alternative ways of coping with their subsistence crisis.

6

The Maya Crisis and the Search for Answers

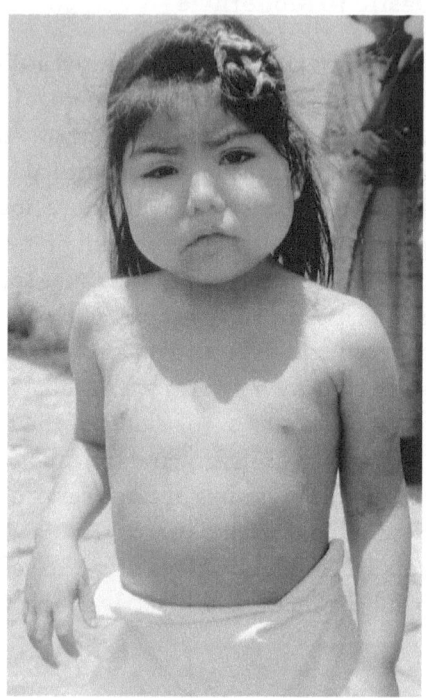

Given the population increase of both the Maya and ladinos, given the intrusion of ladinos with their culturally different legal and political systems into Maya communities, given the Maya loss of land and self-subsistence, given the resulting sickness and high mortality, given the consequent questioning of the Maya covenant, given the search for alternatives to some long-standing strains within the Maya community itself, and given the involvement of Maya in the wage economy, by the mid-twentieth century the traditional Maya culture was in crisis.

An Individual Copes with the External and Internal Pressures

An autobiographical account by Rudolfo Ruíz Santis from Magdalenas, Chiapas, puts flesh-and-blood on the abstract description of the various pressures confronting the Maya during this time (Chojnacki 2004: 21–64). Rudolfo's father was a Maya campesino who had three sons and two daughters. He had inherited a small amount of land, about two and a half acres, from which it was impossible to obtain yearly subsistence for his family. Consequently, his father rented additional land. Still unable to meet the needs of his family, he had to borrow money from labor contractors, to be paid off by working two to three months of the year on coffee plantations for five dollars a day. This amount was frequently reduced for his liquor expenses, as he was a heavy drinker. Rudolfo's father was deeply involved in the civil-religious hierarchy of his community, performing and financing rituals in eight positions, including that of the highly prestigious and expensive alférez.

The oldest of his three sons, Rudolfo, was born in 1959. Rudolfo attended the primary school in town. Every day after school he went to the milpa to help his father. After repeating several grades, he graduated from the sixth grade at age seventeen. During school vacations, he walked several days with his father to work on the plantations. He remembers the taunts as they passed through ladino towns in their traditional dress with their machetes and tortilla bags in hand. Life on the plantations was difficult. They slept on wooden planks in dirty dormitories, rose before dawn, ate some beans and tortillas, started weeding and cleaning the fields at dawn, and worked until mid-afternoon.

> I prayed to God. Why do I suffer like this? Sometimes I cried while I worked. There was so much sickness at home, too. My father drank because of his cargos, and he had lots of debts. We had to buy corn and medicine and pay the *curandero* [shaman], and the debts piled up. My father didn't think, he only drank. Why did I have to leave home, leave my mother? Some of it was my father's fault, but some of it had to do with sickness. It was because of *"la pobreza,"* the poverty. Anyway that's how I explained it to my father. I was the oldest son, so I spoke to him to try to get him to stop drinking. When he was drunk he beat my mother and threatened and hurt people with his machete. (Chojnacki 2004: 25)

The loss of land to ladinos, cargo expenses, drinking with its web of destruction, debts to labor contractors, and poor health were all tied together. Rudolfo, like many Maya, wondered, "Why do I suffer like this? What can I do about it?"

Shortly after graduation, Rudolfo's father arranged for his marriage to Margarita. His father gave him seven or eight cuerdas of land (0.83 to 2.3

acres) as his inheritance. To provide for his growing family of three girls, Rudolfo, like his father, had to work on the coffee plantations. Rudolfo spent a year performing community service in the lowest religious cargo, as a mayordomo in charge of church maintenance. Later, he performed service in the two lowest civil positions, as policeman and regidor. He was poor, struggling. While it may appear that he was following the traditional pattern of community life, other influences were at work. These will be seen in chapter 15.

The Extent of the Maya Crisis

The crisis is frequently referred to as "Maya poverty" or "a subsistence crisis." This implies that the crisis was essentially an economic crisis, a lack of resources to acquire physical goods, especially food. While the Maya crisis had its taproot in their loss of land and the self-sufficiency the land provided, the crisis was much more embracing. It was a crisis of physical survival, of the legitimacy of the Maya's social organization, and of the validity of their worldview from which they drew life's meaning, morality, and their self-identity. In brief, it was the crisis of a culture.

Crisis of Physical Survival

Land loss, abusive labor practices, low wages, alcoholic excess, population increase: the cumulative results of these and additional problems translated into high mortality for Maya families. A life table is a powerful summary, not only of health conditions within a community but of the impact on the community of the total social system in which it is embedded. As shown in table 6.1, the Maya life expectancy at birth (e_0) in Santiago Atitlán in the 1940s and 1950s was 30 years of age. This figure is confirmed by the slightly higher life expectancy, 34.5 years, for the department of Sololá, of which Santiago Atitlán is a part. Sololá as a whole includes a small but better-off ladino population, which slightly elevates the departmental average compared to Santiago Atitlán (Early 1982: 98, 115).

An inspection of the table for the 1950s and 1960s shows that the high infant and child mortality rates (the first four years of life, $_4q_0$) are most responsible for this low life expectancy. From a group of children born in any single year, 13 percent die in the first year of life. The high mortality continues for the next four years. By the time the group reaches five years of age, about 40 percent of these children have perished, only 60 percent surviving (l_5). The high mortality continues, so that by the age of twenty,

Table 6.1. Life Table Functions, Santiago Atitlán, 1940–1969

	1940–49			1950–59			1960–69		
Age	q_x	l_x	e_x	q_x	l_x	e_x	q_x	l_x	e_x
0 (at birth)	0.131	100	29.9	0.130	100	29.5	0.111	100	41.6
1	0.108	87	33.3	0.116	87	32.9	0.054	89	45.8
2	0.083	78	36.3	0.098	77	36.2	0.032	87	47.4
3	0.070	71	38.6	0.071	69	39.0	0.041	80	48.7
4	0.050	66	40.4	0.057	65	41.0	0.028	77	49.8
5–9	0.136	63	41.5	0.128	61	42.4	0.078	75	50.2
10–19	0.059	54	42.6	0.061	53	43.2	0.035	69	49.2
20–29	0.079	51	34.9	0.086	50	35.7	0.056	67	40.8
30–39	0.127	47	27.4	0.135	46	28.5	0.086	63	32.9
40–49	0.196	41	20.6	0.194	39	22.1	0.125	57	25.4
50–59	0.310	33	14.3	0.339	32	16.1	0.256	50	18.3
60–69	0.693	23	8.3	0.508	21	11.5	0.461	37	12.7
70+	1.000	7	5.9	1.000	10	8.1	1.000	20	9.0

Source: Condensed from Early 1982: 98.

Notes: q_x is the probability of dying within the age interval.
l_x is the number (or percentage, if read as decimals) surviving since birth at the beginning of the age interval.
e_x is the years of life expectancy beyond the beginning of the age interval.

50 percent of the original birth group are dead. For those who survive, their weakened immune systems and their strenuous work routines also take a toll. By age fifty, only 32 percent of the original birth group are still alive, and by age sixty, 20 percent.

What is the cause of so much early death? The underlying biological cause is malnutrition. A study of the dietary intake of both Maya and ladino children aged 0–4 from lower socioeconomic families in rural Guatemala found a 58 percent deficiency of calories (Early 1982: 107). For Maya groups, the percentage would be higher. In Chiapas, an estimated 90 percent of the children were malnourished (Benjamin 1996: 231). Figures 6.1 and 6.2 show in greater detail the results of malnutrition, seen in the frontispiece. It is a case of marasmus due to protein and caloric deficiency. The body begins wasting away as it turns in on itself and feeds off of its own body fat and muscles. Bone growth is retarded. Many such cases end in death. This was not a rare case in the Maya area. The boy is held by his older sister, through whose devoted care and that of the Santiago well-baby program, he managed to survive. Many do not.

The introductory photograph to this chapter shows a girl suffering from another type of malnutrition, kwashiorkor. This is due to protein deficiency with barely sufficient caloric intake. Kwasiorkor upsets the body's

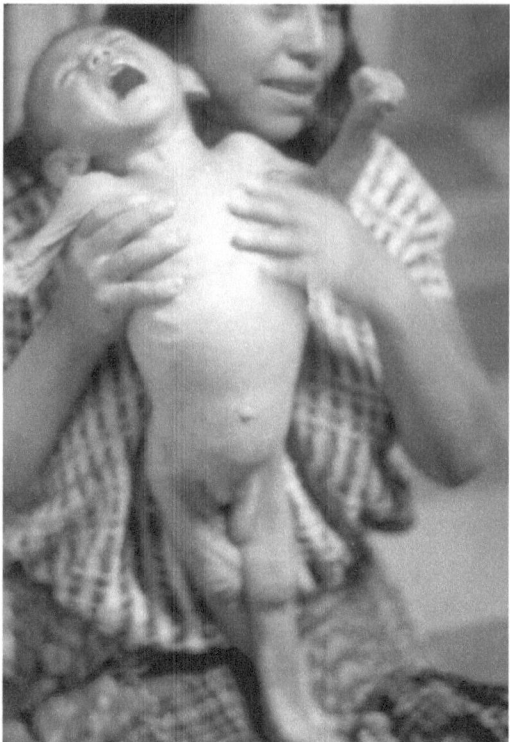 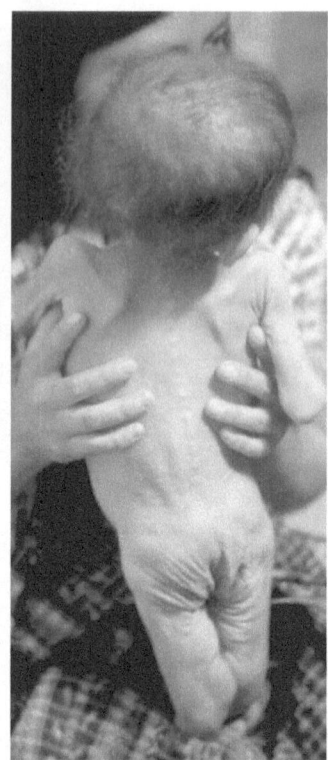

Figure 6.1. A frontal view of the child in the frontispiece shows the results of marasmus.

Figure 6.2. A rear view of the same child showing the ravages of malnutrition.

metabolism, resulting in excess fluids that build up under the skin. This creates the puffiness seen in the girl's face and body. The fluid can contain blood that appears as red blotches, barely visible as darkened areas on the girl's upper left arm. This girl had just returned to Santiago Atitlán with her family after working on a South Coast plantation. Figure 6.3 shows another kwashiorkor case. In bloated areas, the skin sometimes breaks from the weight of the patient's body on the bed. Patients are wrapped in bandages to absorb the seeping liquid, seen here as the dark areas of the bandages. Such patients require extensive care due to the danger of infection where the skin breaks. Figure 6.4 shows recuperating malnutrition patients. Notice the bloating in the patient at the right.

Many malnutrition cases do not die of malnutrition alone. Malnutrition so weakens the immune system that a person easily falls victim to any infectious disease to which he or she is exposed. There is a lethal synergistic interaction between malnutrition and infectious disease (Early 1982:

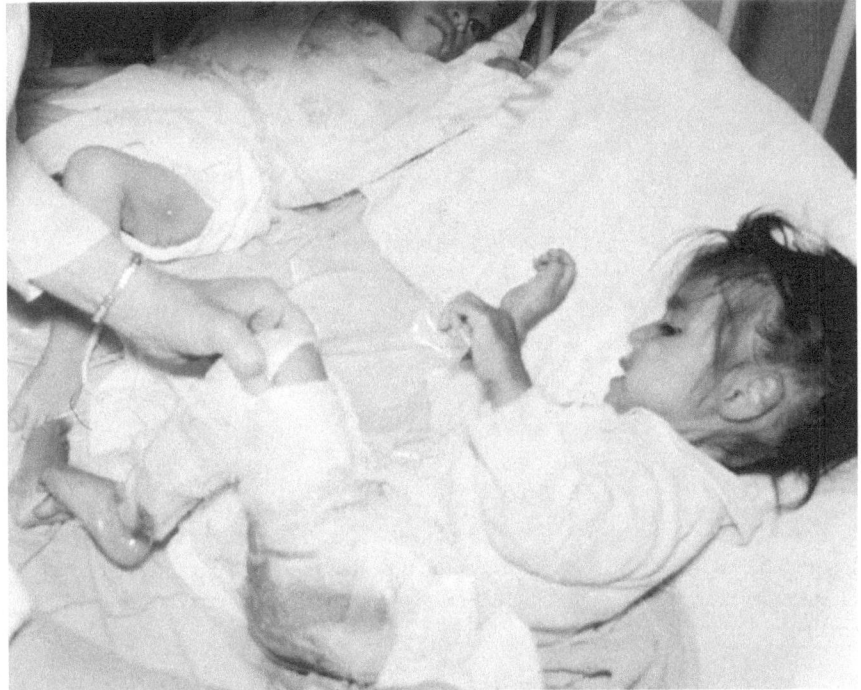

Figure 6.3. A malnutrition patient in the hospital in Huehuetenango.

Figure 6.4. Recuperating malnutrition patients at the hospital in Huehuetenango.

Figure 6.5. For those who do not survive the ordeal of malnutrition and infectious disease, the procession to the cemetery follows. Frequently, as here, the casket is a small one.

Figure 6.6. Early death also overtakes many adults.

101–7). As synergistic, the two together create a multiplier-effect that is greater than the additive-effect, if each one acted independently of the other. Measles, whooping cough, and similar childhood diseases are lethal in these conditions. As seen in the life table, particularly hard hit are infants in the first year of life and young children from the second to the fourth year of life. Clinical data chart the course of the pathology.

> The effect of the common communicable diseases of childhood—measles, whooping cough, mumps, rubella and chickenpox—on the nutritional status of the patient scarcely rests in the episode itself, but rather in the extent to which the attack is part of a sequence of repeated, often nonspecific infections, most often the ordinary infectious syndrome of intestinal and upper respiratory tracts. The effect is additive and cumulative. Conversely, these usually benign diseases derive their enhanced fatality, greater frequency of complications and exaggerated clinical course through attack on the less resistant host, to which nutritional deficiency contributes importantly. To know the full potentiality of these diseases . . . information is necessary on preceding numbers and duration of these other infections, so largely the diarrheas and common colds. (Salomon, Mata and Gordon 1968)

The results are shown in figures 6.5 and 6.6, which depict the daily funeral processions in Maya communities, especially those with small coffins. They illustrate the problem of physical survival with which the Maya had to cope.

The above photographs visualize the meaning of the abstract words "poverty" and "subsistence crisis." Ultimately, the problem was not nutritional nor biological. It was the result of being embedded in a political-economic system marked by systemic injustice that had been responsible for the seizure of large amounts of Maya land and forced them into an unjust wage market. These external forces were reinforced by Maya internal problems of population increase, excessive ritual expenses, and alcohol abuse—the latter two seen as imperatives of the traditional Maya worldview.

Crisis of Doubt about the Worldview

For the Maya, the loss of land was not simply the loss of their primary source of food and livelihood. As seen in chapter 2, the traditional worldview saw the fertility of land as a power, a force, a divine emanation handed down from ancestors to fulfill covenant obligations.

> To the land which nourishes him, which was the home of his ancestors, including his father, and the place which they still frequent in spirit, the Indian is attached by feelings of passionate intensity. The love of the land, of each man for his own piece of land, is one of the deepest emotions which he feels; it is at the root of family life and social structure; it is the basis alike of strongest attachments and bitterest enmities. In gathering information about sorcery, land quarrels, especially quarrels between brothers, were the most frequently mentioned cause of discord which led either to angry visitations from the ancestors or to recourse to black magic by the one who felt himself cheated.
>
> Land is conceived as belonging to the ancestors; one lives upon it by their grace. One does not own land, it is merely loaned to one as a lodging in the world, and for it one must continually make payments in the form of candles, incense and roses to the ancestors who are the real owners. (Bunzel 1959: 17–18)

A ritual ceremony renewed the cyclical power of the land to produce a crop. "From the stories of the elders, one gets the impression that the ancestral patrimony and its residents were united in a single religious community in which prayer to the dead ancestors perpetuated the fertility of the land obtained by the living through inheritance" (Davis 1997: 65).

But by the mid-twentieth century, many Maya had little or no land, despite the fact that many had served in cargo/cofradía roles and performed the required rituals, and that some had lived their lives in conformity with the traditional moral code. Yet the gods had not protected them, had allowed their land to be taken from them, had not provided the food necessary for their survival. In this way, the subsistence crisis struck at the very heart of the traditional worldview and self-identity described in chapter 2. The crisis of worldview grew with the increasing immersion of the Maya in the national economy. Land had been turned into a commodity, to be obtained at a price determined by the blind law of supply and demand or by legal or illegal manipulation of national laws. The logics of these institutions contradicted the Maya worldview. Land lost its sacred character and its relation to the Maya covenant. What did this imply about the Maya covenant itself and the sacredness of nature?

There was a similar problem with regard to labor. Traditional Maya self-identity was that of a producer of maize from the land to fulfill the covenant. Hence, work was something sacred. Work assistance to others was guided by reciprocity principles, a system of mutual sharing. Now labor, and consequently human beings, had also been turned into a market commodity, to be exchanged for a wage determined by a system of structured injustice. What did this imply about the role of human beings in the Maya covenant? Among the Q'eqchi' of Alta Verapaz, Wilson (1995: 158) found the tone of this sixteenth-century lament continually repeated.

> O God, how painful it is to hear you say that what we thought isn't true, rather there is another truth.
>
> O God, don't you hear our ancestors tell us that the *tzuultaq'a* [Maya gods] give us our corn, the water, the rain, the fire? O God, don't you hear that our kin offered *pom* [incense] and hit their legs [with branches as a cure] at the cross roads? And didn't that get rid of the pain?
>
> Don't you know that it got rid of our fevers and chills? O God, how hard it is for us to believe what you say—that there is another truth, O God.
>
> O God, how hard it is to believe that there is no *tzuultaq'a*, that there never was, nor will there ever be, not over there in the sinkhole, nor in the mountains when we hear the thunder and the lightning flashes in the sky.

In spite of doubts, there was fear that the ancestral and saint gods would be enraged by failure to continue the feeding and other rituals. Additional sickness, death, and natural disasters would be the expected punishments. For many Maya, it was a period of fear, confusion.

Crisis of Legitimacy of the Social Organization

As ladinos moved into Maya towns, since they were familiar with the national law mandating the governance of municipios, some became political bosses or took over the positions of civil authority. The lack of resources made it difficult for many Maya to maintain all the religious positions of the hierarchy because of the expense required for ritual items. As the market economy expanded, Maya merchants' need for capital competed with the ritual expenditures of the cargo system. They no longer wished to serve. The result was the undermining of the civil-religious hierarchies of Maya towns and of the role of the principales as the ultimate authority in all matters. The hierarchy, or some variant of it, was the skeleton of the social organization of every Maya community and the protector of its traditional customs, the *costumbre*. The higher officials of this structure, especially the principales, embodied the wisdom of the ancestors. How could this be so if it no longer functioned?

A Crisis of Culture and Self-Identity

The words "poverty," "subsistence crisis," and similar phrases are highly inadequate labels to describe what was happening to the Maya. In varying degrees for different individuals, the crisis involved all the main elements of their culture: subsistence; the morality of their social organization; and the legitimacy of their worldview. It is only with this understanding that one can appreciate the depth of the crisis and the consequent search for survival, community, and identity. "Subsistence crisis," or "physical survival," describe the most immediate problem that had to be faced. As such, they are useful labels, and we will use them a number of times in the upcoming chapters. But it must be remembered that they are linked to problems of social organization, and especially to the defining element of a culture, its worldview.

The ethnographer Moksnes (2003: 22) has expressed the difficulty many have for clearly understanding the depth of this crisis because of its inadequate treatment in the ethnographic literature.

> I felt strangely unprepared in spite of my previous periods of living in San Cristóbal and the ethnographies I had read at the time from the area, for the poverty and vulnerability Pedranos and other Mayas have to cope with. I sensed this was due to disturbing concealments in these texts, perhaps from misplaced politeness or an attempt not to be ethnocentric. However, I felt it was misdirected since Maya peasants themselves point at their poverty, at how hard it is to live under such conditions and at how much they want to find ways to a better living. During my stay I also came to sense that it was this frustration over their poverty and inability to offer their children secure access to food and, if sick, a cure which propelled Pedranos' engagement in the various religious and political groups. For these reasons it became important for me to understand Pedranos' perceptions of themselves as poor and exposed to circumstances beyond their control.

All aspects of the crisis described above resulted in psychological turmoil for individual Maya. It fractured their sense of self-identity, an experience shared by many in contemporary developed societies who have found theological expressions meaningless for their lives and at the same time cannot find ultimate meaning in philosophical, scientific, or other explanation systems. But for individual Maya there was an added note to shattered self-identity. For some, there was a sense of powerlessness and a loss of self-respect, one result of being unable to resist the encroachment and subordination of the dominant group, of being ground down under it and at times psychologically absorbing the ladino stereotype of being a subhuman without rights to resist their unjust demands. A frequent characteristic of the oppressed's worldview is self-depreciation leading to extreme lethargy and a fatalistic outlook (Freire 2001 [1970]: 30–33).

What Is to Be Done? Intensification versus Revitalization

For many Maya, their consciousness was taken up by their immediate needs: curing their illnesses and obtaining sufficient food. What would alleviate these immediate problems? Some principales thought that the capricious gods were not satisfied with their ritual offerings; they wanted more. The traditional rituals must be maintained with increased ostentatiousness. The personal motives of the principales for their strong opposition to any change were based either on values or self-interest. Values were their convictions that it was the only right thing to do. Religious worldviews (also philosophical and scientific worldviews) are often seen by their adherents as all-embracing explanations of reality, as *the* truth based on an epistemology of exaggerated realism, often called fundamentalism. Reform is difficult to envision because it means some part or all of the worldview is not true. Those who opposed any change from self-interest feared loss of prestige owing to their position as village elders. For these reasons, any attempt to alleviate the crisis that involved substantial change of the traditional ways usually resulted in intense conflict with any faction proposing such changes.

Given a deeper understanding of the Maya worldview, some saw one alternative to intensification in revitalization by reform. The contemporary crisis was the death of the present cycle of Maya existence. Its possible rebirth would be in a form that would revitalize it by incorporating changes. The Maya had gone through a similar crisis in the sixteenth century. They had fallen in battle to the Spaniards. The anticipated protection of their covenant gods was absent in spite of Maya efforts by both ritual performance and military resistance. The conquering saints of the Spaniards were incorporated into the Maya covenant as a cyclical revitalization of it (Early 2006). Now, four centuries later, the Maya subsistence base had been seriously undermined by ladinos, the national society, and their own internal problems. Prolonged sickness and early death were the conquerors. Although this crisis was not as dramatic and was not compressed into as brief a period of time, it was nonetheless just as devastating.

Some Maya would embrace any significant change in the contemporary cycle that could be seen as revitalization. They realized the benefits of, and they wanted, medical and educational assistance to help overcome the crisis. Some shamans were pragmatic. They readily conceded that medical assistance could cure some ailments that they could not. Maya communities could appeal to their national governments to provide health facilities, public service programs, and to safeguard their rights as citizens who contributed to the national economy. But the Guatemalan and Mexican governments, while paying lip-service to the needs of the Maya at election times, had little interest in Maya communities beyond their ability to provide cheap agricultural labor. The extent of their illnesses and mortality was, for the most part, unknown, and there was little interest in knowing about them. (As far as this writer knows, he produced the first community Maya life tables in the late 1960s [Early 1970a, 1970b]).

Summary

This chapter has sketched the crisis facing the Maya at mid-twentieth century. It was a crisis of physical survival, worldview, social organization, and psychological identity, all tied together as a single organic entity. In brief, it was a crisis of culture. It was the end result of a series of accumulating problems for a traditional agricultural society in transition to a multicultural society dominated by unjust market mechanisms. It led to a questioning of the Maya covenant and its place in the traditional worldview. The questions were formulated in individual minds in many ways. They would lead to a search during the coming years for answers, a search that would take Maya to worlds far removed from their traditional culture. The remainder of this book looks at the role played by the Catholic Church in this search of the Maya in their time of crisis.

III

Renewed Efforts of the Catholic Church in Maya Communities

During the nineteenth century, suppressions by political liberals reduced the already infrequent appearances of Catholic priests in Maya communities to intervals measured in years. With changes in national policies around the mid-twentieth century, the Catholic Church in Guatemala and Chiapas began to reassemble its ecclesiastical structure. With foreign assistance, the church became deeply involved in the crises of Maya communities and their search for alternatives. This is the focus of the remaining chapters of this book. The late twentieth century was a fifty-year period of rapid change for the world, for Central America, for the Maya, and for the Catholic Church. The theological worldview of Catholicism was evolving, and this was reflected in the types of Catholicism with which the Maya interacted during this period. For the purposes of this study, four main types are considered: Tridentine Catholicism, Catholic Action, Liberation Catholicism, and Maya Christianity. Catholic Action and Liberation Catholicism will sometimes be called "Action Catholicism." The last type, Maya Christianity, began to emerge at the end of the period. All four types are distinguished by their theological emphases and their roles in community organization. Part III examines the first type, Tridentine Catholicism.

7

Worldview of Tridentine Catholicism

The reappearance of the church in Maya communities was not occasioned by the Maya crisis, but by the end of the liberal suppressions. The initial aim of the renewed effort was to form orthodox Catholic communities based on the theological worldview of the sixteenth-century Council of Trent that set the Catholic agenda for the next four hundred years. Some knowledge of the historical development of Christian theology up to that era, and of the circumstances surrounding the Council, help to understand the mindset of the church at that time, its formulation of the Catholic worldview, and the resulting pedagogical catechesis that resulted. Books have been written about these topics. Here, only a skeletal summary is given.

Historical Background of Christian Theology

From an anthropological perspective, theology is a worldview, a paradigm, whose purpose should be to help people live their everyday lives amid the problematic human condition. It uses the categories of a culture in certain historical periods to explain the moral aspect of the human condition and its relationship to a supreme or higher being(s). There can be evolution in theological worldviews. Internal evolution takes place when its categories undergo a change of meaning over time. This often takes place in theological worldviews that define themselves as founded in a certain historical period within a specific culture and then attempt to become normative for future time periods. External evolution may take place if a theological worldview from a different culture is adopted by another culture. Usually, this involves some changes in meaning. There are few, if any categories used for religious or philosophical expression that always retain the exact same meaning across all cultures.

Seen in this perspective, Christianity is a historically based religion that has used the categories of different historical periods and cultures in which it has existed to explain the moral implications of the life of Jesus. The history of Christianity from its origin to the sixteenth century shows it existed within two main cultural domains (Küng 1995). Christianity began among Jewish groups who defined themselves as a reformed Judaism centered on the teachings of Jesus. In other words, Christianity began as a religion inculturated in Hebraic culture. The first Christian groups often used the categories of the Hebrew Bible to explain the significance of Jesus, as recorded in the books of the New Testament. Saint Paul, over the initial opposition of Saint Peter, broke with an exclusive Judaic application of early Christianity and gave it a more universal meaning. Against its Roman and Judaic persecutors, Christianity's early apologists used the intellectual categories of their times, mostly drawn from Greek intellectual culture.

In the fourth century a radical change took place as Christianity became the established religion of the Roman Empire. Institutionally, Christianity took on the political structure of the empire, leading to the emergence of the papacy, which was later carried over to the medieval world of divine-right kings and popes. Any person with authority possessed that authority because of divine endowment of that person. There was a unity of culture in which church and civil authority were one and the same thing. The progression was: God to Saint Peter, to the popes of the church, to the rulers of the Roman Empire, to the kings and queens of various European countries.

A worldview emerged that saw Christianity as belief in a set of literally interpreted biblical statements expressed in the Latin language, although the original biblical languages were Hebrew and Koine Greek. There were problems in harmonizing differing biblical statements into a coherent system. Consequently, early Christian theologians and later the Fathers of the Church undertook the task of rational explanation, usually employing categories of philosophical systems inherited from the Greeks, especially Plato and Aristotle—again inculturation in Greek intellectual culture. Biblical statements regarding God the father, Jesus as the son of God, and the Holy Spirit, were rationalized in the doctrine of the Trinity, which did not become a church dogma until the fourth century (R. Richard and W. Hill 2003: 189–201). The literal biblical interpretations were contested by some in favor of symbolic interpretations, but rejected by church authorities of the period. In the medieval period, Greek-influenced scholastic philosophy became the tool of theological explanation.

A Catholic scholar has noted an important change in Christianity due to this shift from its original inculturation: "In accordance with its Hebraic origins, the truth of Christianity was not to be seen or theorized on; rather, it was to be done, practiced.... The Christian concept of truth was originally not contemplative and theoretical like the Greek concept, but operative and practical" (Küng 2001: 26). The emphasis shifted from commitment shown in everyday moral action to belief in orthodox theological propositions as a necessary precondition to salvific action. This became one of the defining marks of later Christianity. It was reflected in the distinction between dogmatic (systematic) and moral-ascetical theology, with a primacy attributed to dogmatic theology.

Europe prior to the Council of Trent

At the beginning of the sixteenth century, Christianity was encased in a declining European medieval culture battered by a series of natural and social crises. There was a period of poor harvests, with a total crop failure in Germany in 1500. In many places, the peasants reacted by pillaging and looting. Destructive wars broke out in Switzerland, Swabia, and the Tyrol. The greatest threat was the outbreak of contagious disease. Following the Black Death in the previous century, the plague struck again, decimating the populations of many European regions between 1499 and 1502. Another deadly scourge, syphilis, made its appearance, taking a heavy toll. Populations became obsessed with death and questions of individual salvation. In an attempt to make sense of these deadly onslaughts, people sought religious safeguards and turned to their older traditions to find an answer. These traditions, or "popular piety," dominated the consciousness of most Christians of this period. The church with its saints, relics, blessed objects, and sacraments was looked upon as an important source of this piety, although church authorities rejected some usages. In these troubled times, there was a searching out of biblical prophetic ideas about a present period of disaster to be followed by a period of hope—perhaps it was the eschatological period of the Last Judgment to be followed by the final millennium. It was a period of looking for scapegoats for the natural and social crises—Jews, witches, and the notoriously corrupt church itself. "Concrete fears and very real dilemmas determined the structure of men's thought, although in the sixteenth century the meaning of life and the final answers about what life holds in store for men were cast within a Christian context, and it is the tone of Christian life at the turn of the century which provides the indispensable background to the rise of Protestantism and the end of the *Respublica Christiana* of the Middle Ages. . . . Historians have been struck by this singular ferment within Christianity, the heightened religious sensibilities, the almost obsessive preoccupation with death, salvation and the future of man" (Koenigsberger, Mossee, and Bowler 1989: 127–28).

Under these circumstances, Luther, Zwingli, and Calvin rejected some traditional church teachings, especially about justification of the individual person. There was a tendency by the reformers to seek a simpler Christianity that bypassed the complexities of theological explanation and ecclesiastical bureaucracy. Faith alone brought salvation, a position proved by the one source of certainty, "sola scriptura," the Bible alone.

From the Catholic viewpoint, there was a need for clarity in teaching *the truth*, to dispel the confusion created by the teachings of the heretics and to protect the existing European social order from internal and external threats. Disagreements among the dissenters themselves added more confusion for the religious commitments of an age that was looking to religion for answers to its pressing questions. Within the church itself, a serious problem was the moral conduct of its clergy—the popes, cardinals, bishops, and priests. Concubinage was common among the supposedly celibate clergy. But the most serious problems involved justice—as evidenced by the great variety of unjust taxes, the imposition of fees, the selling of ecclesiastical offices, and, especially, the selling of indulgences.[1] Indulgences were important to individuals obsessed with the fear of death and belief in the last judgment. Denying their very existence was the initial breaking point for both Luther and Calvin. The moral crisis of the clergy raised the theological question: How could anyone accept the teachings of a church that was so corrupt?

Authorities were faced with additional problems that called for the truth and unity of church doctrine as a rallying point against the undoing of the medieval period from within and without. The Holy Roman Empire had a fragile existence, in theory only, as the various European kings and princes did not in fact recognize the authority of the emperor, Charles V (King Charles I of Spain). In the continuing medieval concept of the unity of church and kingdom, Charles considered himself the protector of Christianity and wanted clarity about the one true faith to help strengthen the empire in this period of uncertainty.

Adding to the problems within Europe was the external threat of Islam. In the east, it had reached the outskirts of Vienna in 1532. To the south, it had occupied all of North Africa and from there threatened southern Europe. In the west, it had occupied most of the Iberian Peninsula for five hundred years, and parts of it for two hundred more, until the last enclave was finally defeated in 1492. In an effort to unify Spain as a nation under the king and queen, all Muslims were given the choice of conversion to Catholicism or exile. Many remained in Spain as the converted, *conversos*, but the sincerity of many was doubted. This led to the formation of the Spanish Inquisition, which became the model for the later Roman Inquisition. Jews were treated in the same way. For the purpose of conversion, clarity of the true doctrine and conformity to it was seen as a necessity. It was a time of crisis for Christianity, a crisis that had much in common with the Maya crisis of the twentieth century—crop failure, sickness and high mortality, questioning of the theological paradigm of the period, a search for answers.

The Council of Trent

Against this background the church convened a Council at Trent, in northern Italy. A council is a meeting of all Catholic bishops to discuss matters of deep concern to the church. Due to political infighting and European wars, the council met intermittently in twenty-five sessions over an extended period, 1545–63. The disciplinary problem of the clergy's morals was a concern of the Council. But the main purpose was to clarify true Catholic teaching that would refute the theological positions of the various Protestant reformers and become a rallying point for a disintegrating Holy Roman Empire. The Council published a number of decrees that influenced the formulation of Catholicism for the next four hundred years. Issued in the polemical context of the Reformation, they were not a balanced presentation of Christianity, but emphasized aspects pertinent to the polemic. Preliminary statements established the church as the only authority to interpret the Bible. This meant that the church's theological interpretations of the Bible could be considered a source of revelation in addition to the Bible itself. This was done to refute Luther's "sola scriptura" position.

However, in later polemical writings church authorities often sought to legitimate their positions by questionable appeals to the Bible, in an attempt to refute its Protestant adversaries on their own grounds.

Given the concerns of the historical period and the various theological responses to them, the first decrees of the Council dealt with original sin and the consequent need for justification: how individuals attain the state of grace in this life so that at death they attain eternal salvation. The core of the Council's paradigm emphasized the means to attain individual salvation: belief in holiness and justice as the original human condition before Adam's sin and its consequent loss; belief that humanity is redeemed from this sinful state by the merits of Jesus Christ earned by the shedding of his blood; belief that a person receives these merits and is united with Jesus through the sacramental system administered by the church.[2] The majority of the decrees of the Council dealt with the importance and absolute necessity of participation in the sacraments (Denzinger 1955: 262-98, #844-#982). Attached to the decrees were a series of ringing anathemas condemning the doctrines of the Protestant reformers. Following the Council, the Pope issued an oath of allegiance to the teachings of the Council that summarized its principle points. The oath was to be taken by professors of Catholic doctrine and other high church officials (302-4, #994-#1000).

The intellectual outlook of the Council of Trent was that of medieval Europe. The printing press had been invented in the previous century, but books were still relatively scarce, resulting in a sketchy knowledge of history and of differing world cultures. Most learning was ethnocentric in its confinement to European worldviews. Consequently, the statements of the Council, and the later oath, were based on mental constructs fused together from differing cultural worldviews used by Christian theology up to this period. They included Hebrew culture (scriptural narrative without regard for Hebrew genres), Greek intellectual culture (philosophical terms such as "consubstantial" and "proceeds from" in describing the Trinity; "substantially present" and "transubstantiation" regarding the Eucharist), and the Roman institutional title "Pontiff," a shortened form of "Pontifex Maximus," the head of the Roman priesthood. Catholic Christianity can be seen as a syncretic religion, as it gathered bits and pieces from various cultures and historical periods that had been part of its history. It reinterpreted and rationalized them as Western Christianity. This was presented as a worldview given by Christ, independent of any cultural genres, logically elaborated by the fathers of the church, and remaining essentially unchanged down through the centuries.

The Tridentine worldview dominated Catholicism for four hundred years. In the nineteenth century, it was greatly threatened by evolving European worldviews—the scientific revolution; various Enlightenment philosophies; and biblical studies based on linguistic, archeological, and historical methods. Against these threats, an extreme reactionary response further solidified the Tridentine worldview among Catholics, especially during the papacy of Pius IX from 1846 to 1878. In 1864, the Pope issued the Syllabus of Modern Errors, a compendium of statements from the evolving worldviews (Denzinger 1955: 433–42, #1700–#1780). He summoned the world's bishops for Vatican Council I, 1869–70, to reaffirm the Tridentine position, to strengthen the authority of the Vatican over local bishops, and to make the infallibility of the pope a dogma of the Church (442–57, #1781–#1840). The dogma contains several highly restrictive conditions, so that it has seldom been invoked, and never in matters of morality. Dissenting Catholics were labeled "modernists" and often censured. The papacy of Pius X, 1903–14, reaffirmed the Tridentine world, especially against the continuing biblical scholarship. It drew up "The Oath Against the Errors of Modernism," an updated version of the Tridentine oath (549, #2145–#2147). Prominent Catholic theologians who dissented were persecuted and excommunicated from the church.

Some Characteristics of the Tridentine Worldview

Because of the polemical context of the Council of Trent, Tridentine Catholicism took on some distinctive characteristics.

"Will of God" Complex

Since the fourth century and until the time of the Reformation, Christianity was the cultural and state religion of western Europe. By its claim to divine foundation and authority, it considered itself to be the sole keeper of the one, true "deposit of faith." Belief in it was a necessity for eternal salvation. The Council members, in conjunction with the Pope, looked upon themselves as the inheritor of this tradition—the guardians and the only official interpreters of the Bible as an exact reflection of God's will. Correct belief became paramount to all else. With magisterial confidence in its infallibility, the Council issued doctrinal decrees of correct belief, each followed by a harsh condemnation of anyone who held contrary opinions. It was a dichotomous world of fundamentalism, a world of one truth, and anything to the contrary, wrong. The combination of being God's will and of legitimating the ecclesiastical and kingly power of the medieval Respublica Christiana, gave Tridentine Catholicism a tone of arrogance. It was highly ethnocentric, as were most other world and tribal religions at that time—and many continue to be. The last two paragraphs of the post-council oath illustrate this ethnocentrism.

> I acknowledge the holy Catholic and apostolic Church as the mother and teacher of all churches, and to the Roman Pontiff, the successor of the blessed Peter, chief of the Apostles and the vicar of Jesus Christ, I promise and swear true obedience.
>
> Also all the other things taught, defined and declared by the sacred canons and ecumenical Councils, and especially by the sacred and holy Synod of Trent, I without hesitation accept and profess; and at the same time all things contrary thereto, and whatever heresies have been condemned, and rejected, and anathematized by the Church, I likewise condemn, reject, and anathematize. This true Catholic faith, outside of which no one can be saved, (and) which of my own accord I now profess and truly hold, I, N. [name of the one taking the oath], do promise, vow and swear that I will, with the help of God, most faithfully retain and profess the same to the last breath of life as pure and inviolable, and that I will take care as far as lies in my power that it be held, taught, and preached by my subjects or by those over whom by virtue of my office I have charge, so help me God, and these holy Gospels of God. (Denzinger 1955: 303–4, #999–#1000)

The logical conclusion was the claim "This true Catholic faith, outside of which no one can be saved." This became a famous theological phrase that later Catholicism was constantly forced to reinterpret, so that its meaning became quite different from what it appeared to be saying. This triumphal manner of sixteenth-century Catholicism, its emphasis on correct belief interpreted by the church as a prerequisite for salvation, was dominant into twentieth-century Catholicism and currently continues within numerous sectors of conservative Catholicism.

Emphasis on the Individual Person in a Life to Come

Given the high mortality of the sixteenth century, from contagious diseases, poor harvests, and destructive wars; given the resulting obsession with death; and given the consequent theological controversies with

Protestants about personal justification as preparation for a life after death—it is no wonder that the most important decrees of the Council were concerned with individual salvation and its attainment by individual belief and participation in the sacramental rituals. This concern for the individual in a future life downgraded by omission the importance of the present life in a human community. The traditional Ten Commandments were usually considered in the context of individual behavior or conduct between two individuals within a taken-for-granted social system seen as the will of God. These silences of Tridentine Catholicism about community in the present life were conducive to a passive, individualistic, pietistic spirituality incapable of confronting the problems of an evolving social order, much less encounters with non-Western cultures. In the sixteenth century, the ordering of society was to be left in the hands of God's appointed elite, the educated royal and ecclesiastical orders. Silent obedience was expected of the illiterate lower orders. Within this social order, the great sin of the heretics was not their theological positions, but their refusal to retract them and obey divinely bestowed authority when so ordered. A passive, pietistic asceticism for the peasants was a perfect fit for such a social system. No biblical prophets were wanted.

A Ritualistic System

With the emphasis on a future life after death, and with the concerns of the present life seen primarily in an individualistic sense, a question remained: What positively should be done to maintain the state of grace? The answer of the paradigm was the absolute necessity of participation in the sacramental rituals, where the encounter with Jesus and the life of grace were to take place. The absolute necessity was seen as arising from the fact of divine foundation; they were instituted by Christ. The Catholic position was that God could have done it another way, but that the seven sacraments were the decreed way. In addition, the Council said in one of its anathemas, "If anyone shall say that by the said sacraments of the New Law, grace is not conferred from the work which has been worked [ex opere operato], but that faith alone in the divine promise suffices to obtain grace, let him be anathema" (Denzinger 1955: 263, #851). This statement was intended to refute Luther's rejection of the necessity of the sacraments. However, some Catholic theologians also interpreted it as setting a minimal condition for the validity of the sacraments, namely, a mere mechanical performance regardless of all other circumstances. This became a widely held position (Jedin 2003: 170; Kilmartin 2003: 501). In this way, ritual performance itself assumed paramount importance to the subordination or neglect of moral action for which it was supposed to be an empowerment.

The function of the priesthood was a corollary to the theology of the sacraments. The priest is one set aside from the rest of men to be the minister of sacramental rituals. The supreme importance of ritual, coupled with correcting the moral problems of the clergy, gave rise to the withdrawn "sanctuary priest," the "mass-saying priest." In this context, many priests became distant from the communities they were supposed to serve, unable to empathize with the practical and moral problems of their church members' everyday lives. The sacraments were stripped of their function as passage rites, no longer signifying the transition into a new stage of communal life with its characteristic problems and its responsibilities. In this respect, Catholic ritualism had much in common with the problems of Maya ritualism and its place in the Maya crisis. This shortcoming was belatedly acknowledged in 1992 at the Latin American Bishop's conference in Santo Domingo: "The liturgy should express more clearly the moral commitments it entails" (Hennelly 1993c: 138, #240). Some Catholics understood this problem and participated in the liturgical movement in an effort to counteract the degeneration of the sacramental system into ritualism, but the movement had little influence in the Maya areas.

The introductory photograph at the beginning of the chapter shows important Maya and Catholic ritual officials. Taken in 1963 between sessions of Vatican Council II, it shows Bishop Samuel Ruíz, in his Tridentine phase, arriving in Zinacantán to administer confirmation. He is greeted as an important shaman by the two elderly Zinacanteco shamans on each side of the bishop. The bishop's hat and shoulders retain some of the shower of flower petals with which he was initially greeted, a traditional Maya custom for welcoming important persons to their communities.

The Catechisms: Tridentine Catechesis among the Maya

The Tridentine teachings were contained in booklets called catechisms. They contained prayer formulas and summary statements of orthodox Catholic faith and morality. The early colonial catechisms were discursive, composed before or during the Council (Early 2006: 124–30). Catechisms were seldom used before the sixteenth century. Medieval Europe was an oral and visual culture, with communication of the religious worldview dependent on sermons and on visual media such as sculptures, paintings, and the stained-glass windows of the churches. Luther appears to have composed the first catechisms (Jegen 2003: 232). He divided them into three sections: the Commandments; the Creed; and Prayer and the Sacraments. The format was a series of questions and answers that were to be memorized by the student. Catholics quickly followed with their own catechisms. They changed the order of presentation to Creed, Commandments, and Sacraments, reflecting the emphasis on correct belief as a prerequisite to moral action. They retained the format of a series of questions and answers to be memorized. The answers often used abstract theological language and interpreted the Bible according to the Tridentine paradigm presented as an all-embracing theological worldview. The end result was the tendency to use the catechisms in place of the Bible, to substitute abstract language for the concrete expression of the biblical stories.

The content of these catechisms (*doctrina*), with minor revisions, remained in use until Council Vatican II in the 1960s. "During the same period [1560s] Gaspar Astete and Juan Martínez de Ripalda, Jesuits, wrote catechisms which are still in use in twentieth century Spain" (Jegen 2003: 230), and in Chiapas (Maurer 1984: 448). Since many Maya did not know Spanish, and those who spoke it were often unable to read it, the catechisms were initially used by the priests as manuals for their oral instruction that would be translated by Maya catechists.

During the liberal suppression, ladina women gave catechetical lessons for any ladino and Maya children who would attend. "Ladinas had not stressed understanding of the teachings for Indians who were sometimes unable to comprehend the recitation of the catechism in Spanish and more often unable to read it. Lessons were memorized and there was little explanation of their content. As one Maya commented, 'although we could repeat the catechism to all the world, we did not understand it'" (Warren 1978: 100). The first part of the catechisms reproduced the traditional creed and Catholic prayer formulas: the Our Father, the Hail Mary, the Hail Holy Queen, the General Confession, and the Act of Contrition. Based on the ethnocentric assumption that the categories of the prayers were "natural," and therefore understood across all cultures, the ability to repeat the memorized formulas indicated to the priests that the Maya understood their meaning and assented to them. Frequently, instruction did not go beyond memorizing the introductory prayers (Early 2006: 147–77).

Summary

Following the liberal suppressions in the mid-twentieth century, the Catholic Church renewed its efforts of evangelization among the Maya. The effort was to establish orthodox Catholic communities whose worldview had been defined by the Council of Trent in the sixteenth century, a highly ritualistic, individualistic theology. For over four centuries, the Council, with its decrees and pedagogical catechisms, formed the main content of the preachings and pastoral instructions for the Catholic faithful.

8

Presentation and Maya Reception of the Tridentine Worldview

To understand the various ways in which Tridentine Catholicism was presented to the Maya during the reentry period and the various stages in how it was understood, this chapter presents several case histories. The attempt to convert Maya communities to orthodox Tridentine Catholic communities failed to understand the history of the previous four hundred years. The Maya had already defined the Tridentine catechesis and absorbed it into their own worldview during the sixteenth century (Early 2006). As LaFarge (1947: 81) remarked, "The task confronting a priest who wishes to revive true [orthodox] Christianity among the people is made almost overwhelming by the fact that they are not merely non-Christians,

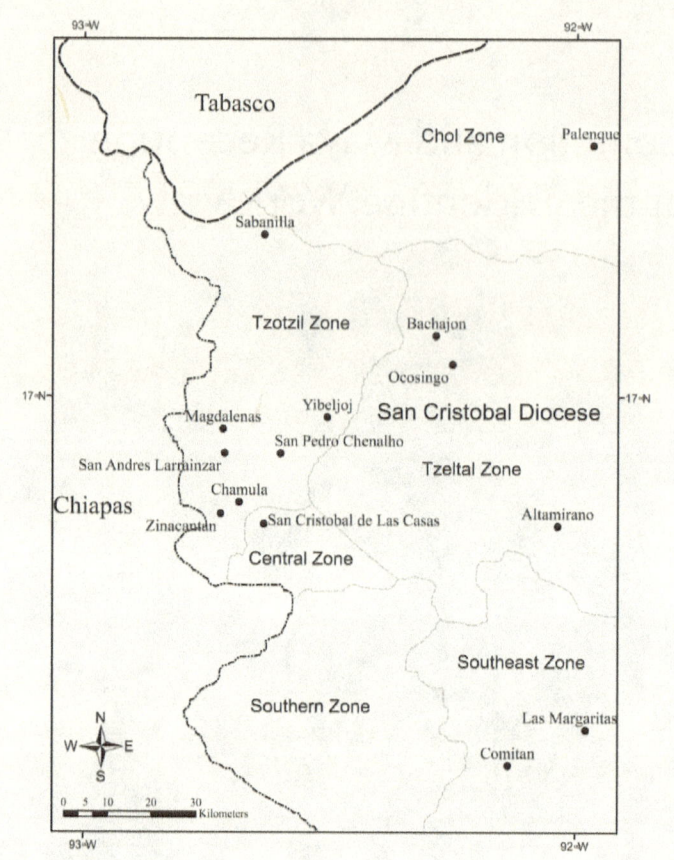

Map 2. Diocese of San Cristóbal, with Its Pastoral Zones and Some of Its Communities

but non-Christians who believe themselves to be the only maintainers of pure Christianity." The priests, without any understanding of the Maya worldview, were unaware of its role in interpreting their renewed efforts. The result was a period of confusion before the situation was clarified.

Early Efforts

In the San Cristóbal Diocese, there were a few catechetical efforts under Bishop Lucio Torreblanca in the 1950s. Some Franciscans initiated a catechetical program in the Chol region (Meyer 2000: 60). A diocesan priest in the Bachajón area instructed some catechists, who zealously established catechetical groups in the Tzeltal area. The pastor of the neighboring

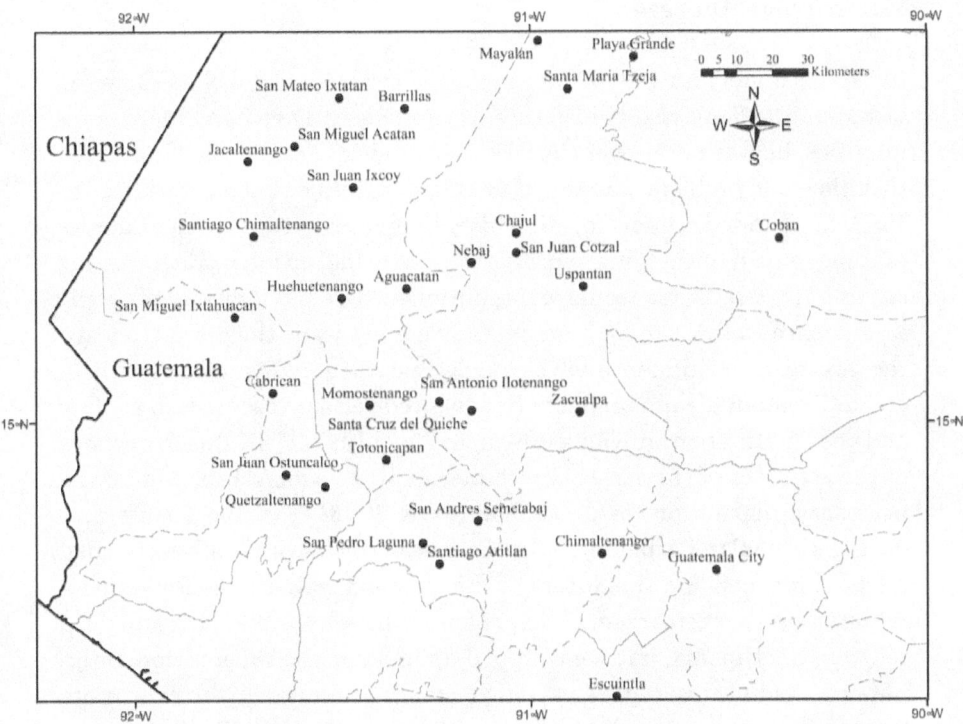

Map 3. Communities in the Western Highlands of Guatemala

Ocosingo parish duplicated the program in that area. These catechists gathered together acquaintances in community chapels to teach them songs, prayers, and elements of the Tridentine question-and-answer catechism. In the later 1950s and early 1960s, when the Jesuits and Dominicans took over the Bachajón and Ocosingo parishes, respectively, they found an already established network of catechists (Vos 2002: 219–200; Iribarren Pascal 1985: 48; Maurer 1996: 61).

In Guatemala, the reintroduction took place earlier than in Mexico. During the 1930s, a visiting priest from Sololá gave catechetical lessons in Santiago Atitlán. In 1937, two Maya catechists from Totonicapán initiated lessons in San Antonio Ilotenango. In 1945, Bishop González initiated catechetical programs throughout the Western Highlands. In 1951, catechists from Tecpán introduced the catechesis in San Andrés Semetabaj. In the early 1950s, a priest began recruiting catechists in San Bartolomé Jocotenango. In 1952, a Maryknoll priest from Chiantla formed catechetical groups in both sections of the Aguacatán community.

San Cristóbal Diocese

In 1960, Samuel Ruíz García (often referred to as Don Samuel) became the bishop of the San Cristóbal Diocese. During his early visits to Maya communities, he was upset over the lack of orthodox Catholicism. He decided that the root problem was the inadequate training of the catechists. In 1963, he founded schools for catechists that conducted intensive courses of three to six months' duration. "We want to install an *internado* [boarding school], modest as one would want, but one where the indigenous people would learn to live a more human life. We don't want them only to learn the catechism; its influence will be much greater if a well-studied catechism and a Christian life are united with advanced secular instruction, as time and possibilities permit" (Ruíz García, in Floyd 1997: 103). This has echoes of the attitudes of the early Spanish period, that civilized urban life was a necessary preparation for conversion (Early 2006: 135). The Jesuits conducted a school in Bachajón. In San Cristóbal, the Marist Brothers opened a school for men and the Sisters of the Divine Shepherd one for women. A Marist brother remarked, "The indigenous have a simple spirit and deep religious sentiments, but are wrapped up in a haze of superstition, of ignorance, and of misery that is going to mark, profoundly, the slow route toward civilization and the Christian religion" (Floyd 1997: 104). Young men and women were selected by priests and nuns to attend the schools. In addition to the catechesis, the students learned or perfected their ability to speak, read, and write Spanish. Because of their schooling, people thought the catechists to be knowledgeable about health, first aid, civics, and agriculture. Therefore, the curriculum included introductions to these subjects (Floyd 1997: 106). Because of their acquired literacy, many catechists later assumed leadership roles in various community organizations (Leyva Solano 1998: 395–98). The exclusiveness of Tridentine Catholicism meant there was no serious dialog with the Maya about their preexisting views, their spirituality and its theological foundations. The instruction was in the manner of an active teacher and a passive student.

A lasting contribution of the schools was establishing a tradition of service to announce the Word of God at the cost of any needed sacrifice (Iribarren 1985: 48–49). A Chamula catechist recalled, twenty five years later: "I will be a catechist for my entire life. Some people say that they need a rest and will stop preaching the Word of God when they are older. But a catechist is a job for life; there is no rest because there is a lot of work to be done" (Kovic 2005: 80). The catechist commitment was the Catholic implementation of the Maya tradition of community service. "Juan's faith and affiliation with the San Cristóbal Diocese is linked at once to his identity as a Tzotzil Maya and to the structural violence (poverty, racism, political oppression) that he resists in daily life" (Kovic 2004: 191). In ten years, these schools placed seven hundred catechists in the Maya communities (Floyd 1997: 105). After catechists finished the course at the catechetical schools, their teachers visited them in their home communities to provide assistance.

The shortage of priests resulted in their still infrequent appearances in Maya communities to administer the sacraments, as called for by the Tridentine catechesis. Most priests were ladinos. Many retained the negative stereotypes of indigenous peoples typical of ladino culture. Realizing this problem and the church's lack of familiarity with indigenous culture, Don Samuel was instrumental in organizing the 1970 Xicotepec Conference. Its purpose was to prepare a pastoral plan for Mexico's indigenous populations. The conference was attended by bishops and priests working in indigenous areas, along with indigenous catechists, including some Maya. In the final session, the catechists presented a summary of their complaints about the relationship between the ladino clergy and their communities. "1. It is indispensible that there be more contact of priests with indigenous communities. 2. It is urgent that the priests' visits to the communities [where there is contact] be more frequent. 3. The priests must spend time in the communities and try to understand their necessary needs and customs. 4. The priests must work out a schedule for themselves so that there is sufficient time to devote to the needs of the communities and not travel about occupied with businesses that have nothing to do with their priestly role. 5. The priests must not directly attack the customs of the community. 6. The priests must take responsibility to implement these points" (Centro Nacional de Pastoral Indigena 1970: 63–64). This was a strong indictment of many ladino priests and their attitudes toward the Maya and other indigenous groups. There remained considerable social and cultural distance between Maya communities and the Catholic Church. This raises the question of how the Maya understood the Tridentine catechesis transmitted by these priests and recently instructed catechists.

Santiago Atitlán

Maya communities welcomed the increased presence of the priests. They continued to interpret Catholicism in terms of their own covenant with the emanated gods. The increased presence meant more-frequent masses would be celebrated for the gods of the covenant, making them more amenable to continuing their protection of the community. It would decrease the time between the birth of a child and its baptism, thus giving earlier protection against sickness.

The phased history of the reentry in Santiago Atitlán contains many characteristics of similar histories in other communities. In the 1940s, a young Dutch priest, Padre Lias, periodically visited Santiago Atitlán. He initiated catechetical instructions for the children of the community. The community's elite—principales, cofradía officials, shamans—ordered their sons to attend. Why would these upholders of traditionalism want their sons instructed in the Catholic worldview that opposed Maya traditionalism? From the Maya viewpoint, the efficacy of the rituals depended on the exact recitation of the prayer formulas. The Catholic priest knew these prayer formulas. Some were included in the first part of the catechisms. Therefore, the traditional elite wanted their sons to learn them and gain greater access to the gods of the covenant.

The students began to realize that the priest's teachings regarding Christ, the mass, the saints, and moral behavior, especially drinking, was different than the costumbre of their elders. Lias did not directly attack the costumbre, as he knew little about it. But the experience left doubts in the minds of his young pupils. They had now been exposed to an outside influence that broke through the certainty of the community-accepted worldview. These doubts were helped by the personality of Padre Lias. After the catechetical sessions, the young priest remained and played soccer with his students, developing a camaraderie with them. This was in stark contrast to the formality and social distance that usually characterized visits by priests. In the 1960s, this writer interviewed the men who were his students and had become catechists. They had a revered memory of Padre Lias. At that time, he was in retirement in Quetzaltenango, where the writer talked with a kindly, outgoing man who also had happy memories of his Atiteco students and their soccer games. It was clear why this humanistic priest and his teachings had left a lasting impression and sown seeds of doubt about then current cofradía practices and worldview.

Other priests succeeded Padre Lias. They gave catechetical instructions to interested people of all ages, with the hope of recruiting some to become catechists. In accordance with the emphasis of Trent, the main objective of the priests was to convert Maya communities to correct Tridentine belief. They were concerned about moral action, but this was dependent on correct belief. They knew little about everyday life in Maya communities. They observed drunkenness and condemned it, but did not realize its connection with the worldview.

A Period of Confusion

Problems arose as the catechists attempted to communicate the Tridentine catechesis. There were ambiguities on both sides of this encounter of worldviews, which caused confusion and initially prevented the ambiguity from being perceived as contradiction. The confusion was rooted in the use of some of the same sacred symbols by both worldviews, but with different interpretations. The sermons of the priests and their catechetical instructions were explanations of the more-important sacraments, the significance of the saint whose feast day was being celebrated, and moral condemnations about some commonly observed behaviors, especially drunkenness. Since the priests knew little or nothing about traditional Maya culture, there was no systematic mention of it in their sermons. Since the sermons had to be translated from Spanish into the local Maya language, there were probably significant differences between what the priests thought they had said and what the translators interpreted them to have said. Catechetical instruction usually involved learning the traditional Catholic prayers in Spanish, on the ethnocentric assumption that the contents were self-explanatory. They were learned by continually repeating the prayers after the teacher, as most Maya were illiterate in Spanish. If the questions and answers were also covered, they were learned in the same way. Catechists in Santiago Atitlán recalled that although initially they realized there were some differences between the priest's instruction and their traditional worldview, they did not understand that it was a different religion. The differences were due to the logical consequences between the immanent pantheistic One and the transcendent God of the Judeo-Christian tradition. These differences were not easily understood. A catechist said that he began to realize the differences only later, when he was required to teach the Catholic view of the saint images and the rituals connected with them. But he preferred not to make a point of it and create difficulties with the principales, including his father.

Ambiguities also existed about the content of the traditional colonial Maya worldview. As seen in chapter 2, the great tradition of the Maya had broken down over the years due to the subsistence problems faced by all Maya, including the priests. There was no longer an established priesthood or a structure for maintaining and teaching the worldview. It continued, diffused among various myths, stories, and rituals, without any effort at coherence or systematization. Some parts were expressed by variant stories that contradicted each other in details but were in conformity about a general theme. Some individuals retained a deep knowledge of the tradition, but they had no official position in the community as teachers. A person became a principal because of past community service that did not require in-depth knowledge of the tradition. The presentation in chapter 2 may be deceiving in its attempt at clarity. It describes a basic logic of the traditional worldview, but it did not exist in this abstract and orderly manner in the minds of many individuals. The worldview was a collection of disjointed parts. It was not perceived as a system that could be contradicted by other systems, as previously there was no experience of other explanatory systems. It is typical of the way the cognitive contents of religious systems exist in the minds of adherents of culturally accepted worldviews. People embrace them because they were brought up in them and are unaware of any alternatives. As long as Maya communities remained relatively isolated, the system was not challenged nor brought into critical consciousness.

Tridentine Catechesis Seen as Revitalization of the Maya Covenant

Because of the ambiguities in the teachings of the catechists and in the mindsets of the Atitecos, the Tridentine catechesis was initially understood to be a continuation of the traditional colonial covenant. It was perceived as teaching the correct prayers to the already existing emanated gods, or perhaps to a new set of gods. The inability of shamans and cofrades to provide relief from the Maya crisis, especially the sicknesses and early deaths described in chapter 6, were occasions for Atitecos to take an interest in the Tridentine catechesis. In the late 1940s, a rain storm in Atitlán lasted three or four days. The watershed off the surrounding mountains flooded the town. Atitecos saw the flood as a punishment of the gods because of their failures to uphold their covenant obligations. In desperation, they asked the catechists to offer prayers to the gods they talked about. They claimed the cofrades did not know how to pray and the shamans appeared helpless. The catechists recited the catechism prayers. The flood subsided. The prayers of the catechists and their gods were proven to have power. Increased requests for catechetical instruction followed (Early 1969).

The Maximon Saga: A Breaking Point Is Reached

Later events made it evident that the Tridentine message required reform: a rejection of the current manner of conducting traditional rituals, of the moral behavior connected with them, and of the worldview that legitimated them. In Santiago Atitlán, the three-year saga of Maximon was the breaking point, occurring about ten years after Padre Lias had sown the seeds of doubt. Maximon is a small human figure made by the Atitecos and revered as one of the most powerful gods of the local pantheon. It is kept in the cofradía Santa Cruz along with an image of Christ-in-the-tomb. Maximon "consists of a flat piece of wood with two legs and a head attached to the main trunk. A carved wooden mask tied around the head serves as its face, an ever-present cigar inserted in the mouth. His costume is a disordered mixture of traditional Tz'utujil pants and non-Maya boots, scarves, suit coat, and not one, but two Stetson hats. The Mam's eclectic taste in fashion reflects his chaotic nature as a deity that violates the order of nature as well as the social norms of Santiago Atitlán" (Christenson 2001: 178; and see the introductory photograph to chapter 5).

The various parts of the figure are kept in the rafters of the cofradía house. They are assembled and dressed for the festivals of San Miguel, of San Andrés, and on Tuesday night of Holy Week. On Wednesday morning of that week, the figure is carried in solemn procession to the town hall, where it presides over the palms and fruits brought from the tropical south coast for the Holy Week decorations in the church. Then it is taken in solemn procession to the front porch of the church, where it is enthroned and lavished with candles, flowers, and cane liquor. A priest arrives and celebrates the Holy Thursday and Good Friday masses. On Good Friday, inside the church a life-size Christ image is hung on a cross by the cofrades, where it is venerated by throngs of Atitecos. After three hours, the Christ figure is carefully lowered into a decorated casket, anointed with shaving lotions, and paraded in solemn procession around the town until late at night, when it is returned to the church. Maximon briefly joins the procession (see the introductory photograph) and then retires to his home in cofradía Santa Cruz. (For an extended account of the Holy Week ceremonies, see Prechtel 1999: 271–327.)

In 1949, a Spanish priest, Padre Alfredo Recinos, began visiting Atitlán to perform the rituals of the major festivals. During Holy Week he observed the Maximon cult for the first time and had the catechists explain it to him (Mendelson 1965: 65–80; Tarn 1997: 1–3, 15–28, 34–47, 60–69). A year later, the Padre again came to Atitlán for the Holy Week masses. Arriving at the church, he saw Maximon enthroned on the church porch amid swarms of Atitecos praying and offering candles, flowers, and cane liquor. Accounts differ slightly about what happened next. Essentially, the Padre went into a rage, grabbed the image, and threw it down the front steps of the church. Infuriated, the Atitecos threatened the priest with stones and clubs. The Padre, anticipating this reaction, went to his room, returned with a pistol, and fired some shots in the air or at Maximon, or both.

The Atitecos again enthroned the image on the church porch. The Padre stalked away from the church to his boat at the lake's edge. The town mayor pleaded with him to return and celebrate the Holy Thursday ceremony and mass. He consented only after a promise that the townspeople would not offer gifts to Maximon while he was in town. That night, the cofrades guarded Maximon with long heavy poles, while the catechists stayed with the priest to protect him. The priest performed the Holy Thursday ceremonies and then left for San Pedro La Laguna. The Atitecos then proceeded to smother Maximon with flowers, candles, and cane liquor.

The Padre was determined to have his way. A catechist (and later a well-known artist), Juan Sisay, tells what happened next.

> One day, not long after that Holy Week in 1950, I met the priest at a catechist's wedding in San Pedro. The priest had told me to tell Atitecos that they should evacuate Maximon from Cofradía Santa Cruz and adore him, if they wanted to with all their customs, as long as it was not done in the same house with the Holy Burial, the Santo Entierro. You know that cofradía Santa Cruz has Crosses, a Holy Burial—a dead Christ in a glass case—and the Maximon up in the trellis. If they did not obey . . . he would be obligated to notify the Archbishop of Quetzaltenango who would come and destroy the Maximon.
> The Archbishop had already written to me, asking for details, but I was worried about acting without authority and suggested holding a session of principales. At this meeting a sacristan assaulted me and almost hit me, saying that I was a criminal to do away with such old customs. . . .
>
> The Archbishop was prevented from getting here by a violent rainstorm so that only three priests arrived: a Father superior, Father Recinos and a third man. I met them on arrival and they went to see the Town Mayor. The Mayor happened to be away and they spoke to a senior officer, the First Regidor. . . . He was violent with the priests and refused them any aid in their plans to burn Maximon in the market place.
>
> Finally they rushed over to the cofradía Santa Cruz taking with them a group of junior officers, the *alguaciles* [policemen]. They prayed for a while and exorcised the house. I stood at the door, not wishing to interfere in such delicate matters. Then they looked for the head and masks of Maximon and took them away.

> Later there was a great outcry in the village and people wanted to kill the priests. But it was too late. Their anger did not turn against me until some time later. (Tarn 1997: 18)

According to other accounts, it was Juan Sisay himself who climbed up the ladder, got a head and mask of the disassembled Maximon and gave them to a policeman. He took another mask and gave it to the priests (Tarn 1997: 19).

The first interpretation of these events by the Atitecos was that the Pope had sent for Maximon to pay homage to him. The principales knew differently. This was an outrageous denunciation of a powerful god of the traditional religion. A breaking point with the priests and their catechesis had been reached. The principales took the rest of the image and put it in one of their houses for safekeeping. They sent a delegation to the governor of Sololá to redress the injustice. He listened politely but did nothing. The mayor, who was sympathetic to the priests and the catechists, issued an edict against the worship of Maximon. The principales agitated the people to protest the edict as an outrage against the traditions of the town.

In 1951, Padre Recinos again visited Atitlán. To thwart any repetition of the Padre's efforts during the coming Holy Week, the most influential shaman of the village sent a telegram to the president of Guatemala, Jacobo Arbenz (Tarn 1997: 21–22).

> Urgent. Santiago Atitlán 19/3/51 14 hours. To the Constitutional President of the Republic: Respectfully we request your orders to continue with the investiture of Saint Simon [Maximon] and the celebration of Holy Week; the parish fathers have forbidden us to perform our ceremonies. We ask you to intervene. We greet you. Baltazar Ajcot by Pascual Pakay.

A reply was received the next day.

> Pascual Pakay and associates, Santiago Atitlán. From the National Palace 20/3/51. The Departmental Government has instructions to intervene in your case. Respectfully Jacobo Arbenz.

An hour later, another telegram arrived from the departmental governor giving permission to proceed with the traditional Holy Week ceremony for Maximon. But to dampen the conflict, the enthronement of Maximon was to take place in a small unused chapel near the church. During Holy Week, Padre Recinos performed the usual ceremonies without incident. At this time, there were approximately eighty Atitecos who were followers of the catechists. Holy Week of 1952 also proceeded without major incident. Padre Recinos complained about the loud prayers and drunken singing from Maximon's chapel near his quarters. The image was moved to the marketplace on the main square.

In 1953, the cofrades decided to return Maximon to his place on the church porch during Holy Week. Padre Recinos learned of it and sent a letter to the mayor forbidding its placement there. It appeared that the cofrades were defeated. But at this point the town's small but influential Protestant faction entered the dispute. In a very unusual alliance, they contacted the departmental governor requesting his support of the cofrades and their image. He replied in the affirmative and the Padre was outmaneuvered. National political parties were beginning to operate in Atitlán at that time, with members from all the town's factions. Political maneuvering was behind the unusual Protestant intervention in defense of a traditional idol. However, the showdown with Padre Recinos did not take place. He was replaced by another priest who simply performed the Catholic ceremonies and ignored Maximon's enthronement on the church porch.

The Maximon saga lasted three years, from 1950 to 1953, occurring about ten years after the return of the priests. The dispute made it clear to all that orthodox Tridentine Catholicism was something distinct from traditional colonial Atiteco costumbre, not a part of it. With Maximon as the example, Tridentine Catholicism rejected the traditional Maya interpretation of the saints, cofradía rituals, and the place of alcohol in them. The catechists suspended their lessons for a short period to let the hostility subside. Juan and Diego, the two catechists who had helped the priests take the mask, were subjected to constant verbal attacks and predictions of their soon-to-be deaths as punishment from the powerful Maximon. But for a number of Atitecos, the fact that Juan and Diego did not die, that there was no punishment from Maximon or any of the gods, that they continued their lives without incident, validated the position of the catechists. This was a frequent outcome of disputes between traditionalists and the priests or catechists. The saga was an important factor in increasing the number wishing to take catechetical lessons. (The stolen mask found its way to a museum in Paris and was later returned to Atitlán [Tarn 1997: 156–83].)

The conflict involved all the elements typical of the clashes between Tridentine Catholics and traditionalists. In the long absence of the priests from the communities, the local authorities had taken over jurisdiction of the churches from the dioceses and their bishops. The local civil authorities then recognized the elders of the cofradías and the shamans as being authorities responsible for the church without denying the authority of the absent priests. Later, the priests, relying on the Spanish history of the churches, assumed that the churches were still under their full jurisdiction as the representatives of the bishop of the diocese. When taking up residence in a Maya town, a priest was usually surprised to find that his was not the local understanding, but a matter needing clarification. In the Atitlán case, the church porch became the battlefield. Frequently, the disputes could not be settled by the priest and traditionalists themselves, but passed to the town's civil authorities for adjudication. In traditional towns, traditionalists also held the civil positions and would rule against the priest. The priest would then appeal to civil authorities at the departmental or national level, where he would receive a favorable ruling from ladino officials. This case was an exception, where freedom of religion to worship Maximon was invoked by the traditionalists against the priest and upheld by the national government.

The Merchants

A segment of the Santiago community showed a special interest in the teachings of the padres. Many merchants and others with familiarity of large market centers and the national culture wanted reform of traditional cofradía service and behavior. As a result of their travels and experiences, they had broken out of the bonds of local community custom, become economically acculturated to national ways of conducting business, and had begun to criticize aspects of their own community as backward. Many of these men were highly intelligent, with an encyclopedic knowledge of many market centers and price structures. They worked hard to expand their businesses. Many had started as solitary tradesmen, buying a product cheap in one market, then hauling it by foot and tumpline to another where they could sell it at a profit. Gradually, they would buy pack animals and hire assistants. Greater immersion in the national economy took place when they could purchase a truck, enabling them to haul a greater volume of goods and, more importantly, to expand their range of markets.

The merchants needed capital for expansion. But the time and money required to fulfill their community service in ceremonial positions absorbed these needed resources. Merchant accumulation of resources for capital investment in a market economy was not understood in traditional Maya communities dominated by the exchange principles of reciprocity. Savings for investment were seen as personal wealth that should be shared within the community. An important mechanism of sharing was to take the more-expensive positions in the cargo/cofradía system, fulfilling the obligations mandated by the community's covenant. Strong pressure to take these positions was brought upon those merchants perceived to have excess wealth. Against this pressure, the merchants wanted a reform of what they saw as degenerate aspects of traditional Maya religion. They especially criticized the costs connected with the ritual system as greatly excessive, and the abuse of alcohol. They saw the latter as degrading Maya men and opening the way for ladino domination by the debts the men incurred. Their criticism was from both a moral and economic viewpoint. Many were committed to the service of their communities and willing to invest considerable effort for reform that would revitalize them (for a similar view of revitalization among the Iroquois, see note 1 and Wallace 1966: 32–36).[1] They were not simply profit seekers escaping into an anomic individualistic world. The message of the padres fitted their need of reform, and many became ardent catechists.

Similar Confusions, and Reform Efforts in Other Maya Communities

The account from Santiago Atitlán is typical of what happened in a number of Maya communities. In San Andrés Semetabaj (Warren 1978: 95), San Bartolomé Jocotenango (González 2002: 279–81), and San Antonio Ilotenango (Falla 2001 [1978]), the merchants were early and eager converts to the Tridentine catechesis seen as reform and revitalization of their communities. Many were among the first catechists. In Alta Verapaz the early catechists "were usually those with exposure to the outside world through merchant work, military service, or labor migration . . . [and] included a large proportion of entrepreneurs in the village—those who supplemented their subsistence agriculture with work as traveling merchants, makers of musical instruments, or transporters of goods by boat or mules" (Wilson 1995, 177).

The catechesis was misunderstood by many of those who apparently embraced it. In Aguacatán, "What made the early converts willing to run this risk [punishment from the traditional gods] was not that they had become cynical or agnostic about the dead [ancestral gods], but rather that they had understood that the new religions were more powerful than the dead and could protect them against their wrath. . . . The first catechist often told me and so did the missionary priest, that the first converts did not really understand the 'doctrina' correctly because they had thought that the 'doctrina' would prevent them from falling ill. Given the role of the dead in causing illness in traditional belief [ancestral punishment for neglect of the covenant], it is clear that those who converted believed they were going to be protected, just as those who remained faithful to tradition believed that the converts would be punished" (Brintnall 1979: 138).

In San Bartolomé Jocotenango, "The most receptive of the new religion were young men tired of the rigidity of *Costumbre*, and men who migrated with their families to work on the plantations and returned home sick and without any money. They were accustomed to seek a solution to their health and economic problems from the shamans without being able to resolve any of them. . . . they adopted the God of the Church as a new unit within the established order of the ancestors and emanated gods. . . . they talked of conversion in terms of its economic implications; they compared the costs and benefits between the 'old' and 'new' ritual. The principal argument in favor of the new religion was that it did not cost anything. The masses were in Spanish and some parts in Latin that the people did not understand. But the ritual and the symbols used in its celebration allowed the belief that one could ask for all the same things that they usually requested through the mediation of the shaman" (González 2002: 280, 285, 289). There were similar initial reactions to the catechesis in San Antonio Ilotenango, Yibeljoj, and San Francisco (Falla 2001 [1978]: 117–38; Mosknes 2003: 118; Koizumi 1981: 188–204).

In San Andrés Semetabaj, there was considerable lag between the introduction of the catechesis and the realization that it contradicted current customs. The dispute was centered on how the festival rituals should be conducted. The catechists, wishing to reform the performance of rituals, originally participated with the traditionalists in the festivals of the saints in the community chapel. Some served in cofradía positions as part of their reform efforts. They would also conduct their own services on these occasions. But tensions built between the two groups over these efforts. Finally, in 1958, the traditionalists, backed by the town authorities, expelled the catechists and their followers from the community chapel. This became a clear sign of the differences between the two groups, seven years after the initiation of the Tridentine catechesis. The catechists reconstructed the community church that had been ruined by earthquakes. It became their center for instruction and ritual.

These breaking points often occurred over who should control the traditional sacred images and sacred places. Frequently, the disputes were over mundane matters, but of great symbolic importance: control of the church keys with the power to regulate times of admittance; types of rituals allowed in the church; ritual drinking of alcohol in the church; use of ritual items kept in the church; cleaning, painting, or repair of the saint images; removal of some saint statues; painting the church interior that knowingly or unknowingly covered over sacred places of the traditionalists (Brintnall 1979: 121–28; Carmack 1995: 228, 232–33, 237–41; Falla 2001 [1978]: 64, 178–82; Melville and Melville 1971: 52–69; Warren 1978: 101–2). It took time for people to realize the different interpretation of the saints. This resulted in a lag between initial allegiance to the catechists and the breaking point with the traditionalists. It was often occasioned by some dramatic event that made clear the renunciation of the traditional worldview.

The Momostenango Experience

This community was the scene of two extreme types of reentry. Two German priests had been in residence for an extended period of time:

> [They took] an active interest in the native religion. They tolerated the Indians' syncretistic ritual practices, rationalizing them as acts of faith and not, therefore, violations of church doctrine. The priests gave the Indians free access to the church. . . . The priest-shamans burned candles to the ancestors on the church floor, a practice so ancient that half a foot of wax had accumulated there. The cofradía officials conducted ceremonial dressing, drinking, and dancing with the saints, saving their noisiest and most drunken excesses for the finale in the church. The priest-shamans were even allowed to bless the plate used each week to bake the Eucharistic host at one of their "pagan" altars. Father Knittel began to teach Catholic doctrine to a group of Indians around 1940, but most of its adepts continued to practice the traditional religion along with Catholicism. (Carmack 1995: 174, 227)

The German priests left Momostenango, and in 1945 Bishop Rafael González Estrada became the pastor. He not only reorganized the catechetical instruction in Momostenango, but throughout all eight departments that comprised the Los Altos Diocese.[2] The bishop visited many towns and small villages on mule back to further the catechetical effort (Falla 2001 [1978]: 172).

> González learned the K'iche' language and preached in that language attacking the traditionalists as idolaters and repeatedly citing the biblical admonitions against worshipping false gods. There was to be no compromise, and one of his favorite scripture lessons was that you cannot "serve two masters." Traditional ritual was denigrated as

barbarous superstition and its practitioners as "savages." The bishop apparently went so far as to refer to the traditionalists as "coons, coyotes, armadillos" and other beasts.... González taught his followers how to chant prayers, how to read the Bible to combat the evangelicals, and especially how to convert the traditionalists. Converts ceremoniously burned their tz'ité bean bags [used for divination] as a sign of religious change. In those days the activities of the ... [catechists] consisted mainly of religious schooling and missionary work. (Carmack 1995: 232-33)

The combative nature of the bishop's remarks reflected his self-image as a Tridentine warrior defending the one, true faith, as well his ladino background. By 1954, González's catechetical program had 3,000 to 4,000 adherents in eight departments of the Western Highlands (Carmack 1995: 227).

Bitter conflicts between traditionalists and the catechists followed in many communities where the reentry was carried out in this manner. In San Antonio Ilotenango, catechetical efforts began in 1937. There were only three elderly priests in the Department of Quiché at that time who tolerated the ceremonies of the traditionalists. In 1952, the crusading Bishop González visited San Antonio. "Arriving with the bishop was a catechist who preached movingly in the Quiché language. Listening to him were not only Catholic Action members [in the broad Guatemalan usage of the term[3]], but also *cofradía* members and some principales, who had gathered to welcome the bishop. The catechist unleashed an attack against the *zahorines* [shamans], candles, drinking" (Falla 2001 [1978]: 175). The attack was bitterly resented by the traditionalists, and made clear the division of the two groups. The traditionalists rose up against the catechists, "because they wanted to take away the Tradition." The breaking point here occurred fifteen years after the reintroduction of the Tridentine catechesis. Soon afterward, the traditionalists threw stones at the head catechist and beat him with sticks. At that time, there were about thirty members of the Catholic group.

González's catechetical effort became part of Archbishop Mariano Rossell y Arellano's national catechetical plan. Rossell saw the programs of land reform and peasant rights of the Arévalo-Arbenz period (1945-54) as a dangerous drift toward communism and feared that it might find a receptive hearing in the impoverished Maya villages. "Today [the Indian population] is a tame and long suffering lamb, but it is very easy to turn it into a cruel work, or a ravenous lion, or a poisonous snake" (Rossell 1955, in Warren 1978: 89). Like most of the socially distant ladino clergy,

Rossell considered the Maya to be ignorant Catholics. It was hoped that Maya catechists would lead a Tridentine revival and that, as community leaders, they would counteract any possible subversive activity. Adding an anti-communist crusade to catechetical efforts among the Maya had little relevance for most Maya, who knew nothing about communism or even had heard of it. The main objective remained the instruction of the Maya in orthodox Tridentine Catholicism, with their consequent rejection of what were considered to be pagan beliefs and practices. Rossell also wished to encourage education among the Maya and founded schools for Maya boys and girls in the capital. He was also concerned about fending off the increasing inroads of Protestantism (Calder 2004: 90).

Beyond Ambiguity: A Mature Version of the Catechesis

The San Andrés Semetabaj version of the catechesis was recorded by Warren from well-informed, older catechists in 1970–72, twenty years after its introduction. It represents a mature version used by catechists who understood the Tridentine difference from the traditional colonial Maya worldview. The writer also heard this catechesis in Santiago Atitlán. Catechists held regional meetings and exchanged ideas, so that it probably emerged from the input of catechists from a number of towns in the area.

As seen in chapter 2, the main concern of the traditional Maya covenant was the survival of the material body for the length of its allotted cycle in spite of the many dangers encountered during its passage through the life cycle. As long as the covenant was kept, there would be harmonious balance in the community, resulting in material survival and prosperity. The catechesis called harmonious balance by a Tridentine term: the state of grace. But contrary to the traditional worldview, the Tridentine catechesis saw it resulting from individuals giving primacy to the care of the spiritual soul over concerns for the material body. "Jeopardizing grace are the ever present temptations of the World, which upset the balance of the body and soul and lead the Christian into a state of sin. The individual may regain grace through a cycle of rebalancing the stress between body and soul. First, the Christian acknowledges his or her sin by repenting. Then the Christian shows personal understanding of the relation between internal intention and external action by confessing to the priest to gain God's pardon. Finally, the Christian is reconciled with the brothers and sisters in Christ [by receiving the Eucharist] in order to cease being a detrimental model for the Christian community" (Warren 1978: 115; see also 103).

Harmonious balance, the state of grace, is a spiritual gift from God through the merits of Jesus Christ, who gained salvation for humans by himself willingly undergoing the familiar Maya cycle of birth-death-regeneration. As abstractions, the Maya and biblical cycles are similar. The state of grace is maintained by observing the commandments listed in the catechism that were not unlike the moral precepts of the wisdom of the ancestors. It is lost by sin, defined as giving the demands of the body primacy over those of the soul. The devil is associated with the will of the material body in its effort to have the material triumph over the spiritual. The state of grace, redemption from sin, harmonious balance, is regained through the sacraments of penance and the Eucharist. The apostles and the saints were seen as ancestors who were successful in overcoming the tension between soul and body and, as a result, maintained balance in their lives (Warren 1978: 113–16). "These three sets of contrast; body-soul, internal-external, spiritual-material; gave Catholic Action [in the Guatemalan sense; see chapter 9] a single analytic framework within which to discuss the individual as well as religious groups in the community. To be a moral actor was to choose both to recognize these distinctions and to struggle for a balance between them" (105). With this message, "literate converts who had studied the Bible and catechism in depth became catechists with special duties to direct meetings and instruct new converts" (97).

Critique of Cofradía Rituals

The three contrasting categories framed the catechists' rejection of what they considered the degenerate rituals practiced by the cofradías.

> By not distinguishing the soul from the body, the brotherhoods were characterized as celebrating the body through the "drunkenness, fights and impurity" of the dances. The "doctrine" of the brotherhoods became the expression of conflict between individuals and families because Indians awaited the brotherhood dances to bring up problems they were usually unwilling to express. Along with ritual meals, procession and masses, brotherhood-sponsored dances marked high points in the ritual cycle. . . . In the tightly packed dancing hut where the marimba played long into the night. . . . Sons fought with their fathers, neighbors aired old grievances, wives left the dance hut to initiate affairs and ladinos became involved in altercations. To the brotherhoods, who recognized this variation from the social order, it was an important part of the gaiety and excitement of the religious celebration. They went no further than *costumbre* to explain the significance of brotherhood rites. . . . For members of Catholic Action . . .

> this part of the celebration was ... becoming one of the chief enemies of the soul and a place where the Devil tempts man to sin.
>
> The disorder of the brotherhood celebrations was also thought to exemplify an imbalance between internal and external manifestations of belief as well as between spiritual and material aspects of life. The brotherhoods relied on external manifestations of belief such as the celebration of the saints in processions and the decoration of altars and tombs with flowers without paying attention to the internal forms of celebration such as meditation and mental prayer. ... By failing to consider the difference between the spiritual and material domains of life, the brotherhoods emphasized the material pole. They were concerned with the immediate material benefits of the ritual. (Warren 1978: 103–4)

Given the other internal strains in the community as a result of cofradía and shamanic demands, discussed in chapters 4 and 5, this critique for revitalization found a receptive audience among significant numbers. The catechesis was seen by Catholics as a reformed, revitalized morality. It attempted to liberate the community from the consequences of both ritual and secular drinking. Among these consequences was freedom from debt to ladino labor contractors and money lenders that led to forced work on ladino plantations—in short, liberation from ladino domination.

Critique of Ladino Conduct

The critique of local ladinos used the same principles. "Ladinos do occasionally participate in the sacraments; yet their behavior belies the fact that they see no relation between the state of grace and behavior, ... they are talked about in the same terms used for Indian converts who forsake Catholic Action for brotherhood activities. Their intentions are not consonant with the belief system of the Catholic Church. Local Ladinos lack a spiritual concern, whereas Catholic Action is quick to point out that many Ladinos in the capital are practicing orthodox Catholics and have become active in other chapters of Catholic Action" (Warren 1978: 133).

Tridentine Catechesis as Revitalization of the Colonial Maya Theological Worldview

The first generation of catechists in San Andrés did not see themselves as completely rejecting the colonial Maya worldview on which they were raised. Chapter 3 described how the San Andrés worldview had two differing explanations for the origin of cofradía rituals. One saw them as received from the ancestors, and the other as brought by the Spaniards to the Maya living in the wilds as uncivilized people. The catechists saw themselves as k'amöl b'ey, principales known for their knowledge and role in perpetuating these rituals (Warren 1998: 164). Current cofradía beliefs and practices were degenerate because cofradía officials no longer followed the teachings of the catechism and mixed Catholicism with the confusion of the undifferentiated pre-Columbian epoch (Warren 1978: 99). The result has been the Maya crisis described in chapter 6. Therefore, to revitalize colonial Maya culture, there must be a return to the teachings of the Tridentine catechism. For Catholic priests unaware of the Maya metaphysical paradigm, the catechists and their followers were simply orthodox Catholics.

Conclusion

This chapter has examined the reintroduction of Tridentine Catholicism into Maya communities. From the viewpoint of the Catholic priests, its purpose was to convert the Maya to orthodox Catholic belief in place of their current ignorance and superstitions. While the priests were concerned with moral action, in the Tridentine worldview this had to begin with correct belief and participation in sacramental rituals. From the viewpoint of the reform-minded catechists, their primary concern was revitalization of the community by actions that did away with degenerate aspects of the current ritual system, especially the abuse of alcohol and needless ritual expenses. Tridentine belief and ritual were a means to these ends. This difference of emphasis between belief and action was seen in the evolution of Christianity from Hebraic to Greek forms of expression, discussed in the previous chapter.

Due to the pantheistic character of the Maya worldview, it lacks firm systemic boundaries. By its concepts of emanations, epochs, and cycles, it can easily expand and absorb elements from other systems without undergoing systemic change itself. In the mid-twentieth century, the initial reaction to the reintroduction of Tridentine gods was much the same as in the sixteenth century. At that time, the Catholic elements were absorbed into the traditional covenant and seen as stronger prayers or gods that would protect against sickness and early death—a revitalization of the existing covenant. Various forms of further revitalization will be seen in upcoming chapters.

The descriptions in this chapter have been ideal types. In situations of transition, it is difficult to know how individuals may interpret paradigms that are historically or logically distinct. As Christenson (2001: xvi) notes about Santiago Atitlán: "This book does not present a unified view of traditional Tz'utujil-Maya theology, for such a thing does not exist. Researchers . . . have noted that religious beliefs and practices among the Maya may vary from individual to individual. This is because there is no unity of opinion in matters of Tz'utujil faith. Certain core myths are widely known among nearly all segments of the community, but the particulars of these stories are learned primarily through oral tradition and thus are not codified in any single source." This also applies to catechisms and the Bible. They were read to mostly illiterate groups by catechists, and certain core elements were stressed, but their comprehension was always individualistic.

This was the face of Catholicism in the Maya areas of Chiapas and Guatemala during the 1950s and 1960s. The conflict with traditionalists touched many parts of a community's social structure (Falla 2001 [1978]). In the polemic against the traditionalists, many in the next generation of Catholics were taught to have little respect for the traditional culture in any of its forms. Each community had its own history of how the renewed contact with the priests and catechists played out. Each individual had his or her personal psychological history of sorting through the cultural crisis to find self-identity. This chapter has attempted to sketch some of the trends.

IV

Crisis within the Catholic Worldview

As Maya Catholics became more familiar with the Tridentine catechesis, they began to speak out about its failure to confront their continuing subsistence crisis. At a 1968 meeting of the Tzeltal zone of the San Cristóbal Diocese, elder Maya catechists made clear to the bishop a problem that was being discussed in Maya communities.

> The Church and the Word of God have told us things to save our souls but we do not know how to save our bodies. While we work for the salvation of ours and the souls of others, we suffer hunger, sickness, poverty and death. Now we know the Bible, its authors, the names of its books . . . we sing and pray every Sunday. . . . There are catechists, choirs and leaders . . . but hunger and illness and poverty continue without end. (Womack 1999: 29; Kovic 2004: 195; Floyd 1997: 107)

The statement reflected the continuing influence of the Maya covenant, where survival in this life for the duration of one's allotted cycle is its main purpose (see chapter 2).

9

Beyond Tridentine Belief and Ritual

Worldview of Vatican Council II

The problems that the Tridentine catechesis failed to confront were forcefully enunciated at the Indian Congress of 1974 (Ruíz García 2003: 58–63; Ruíz García 1994: 597; Iribarren 1985: 49–50, 61–62; Fazio 1994: 103–5; Morales Bermudez 1995: 305–40). The year of the congress coincided with the 500-year anniversary of the birth of Bartolomé de Las Casas, the Dominican friar who was the first bishop of the diocese and an ardent defender of indigenous rights. The governor of Chiapas wished to take advantage of the anniversary by sponsoring a commemoration that would make Maya leaders amenable to an alliance with the government's political party (the PRI). He appointed an organizing committee and proposed that they bring together a group of experts to deliver scholarly papers about

Las Casas. Bishop Ruíz, a member of the committee, objected, saying that the Maya would receive little benefit from listening to scholarly papers read in Spanish about an unknown person. Instead, he suggested that it would be much more appropriate to have the Maya themselves organize the celebration. The committee agreed, and asked the bishop to implement the suggestion. He consented on one condition, that the Maya themselves have a voice in the celebration. To carry out the proposal, the bishop called on the network of catechists in the Tzotzil, Tzeltal, Tojolabal, and Chol zones of the diocese.

The catechists turned the celebration into a public demand for social justice and assistance in meeting their subsistence crisis. Word of God communities, led by their catechists, came together and discussed their ongoing problems. Over the course of an entire year, the catechists talked with 400,000 people in a thousand communities. Then the catechists in each zone and, later, the representatives from each zone, met to synthesize their discussions. They found that all the communities had four problems in common: lack of land, lack of health facilities, lack of educational resources, and persistent inequities regarding access to and value received in market transactions for their goods. The results were presented at a convocation in the San Cristóbal convention center on the anniversary of the friar's birth. A large banner stating the theme of the convocation, Igualdad en la Justicia (Equality in Matters of Justice), was placed over the entrance to the center. Two thousand Maya came to San Cristóbal to hear presentations in Tzotzil, Tzeltal, Chol, and Tojolabal, with simultaneous translations between these languages and in Spanish. Over three days, members of each zone presented their analysis of the problems, accompanied by sharp denunciations of the role of government corruption in each problem area, and proposed solutions. The governor never returned after the first session.

The congress initiated a Maya civil rights movement, as it brought together different Maya groups and gave them a sense of a common goal thwarted by systemic resistance. The attendees attempted to form a permanent organization, independent of both church and government. They conducted a series of seminars in the various zones over the next few years, the content of which included the history of Mexico and Chiapas, elements of political economy, lessons on how to form organizations, and the dignity of work. But due to internal disagreements over specific goals and strategies, the organization fell apart in 1977. The 1974 congress set an agenda for the pursuit of social justice, a theme of the upcoming chapters.

Extent of the Catholic Crisis

Without the Maya realizing it, the complaint of their elder catechists formulated a serious problem with Tridentine Catholicism, not only in the Maya areas, but for Catholicism worldwide. In the Middle Ages, theological reflection about the problems of Europe dominated Catholic thinking and expression. As a defense against the Protestant reformers, the Council of Trent had focused on the individual, by fortifying belief in traditional theological formulations and by its emphasis on sacramental rituals as necessary for salvation.

By the nineteenth century, this narrow focus had little relevance for Europe's problems. Europe had evolved from a hierarchical, Christian, agrarian society with incipient capitalism into an industrial, capitalistic society heavily influenced by the scientific revolution and the philosophies of the Enlightenment. Western culture was becoming more and more secularized. Many, unable to find meaning in Trent's presentation of Christianity, affirmed atheism or agnosticism along with a skepticism about life's meaning. Some Christians agreed with Marx's indictment of Christianity for its failure to be concerned with contemporary communal life, which was wracked by problems of social justice arising from early industrial capitalism. "The social principles of Christianity preach the need of a dominating class and an oppressed class. And to the latter class they offer only the benevolence of the ruling class. The social principles of Christianity point to heaven as the compensation for all the crimes that are committed on earth. The social principles of Christianity explain all the viciousness of oppressors as a just punishment either for original sin or other sins, or as the trials that the Lord, in infinite wisdom, inflicts on those the Lord has redeemed. The social principles of Christianity preach cowardice, self-hatred, servility, submission, humility—in a word, all the characteristics of a scoundrel" (Marx, in Gutiérrez 1968: 76). A number of Catholics agreed with the indictment. While dissenting from the Marxian position that religion was simply a superstructure that reflected the economic organization of a society, they saw the indictment as indicative of a religious system that had degenerated and needed reform, revitalization.

In addition, European Catholicism was encountering new worldviews and social structures in the non-Western cultures of a world constantly shrinking by the tightening bonds of communication and transportation. Western culture, Christianity, and Catholicism, all were undergoing a crisis of worldview, identity, and social structure not unlike that of Maya culture. They also needed revitalization.

Catholic Reform Efforts

New movements were arising within Catholicism that did not see Christianity as adequately expressed by the Tridentine paradigm. Broad-based coalitions of clergy and laity emerged that insisted on taking the contemporary world seriously and finding meaning in it. They pressured the upper levels of the clerical bureaucracy to realize that it could no longer continue the illusion of living in the Tridentine sixteenth century. Beginning in the late nineteenth century, a loosely structured movement called Catholic Action emerged in Italy. Its main emphasis was the laity acting in secular society on the basis of Christian principles. Originally, it focused on protecting the church. As the movement developed, the emphasis shifted to action that implemented social justice and Christian family life in a society convulsed by the problems of transition from an agricultural society to industrial capitalism. For these purposes, clerical Catholicism began to recognize that it was necessary to meaningfully incorporate the laity as active promoters of the Gospel in a society growing more specialized and complex. Catholic Action encompassed a broad spectrum of activities. At one extreme it included any communal contributions by Catholic laypersons inspired by their faith. At the other extreme the name was restricted to those actions of lay groups mandated by a local bishop (Geaney 2003: 3:275–78). The movement was more interested in action than in theological speculation, definitions, or clerical authority.

Belgian Catholic Action (or JOC, for Jeunesse Ouvrière Chrétienne [Young Christian Workers]) became a template for many in the movement. Lay Catholics who worked or lived in similar surroundings were encouraged to meet in small groups. For their agenda, they employed the triadic formula, "observe-judge-act." Observe meant understanding the dynamics of the particular social milieu in which they participated: university, office, factory, etc. Judging involved assessing the morality of the milieu. Acting involved implementing a specific plan of action to inject Christian social principles. (Some groups substituted "think" for "judge," to avoid the connotation of self-righteousness.) For the most part, they used the Bible rather than catechisms as a stimulus for their reflections. The movement insisted on a strong spiritual foundation as a basis for the activity of the laity. Its spirituality was centered on the life of Christ as a moral teacher and social prophet. "Act" frequently led to involvement in social, economic, and political organizations seeking social justice. These small groups were assisted by insertion into regional organizations with elected officers and boards of directors. While clergy were frequently involved in advisory roles, the laity were the experts about their specific social environments. The "lay apostolate," a term often used for the Catholic Action movement, was recognized in a series of social encyclicals in which the Vatican began to understand some of the problems of industrial societies, such as the Rerum Novarum (1891) of Leo XIII, on the conditions of working men; its reaffirmation, in Quadragesimo Ano (1931), by Pius XI; and Mater e Magistra (1962) and Pacem in Terris (1963), by John XXIII.

The Problem of Change within Catholicism

It is necessary to understand Catholic structure and how change is possible in an institution that has evolved into a divine-right royal bureaucracy. One stereotype sees all Catholics as passive individuals waiting for directions from a divine-right Pope. This view has its basis in an exclusive, top-down view of ecclesiastical structure, with the Pope as the highest and supreme authority, along with his Vatican bureaucracy. The next level of authority consists of the councils, for which all Catholic bishops are called by the Pope to consider matters of serious concern to the church. All their decrees must be approved by the Pope. Below the councils are the regional conferences of bishops, called by their presidents, who are elected by the bishops themselves. Again, all final results must be approved by the Pope. At the lower level are dioceses, presided over by an individual bishop. These are composed of local communities, or parishes, presided over by a pastor.

This structure retains its medieval characteristics, a divine-right kingship with supreme authority over all lower bureaucratic levels. It is a structure that assumes a society largely composed of two dominant tiers—an educated elite of nobility and higher clergy comprising an upper level, and a lower level composed of an uneducated peasantry and local clergy. The latter need guidance from the upper level, to whom authority has been entrusted by God. It is a structure vastly different from the Pauline churches of the early centuries of Christianity (McBrien 2008), and it has had many problems coming to terms with industrial democratic societies.

However, in spite of its emphasis on the top-down royal authoritarian structure, the system also functions within a ground-up process. Recommendations and petitions can ascend the same line of organization, with initiation at any level. Local parishes can petition bishops for changes. Individual bishops may organize movements for change among conferences of bishops, or in the Vatican itself. Creative theologians, like any other group of scholars, do research, speculate, and write about the meanings of Christianity in various historical periods and cultural circumstances.

Movements begin on the lower levels and may eventually seek approval at the top.

The push for changes from below usually creates controversy. The Catholic bureaucracy, like any bureaucracy, is a slow-moving, cumbersome, often self-protective organization, very suspicious of any perceived threat. Past tradition is often seen as a sacred entity, and therefore unchangeable. At the councils and bishops' conferences, committees of bishops are assigned to write position papers about the topics under discussion. The bishops are usually assisted by experts on these topics. The documents produced by the committees are submitted to the council or conference, where they may receive a vote of approval, rejection, or be sent back for revision. There is usually intense politicking within these committees or on the assembly floor between those who support and those who oppose change, or sometimes over the means to accomplish change. Encyclicals of the popes are Vatican pronouncements that go through a similar process but involve a much smaller group.

The controversies result in a series of final documents from the assembly, which are frequently of excessive length, repetitious, and confusing. These documents constitute a unique genre of writing. The Protestant scholar, Robert McAfee Brown, with the help of a Catholic bishop, deciphered how to read the genre.

> An initial reading of the Puebla document is likely to be a letdown, particularly to Protestants unused to episcopal prose. The document is rambling, repetitious, and occasionally contradictory. Sometimes it rises to eloquent heights, on other pages it resembles an accumulation of trivia. One almost wishes there could have been a winnowing process so that the best could have emerged more clearly without being so hard to find. . . . The length and frequent ambiguity need not, however, be cause for despair. For I learned something as a Protestant "observer" at Vatican II about reading Catholic documents that has held me in good stead ever since. I was complaining one night about the constant references to previous popes "of blessed memory" in order to show that the new statements were "what the church has always taught." A bishop (whose name it would be madness to reveal) said to me, "Discount everything in the draft that simply repeats what was said by earlier popes or councils; that is Roman style and it appeases the conservatives. Look instead for the two sentences where something *new* is said, for that has a chance to emerge as the 'teaching' of the council, and it indicates the direction of the future of the Church. Concentrate on the two sentences. We often trade [in

committee or on the assembly floor] three paragraphs for two sentences. It's a good bargain. (Brown 1979: 331)

The key two sentences become acceptable changes, some of which have already taken place on lower levels, amid controversy. Acceptance of change allows the now legitimized movement to go forward, and for possible implementation of the change among a wider audience of Catholics, but still amid controversy and resistance. After a council or conference, the usual comments are, "It did not go far enough," by those pressing for change, who are often labeled as liberals. Or, "It went too far," by those resisting change, labeled as conservatives. This sets the stage for the next protracted cycle of controversy.

Vatican Council II

To shake off the limitations of the Tridentine church, to confront the challenges of the contemporary world, to review the Catholic movements from below pressing for changes, to revitalize a Catholicism in crisis, "to throw open the windows and let in the fresh air," in the words of Pope John XXIII, he unexpectedly convened Vatican Council II in 1962 over the strong opposition of his Vatican bureaucracy with its Tridentine outlook. The Council was held in four sessions between 1962 and 1965. Catholic bishops, predominantly Europeans but including others from Catholic dioceses worldwide, wrestled with the problem of the formulation of Catholicism for the twentieth century. "Therefore, the sacred Synod proclaims the highest destiny of man and champions the godlike seed that has been sown in him. It offers to mankind the honest assistance of the Church in fostering that brotherhood of all men which corresponds to this destiny of theirs. Inspired by no earthly ambition, the Church seeks but a solitary goal: to carry forward the work of Christ himself under the lead of the befriending Spirit. And Christ entered this world to give witness to the truth, to rescue and not sit in judgment, to serve and not be served" (Gaudium et Spes #3). The self-image of the church as servant was radically different from the triumphant Tridentine image of the church as the supreme authority and infallible judge. The Council covered many topics described in volumes dedicated to the history of its deliberations (O'Malley 2008).

Below is a brief review of some themes important for the Catholic presence among the Maya. For the Tzeltal catequists' complaint that the Tridentine catechesis was not doing anything for their bodies, some of Vatican II's discussions were of special importance. The most significant document was the pastoral constitution, *The Church in the Modern World*, the longest and most-detailed of the Council's sixteen documents. It attempts to overcome Trent's unbalanced treatment of Christianity by showing concern for the contemporary world. Throughout the document there is a strong emphasis on social justice as the road to peace in the international community, and on the moral responsibility of Christians to be involved in its pursuit. As with all such bureaucratic documentation, the continuing problem was its implementation by Catholic bishops, priests, and laity. (The introductory photograph to this chapter shows Maya outside the San Cristóbal cathedral waiting to greet Bishop Ruíz upon his return from a session of the Council. The banner reads, "Your children of the Mexicanos neighborhood deposit their hearts at your feet.")

Given the unjust seizure of Maya land by violence or illegal means; given the resulting shortage of land; given the lingering debt servitude on plantations; given the unjust discrimination in the labor and produce markets; given the subsequent impoverishment of Maya communities, especially the high rates of sickness and mortality at an early age; given the denial of health and educational services from the public sector—given all of this, the Council's decrees about the concern of the church for contemporary life were of great importance. "The Council focuses its attention on the world of men, the whole human family along with the sum of those realities in the midst of which that family lives. . . . Therefore while we are warned that it profits a man nothing if he gain the whole world and lose himself, the expectation of a new earth must not weaken but rather stimulate our concern for cultivating this one" (Gaudium et Spes, #39; #2). The document contains extended discussions about human dignity and cultural and economic development.

The Council defined the church as the "People of God," the vast majority of whom are laity. To carry out many of the church's concerns for a world dominated by complex industrial and industrializing societies, a laity dedicated to the principles of Christ is needed. Their function is "evangelization . . . [in which the] announcing of Christ by a living testimony as well as by the spoken word, takes on a specific quality and a special force in that it is carried out in the ordinary surroundings of the world. . . . For where Christianity pervades a whole way of life and ever increasingly transforms it, there will exist both the practice and an excellent school of the lay apostolate [Catholic Action]" (Lumen Gentium, #35). The means to accomplish this depend on the circumstances of a person's life. "By their competence in secular fields and by their personal activity, elevated from within by the grace of Christ, let them labor vigorously so that by human labor, technical skill and civic culture, created goods may be perfected for the benefit of every last man, according to the design of the Creator and the light of His Word" (63, #36). Most of the Maya area by the mid-twentieth century consisted of relatively isolated villages, untouched for the most part by modern technology. Pastoral workers, mostly priests and nuns, were often the only ones present who could initiate such endeavors with a view to empowering the Maya to undertake various projects of their own empowerment.

Terminology: Catholic Action, Liberation Movement, Action Catholicism

This section attempts to clarify the use of these terms in this and the following chapters. Cardinal Saliège, the former archbishop of Toulouse was one of Catholic Action's early advocates. "Less concerned with theory than with contemporary conditions of life that many found unworthy of human beings, [he] viewed Catholic Action in terms of institutional change, having for its task 'to modify social pressure, to direct it, to make it favorable to the spread of the Christian life, to let the Christian life create a climate, an atmosphere in which men can develop their human qualities, can lead a really human life, an atmosphere in which the Christian can breathe easily and stay a Christian.' It would, he said, 'lift up the mass, not a couple of individuals; the mass, prompted and set in motion by a natural leader chosen from the mass and remaining part of the mass'" (Geaney 2003: 3:277). These ideas were reflected in the description of the lay apostolate by Vatican II.

For analytical purposes, this writer prefers to include "competence in secular fields," "technical skills," and "civic culture" as necessary criteria of Catholic Action. Contrary to a frequent usage in Guatemala and elsewhere, the work of the Maya teaching catechism is not Catholic Action, since it is not a secular field. Also, the writer finds it analytically helpful to add an additional criterion. Catholic Action is often focused on local communities. This has precedence, as European Catholic Action was originally composed of small groups, cells, who were concerned with instilling Christian principles in their immediate communities. Also, in earlier concerns about the "underdeveloped world," the focus was on local community development. As Catholic Action worked at this level, it began to realize that problems were not simply local. It began to mature in its outlook and also focus on entire social systems. Here it takes on the characteristics of what will be called the liberation movement, with its systemic outlook, which will be discussed in the following chapter. There are circumstances in which it is difficult to distinguish the two, especially in transitional situations or where there is a lack of detailed information. For these reasons, in the following chapters, the writer has sometimes used the generic term "Action Catholicism" to mean either Catholic Action or the liberation movement, or a combination of both.

In Guatemala, beginning with Archbishop Rossell, the term "Catholic Action" was used to describe the Maya laity acting as catechists. They were substituting for the shortage of priests in a sphere of competence not specific to the Maya laity. This usage has persisted, so that in Guatemala the term "Catholic Action" came to be used indiscriminately for Maya laity giving catechetical instruction, or involved in local community-development programs, or undertaking activities characteristic of the liberation movement. This vague usage led to some deadly results, as will be seen in later chapters.

Summary

This chapter has sketched the attempt of the Catholic Church at Vatican Council II to widen its Tridentine vision, to revitalize Catholicism and restore its concern for the here-and-now human community. Social justice was seen as a principle of world order, and there was a realization that the laity have a distinct religious function both within the church and in the wider community. The window had been thrown open. It remained to be seen how much fresh air would come in, to what degree the words on paper would be translated into action. The Council framed a dynamic of controversy among Catholics for the remainder of the century and beyond, between those now-legitimized movements advocating change versus those fighting to retain a Tridentine church.

10

Crisis and Reaction in Latin America

The Liberation Movement

The statements of Vatican II were necessarily general and abstract as they were intended for Catholics worldwide. Their specification was left to bishops' conferences that would take into account the local circumstances of their dioceses. This was the task of the Second General Conference of Latin American Bishops, held at Medellín, Colombia, in 1968. In the years prior to the conference, there existed within the Latin American church a liberation movement (too-narrowly labeled "liberation theology"). It had anticipated and influenced some orientations of Vatican II. This movement has generated a literature, theological controversy, social action, and consequent Vatican oversight. It influenced the formation of many of the younger pastoral workers active among the Maya in Guatemala and

Chiapas. The purpose of this chapter is to briefly sketch some trends of consequence for Maya communities.

Social Conditions in Latin America

The crisis in Maya communities discussed in Part II was a relatively small example of what was happening in many Latin America communities. It was more severe in indigenous communities, but the rural peasantry and urban poor throughout the region were also in crisis. From the sixteenth to nineteenth centuries, European nations had embraced the philosophy of colonialism, frequently legitimated by their national religions under the guise of civilizing the local labor force. Over four centuries, much of Africa, Latin America, and significant parts of Asia were brought under European control. As colonialism waned in the nineteenth and twentieth centuries, these societies gained political freedom. But they had not been prepared for their newfound freedom. They were left with their economies in shambles, presided over by local oligarchs who had prospered as puppets for the colonial powers. Political independence did not bring economic independence. The international system, in conjunction with local economic and military elites, continued the traditional systems of domination. These elites were determined to maintain control of their countries, to continue their luxurious lifestyles, and to increase their wealth at the expense of the poverty, sickness, and early deaths of the rest of the population. In their worldview, the peasantry existed only to supply cheap agricultural labor. The urban poor, the inhabitants of the immense shantytowns within and surrounding almost all Latin American cities, existed for a similar purpose, to feed incipient industrialization. It was a situation crying out for social justice.

This scenario was especially true in Central America and Mexico. From 1936 to 1979, Nicaragua was under the iron fist of the Somozas and their National Guard. All the presidents of El Salvador from 1944 to 1980, and of Guatemala from 1954 to 1986 (with one insignificant exception), were military generals presiding as dictators. As a part of Mexico, Chiapas was under the dictatorship of a political party (the PRI) that furnished all of Mexico's presidents from 1929 to 2000. The common characteristic under all these governments was loss of land by the rural peasantry and their impoverishment as economic and military elites assembled huge landholdings to supply raw materials for international markets. If the peasants were Indians, so much the easier to take their land, as seen in Part II. In these circumstances, the Latin American church needed its own distinctive response, different from the European formulations that historically had dominated Catholic concerns at higher ecclesiastical levels.

A Catholic Response: The Liberation Movement

A number of liberation movements sprang up around the world in reaction to colonialism and co-opted nationalisms. The Latin American version was an all-embracing religious-political-economic movement of disparate and sometimes contradictory elements. For this reason, the term "liberation movement" is used here instead of the more frequent term "liberation theology." Theology was an important aspect of the movement, but only one of several aspects. Even with regard to theology, Segundo (1990 [1983]: 361) notes, "We are faced here with two different theologies under the same name: different in scope, different in method, different in presuppositions, and different in pastoral consequences."

The essence of the movement can be summarized by the answers to the three questions posed by Catholic Action's triad. The first is liberation from what—that is, what reality gave rise to this movement? In Latin America, it was the recognition that the majority of its people were living in conditions of dire poverty because of systemic oppression that existed at all levels of the social structure: local, national and international. Catholic Action, as an older effort at liberation, had focused on development programs within local communities. Given the time and effort required for educating communities about development programs and implementing them, material and human resources were originally concentrated at this level. As Catholic Action evolved, it realized that its focus on local communities was insufficient, that local problems also stemmed from the larger system in which the local communities were embedded. Local Catholic Action programs were necessary, but in themselves did not get to the root of the problems and often withered for lack of sustained support from the larger society.

The liberation movement placed the problems within a critical systemic view that included regional, national, and international systems. The ordinary Catholic was unaccustomed to this type of thinking. As seen in chapter 7, the Council of Trent had formulated a Catholic worldview that theologically legitimated a social system based on divine-right authority, of both kings and popes. It took this system for granted and emphasized obedience to it. If there was criticism, it was quickly suppressed as either treason or heresy. The decrees of the Council of Trent focused on individuals rather than on a systemic view of society.

The next step of the triad was the moral judgment about these conditions, as well as the individuals and institutions responsible for them. Based on the intrinsic dignity of human beings, no matter on what basis this dignity is affirmed, one could only morally condemn the colonial legacy. For this, no theology was needed. It was simply a matter of common sense (Segundo 1990 [1983]: 355; Ruíz García 2006). But to combat the oppressive use of a theology as an ideology that ignored and, in some cases, legitimated systemic injustice, and that blindly influenced many Catholics, a theology of condemnation was needed. The moral imperative calling for action stemmed from the biblical example, both in words and deeds, of Jesus in his prophetic role of liberator. For guidance and empowerment, Christians also turned to biblical narratives of the oppression and liberation of the Jewish people. This was not a new theology. It was bringing back to life a part of the Judeo-Christian tradition that Tridentine Catholicism had pushed aside.

The final question was what to do about it. The answer depended on the circumstances of the individual, his or her community, the country, and the particular moment in history. But regardless of the circumstances, the movement mandated that there must be some kind of concrete, realistic, liberating action by both the oppressed themselves and others. Resignation, self-pity, despair, blind acceptance of corrupt authority by the oppressed would not do. For those not of the oppressed, there must be actual experience among the poor, a listening to and learning from them to undergo a conversion experience. Only then would people become involved in helping to liberate communities from oppression. Mere empathy, charity at a distance, or well-meaning actions proven inadequate in the past would not do. Realistic action was required—the third phase of Catholic Action's triad. An example is shown in the introductory photograph, which shows Thomas Stafford in Santiago Atitlán teaching soil enrichment for small plots by worm cultivation.

Table 10.1 shows the skeleton of the liberation movement. There are three levels of involvement: professionals, pastoral workers, and the heart of the system, the Word of God communities, also called Christian Base or Catholic Action communities. (Some commentators make careful distinctions between these terms that are not helpful for purposes here.) The table briefly describes some characteristics of each type of involvement. All employ biblical reflection as an aid to their examinations of current realities (see Part V). The flow of activity is not simply from the professionals to the Word of God communities. All in the movement agree that there is creative theological input produced by the individual communities

Table 10.1. Structure of the Liberation Movement

Characteristics of Involvement	Type of Involvement		
	Professional	Pastoral Worker	Popular
Roles	Theologians, professors, teachers	Priests, religious, laity	Any interested person
Method	Social and biblical analysis; theory related to action	Observe, judge, act, aided by biblical reflection	Observe, judge, act, aided by biblical reflection
Logic	Methodical, systemic, dynamic	Related to specific action; prophetic	Related to action in everyday life
Oral Presentations	Conferences, lectures	Sermons, talks, group discussions	Group reflections, celebrations
Written Presentations	Books, articles	Pastoral guidelines	Notes, letters
Discourse	Detailed, rigorous	Practical	Spontaneous, diffuse
Place	Theological institutes	Pastoral institutes	Base communities
Promotion	Theological congresses	Pastoral congresses	Training programs

Source: Adapted from Boff and Boff 1987: 13.

themselves, growing from their examinations of their current situation, and assisted by biblical reflection. This input should be taken into consideration by the professionals.

Origins of the Latin American Liberation Movement

The hard look at the Latin American reality was made by several sectors of Catholics. Some of the earliest insights came from middle-class university students.

> It involved a kind of Christian conversion as far as the social consequences of our faith were concerned. Without taking this context into account, one easily falls into the mistaken notion that liberation theology is a specific branch of theology, recently created and somehow inflated, dealing with "liberation" whatever this term may mean. . . . the university student, using above all the option of the

social function of *ideologies*, had already discovered that our whole culture, whatever the intention in constructing it may have been, was working for the benefit of the ruling classes. It was not, of course, necessary to be a Marxist to make such a common sense discovery, but it is also true that many Christian students at the university were led by their Marxist fellows to this realization and to be concerned with this fact.

Furthermore, Christian students could do nothing except include *theology*—the understanding of Christian faith—in the ideological mechanisms structuring the whole of our culture. . . . I mean by that, that even though ideologies are consciously or unconsciously developed in the ruling classes which benefit from them, they also pervade the whole of society, since they are injected even into the minds of those who are their victims. Unlearned and so incapable of utilizing developed tools of ideological suspicion in a culture considered impartial and the same for all classes, poor and marginalized people were led by their culture to accept distorted and hidden oppressive elements which "'justified" their situation, and, among all these elements, a distorted and oppressive theology.

From Christian students to theologians working with them, this ideological suspicion thus became a source of a new worldview about what theology should become and about how a theologian was supposed to work to unmask the anti-Christian elements hidden in the so-called Christian society. (Segundo 1990 [1983]: 354–55)

Other sectors of Catholics were also undergoing a conversion experience and coming to similar conclusions, especially theologians engaged in pastoral work among poor and marginalized peoples. "Conversion means . . . for many intellectuals, a kind of self abnegation. Instead of teaching, they should learn. And in order to learn from common people, they should incorporate themselves, even mentally, with these common people, and give up the chronic suspicion among intellectuals that common people are always wrong" (Segundo 1990 [1983]: 359). As a result, "Theologians, wanting to be in religious matters the 'organic intellectuals' of poor and uncultivated people, began then to understand their function as one of unifying and structuring peoples' understanding of their faith" (360). In other words, theology and its catechesis should have input from the faithful liberating themselves from the prepackaged formulas used as ideology by both the upper classes and naive clergy—formulas to which the faithful were supposed to conform in order to be orthodox Catholics and gain eternal salvation.

In its origins and early analyses, the movement was concerned with the need for liberation from the oppressive ideologies and social structures of Latin American societies. Indigenous populations were included as part of the oppressed poor in need of liberation. But as a cultural and demographic minority with their own distinctive problems, their problems were not included in the movement's vision until a later date.

It is difficult to give an exact definition of the term "liberation theology." As Segundo noted, it is not a new theology. Berryman (1987: 4) simply states: "Liberation theology is an interpretation of the Christian faith out of the experience of the poor. It is an attempt to read the Bible and key Christian doctrines with the eyes of the poor." As liberation, this theology examines a society guided by the threefold formula, "observe-judge-act." As will be shown in chapter 13, the Bible, with its message of justice and love, helps to throw light on situations in need of liberation and shows the necessity of personal involvement. In many instances, this is not a complicated process. As mentioned above by Segundo and Ruíz, it is common sense. The difficult part is breaking through the social and personal defense mechanisms used to legitimate a situation obviously in need of liberation. Among these defense mechanisms are the ideological use of theological paradigms, specifically the Tridentine theology.

Ecclesiastical Recognition of the Liberation Movement

The movement has been scrutinized by conferences of bishops from all of Latin America. These conferences are supposed to take place every ten years to set guidelines for the pastoral work to be implemented in various dioceses. The ten-year norm is sometimes changed to correlate with the dates of other important events.

The Medellín Conference

Following the close of Vatican II in 1965, the first Latin American post-counciliar conference was the Second General Conference of Latin American Bishops, held in Medellín, Colombia, in 1968. Its title states its purpose: "The Church in the Present-Day Transformation of Latin America in the Light of the Council." The liberation movement had been in existence since the 1950s, and had to be considered in the bishops' deliberations.

Most of the bishops had been trained in the Tridentine mold. This set up the same kind of conflictive dynamic that had prevailed at Vatican Council II.

Medellín opened with selected bishops delivering seven position papers to become the basis of the subsequent discussions and committee work. The final results of the conference were the issuance of documents on sixteen topics: five on Human Promotion, four on Evangelization and Growth in the Faith, and seven on the Visible Church and Its Structures (Colonnese 1970). Clearly stating the problems confronting Latin American society, the pastoral conclusions touched all its sectors. As at Vatican II, they reflected a compromise between conservative and liberal bishops, the latter representing the liberation movement.

Buried within the fourteenth document, titled "Poverty of the Church," in the section on Pastoral Orientations, there appeared these paragraphs:

> The Lord's distinct commandment to "evangelize the poor" ought to bring us to a distribution of resources and apostolic personnel that effectively *gives preference to the poorest and most needy sectors and to those segregated for any cause whatsoever,* animating and accelerating the initiative and studies that are already being made with that goal in mind. . . . This has to be concretized in criticism of injustice and oppression, in the struggle against the intolerable situation which a poor person often has to tolerate, in the willingness to dialogue with the groups responsible for that situation in order to make them understand their obligations. (Colonnese 1970: 2: 216–17, #9–#10; emphasis added)

This paragraph is the equivalent of the previously described "important two sentences" in episcopal documents. It states the "option for the poor" of Medellín, which became the rallying cry of the liberation movement. "The Latin American bishops did not merely apply the teachings of the Council [Vatican II], they offered a bold new interpretation that offered a significant development in Catholic social teachings themselves. Out of Medellín there emerged a new Christian vocabulary: option for the poor, liberation, structural sin" (Cleary 1993: 3).

The Puebla Conference

The forces let loose by the liberation movement and its legitimation by the Medellín conference continued to meet strong opposition from conservative bishops and lay elites, many of whom initially had not realized the implications of some of its documents. The presidency of CELAM, the Latin American Bishops' Conference, came into the hands of a bishop intent on reining in the movement. Latin American governments, national and international corporations, as well as the U.S. government with its spheres of influence, realized that the whole Latin American social order, and their privileged positions within it, were being scrutinized, and in many cases challenged. This concern can be seen in the intense interest shown for the third Latin American Bishops' conference, in 1979, at Puebla, Mexico. "Given the growing theological prominence of the Latin American church and of its publicized conflicts with military governments in eight of the ten South American countries, more than 3,200 journalists and observers flocked to the Puebla conference" (Cleary 1993: 3–4).

The conference produced a massive document, *Evangelization at Present and in the Future of Latin America*, almost two hundred pages in the English version. It followed the "observe-judge-act" triad that was also used at Medellín. The document's fourth part, Missionary Church at the Service of the Evangelization of Latin America, devoted an entire chapter to "A Preferential Option for the Poor," with three subsections: "From Medellín to Puebla" [observe], "Doctrinal Reflection" [judge], and "Pastoral Guidelines" [act] (Eagleson and Scharper 1979: 264–67, #1134–#1165). The preferential option is also mentioned in the section on the obligations of bishops (220, #707). This again legitimated the movement, but did not remove opposition to it. The results of the conference, "confirmed advances, affirmed for the hesitant no turning back, and made a strong step forward. However, enough ambiguities persisted that the conservatives would find justification for cautionary interpretations which they increasingly thrust into the debate" (Cleary 1993:4).

Once again, indigenous groups were included, but not in terms of their specific problems. In the section on the preferred option, they are mentioned in a footnote (Eagleson and Scharper 1979: 264, #1135), and two paragraphs discuss the question of values inherent in their traditional cultures (267, #1164–#1165), but without an extended treatment of the question. They also received passing mention in the section "The Evangelization of Culture" (177–84, #385–#443), which was primarily interested in the shift from rural to urban life.

Vatican Reaction

The liberation movement aroused strong opposition among conservative Catholics, who initiated a prolonged battle against its theological basis. The Congregation for the Doctrine of the Faith is the Vatican's watchdog for theological orthodoxy. Since the liberation movement touched on a number of central theological issues, it came under Vatican scrutiny. Warnings about the movement followed, and some liberation theologians were silenced (Hennelly 1990: 367, 425). The Congregation issued its first instruction in 1984 (393). It reflected a European mindset that was still attempting to understand Latin American realities and reflections. A more-focused treatise was issued in 1986 (461), which was more than twenty years after the priest-guerrilla Camilo Torres had raised the question of liberation for Latin America (see chapter 25), five years after the crisis of the insurgency-counterinsurgency in Guatemala, and eight years prior to the uprising in Chiapas (see Part VI). The treatise gave approval for small groups involved in biblical reflection and consequent action. "The new basic [sic] communities or other groups of Christians which have arisen to be witnesses to this evangelical love are a source of great hope for the church. If they really live in unity with the local church and the universal church, they will be a real expression of communion and a means for constructing a still deeper communion. Their fidelity to their mission will depend on how careful they are to educate their members in the fullness of Christian faith through listening to the word of God, fidelity to the teaching of the magisterium [church authorities], to the hierarchical order of the church and to sacramental life. If this condition is fulfilled, their experience, rooted in a commitment to the complete human liberation, becomes a treasure for the whole church" (483). The approval is Tridentine, conditional, requiring submission to clerical authority. But since a number of communities were coping with clerical indifference to the powers of oppression, if the clerics were not actually in alliance with them, the document did not fully comprehend what "liberation" involved.

The document also discussed the question of liberation and the use of violence. For the most part, it draws on the traditional Catholic position about just and unjust wars.

> Systematic recourse to violence put forward as the necessary path to liberation has to be condemned as a destructive illusion and one that opens the way to new forms of servitude. One must condemn with equal vigor violence exercised by the powerful against the poor, arbitrary action by the police, and any form of violence established as a system of government. . . .
>
> Situations of grave injustice require the courage to make far-reaching reforms and to suppress unjustifiable privileges. But those who discredit the path of reform and favor the myth of revolution not only foster the illusion that the abolition of an evil situation is in

itself sufficient to create a more humane society; they also encourage the setting up of totalitarian regimes. The fight against injustice is meaningless unless it is waged with a view to establishing a new social and political order in conformity with the demands of justice.

...

These principles must be especially applied in the extreme case where there is recourse to armed struggle, which the church's magisterium [official teaching] admits as a last resort to put an end to an obvious and prolonged tyranny gravely damaging the fundamental rights of individuals and the common good. . . . Indeed, because of the continual development of the technology of violence and the increasingly serious dangers implied in its recourse, that which today is termed "passive resistance" shows a way more conformable to moral principles and having no fewer prospects for success. . . .

It is not for the pastors of the church to intervene directly in the political construction and organization of social life. This task forms part of the vocation of the laity acting on their own initiative with their fellow citizens. (Hennelly 1990: 486–87, #76, #78, #79, #80)

These sections of the document raise questions for the upcoming examination of the involvement of some Catholic liberation communities in armed uprisings (Part VI).

The Santo Domingo Conference

The Fourth Conference of Latin American Bishops was held in Santo Domingo in 1992, in conjunction with the 500-year anniversary of Columbus's "discovery" of the New World. Its final document, *Conclusions*, reaffirmed the option for the poor (Hennelly 1993a: 122, #178; 123, #180). The option for the poor was also reaffirmed in the Pope's opening address, which was largely a plea for sacramental observance (50, #16). In recognition of the Columbian anniversary, the final document contains an explicit discussion of indigenous populations and African Americans. This was an advance over their casual inclusion in discussions about poverty in the preceding conferences. But, by 1992, the problem of the treatment of indigenous populations had expanded to a liberation theology that included a *Teología India*. This theology includes the liberation of non-Western cultures from the imposition of Western theological categories, in favor of the inculturation of Christianity using the indigenous's own cultural categories. Teología India has a somewhat different history than the quest for social justice, and will be discussed in Part VIII.

Summary

This chapter has briefly examined the Latin American liberation movement. The movement grew out of Catholic Action groups taking a hard look at the poverty crisis in Latin America, enlarging their vision, and coming to realize its systemic nature. The movement was influential in shaping the documents of the bishops' conferences at Medellín, where it received legitimation in the midst of controversy. This validation was confirmed at the Puebla and Santo Domingo conferences. After much suspicion of the movement, in 1986 the Vatican finally gave partial approval, as long as the movement remained subject to Vatican discipline. In the typical fashion of episcopal documents, the Vatican's response was to ignore the not infrequent situation in Latin America, where the church's theology and church officials themselves were the oppressors—concerns raised by Segundo and other liberation theologians. At a later date, the liberation movement began to concern itself with "inculturation"—the right of non-Western cultures to express a Christian theology using their own theological categories instead of those imposed by Western clergy.

11

Maya Dioceses Reorganize for Action Catholicism

With the impetus of Vatican Council II and Medellín, bishops and/or pastors in Maya areas began to restructure their dioceses. In Guatemala, the large western diocese of Los Altos had been divided into individual dioceses, increasing the number of bishops (map 1). Due to the lack of local clergy, most of the dioceses were headed by foreign bishops, who recruited in Europe and the United States for priests, nuns, and lay volunteers to be diocesan pastoral workers. This also gave access to foreign funds to initiate and maintain Catholic Action programs.

In Mexico, Chiapas had been divided into three dioceses: Tuxtla Gutiérrez and Tapachula, both with largely mestizo and acculturated Maya populations, and the San Cristóbal Diocese, with a large traditional Maya population, many living in semi-isolated villages or as impoverished workers on agricultural plantations (map 1). The diocese had only thirteen Mexican priests in the early 1960s, including a Mexican congregation based in Tenejapa and Jesuits based in Bachajón, who had worked in the municipios of Chilón and Sitalá since 1958. The bishop invited several congregations of Mexican nuns to assist with catechetical work. In 1963, American Dominican priests arrived to administer the large Ocosingo parish that included Palenque, Altamirano, and most of the sparsely populated but changing Lacandón (maps 1, 2, and 4). A few years later, they were succeeded by Mexican Dominicans.

Objectives and Reorganization

The option for the poor seeking social justice became the agenda of a number of these dioceses. They wanted to turn passive Tridentine catechetical groups into energized Action Catholic groups led by retrained or new catechists in each Maya community. The catechists were to form groups to pray together, and with the assistance of the Bible, to reflect on the crisis conditions of their communities using the "observe-judge-act" formula. All this was in contrast to Tridentine Catholicism, as commented upon by Rigoberta Menchú (1984: 121): "It's not to undervalue the good things they [priests] have done for us, but they also taught us to accept many things, to be passive, to be a dormant people. Their religion told us it was a sin to kill while we were being killed. They told us that God is up there and that God had a kingdom for the poor. This confused me because I'd been a catechist since I was a child and had had a lot of ideas put into my head. It prevents us from seeing the real truth of how our people live."

A diocese is a subculture with its own worldview and social organization under the authority of a bishop. The bishop, with a group of advisors, draws up a pastoral plan that details the specific goals and means to implement the worldview of the church among the people in his area of responsibility. At the next level are the pastoral workers—the priests, nuns, and laity who bring the plan of the diocese to the catechists, key representatives of the Maya communities.

In Chiapas, Bishop Ruíz established an annual diocesan assembly of pastoral workers and expert advisors to make and vote on diocesan plans and policies. The diocesan assemblies went through several phases (Iribarren 1985: 6–47). The assemblies of 1975 and 1976 were concerned with diocesan options. The bishop insisted that there needed to be changes in the diocese's pastoral work, that the older theology and the structures that implemented it needed to be changed to focus on the option for the poor since they composed the vast majority of the diocese. A theology of liberation, helped by biblical reflection, should consider the problems cited by the Indian Congress of 1974.

The assemblies from 1977 to 1980 were mainly concerned with the diocesan structures needed to implement these goals. Since this involved a radical departure from past experience, there was a period of trial and error. Problems arose about the ability of committees and the zonal groupings to attain their goals, and there was poor coordination between the zones. The discussions showed the need for a greater understanding of Maya culture in order to implement meaningful action.

The Heart of Action Catholicism: The Sunday Service

The transmission of the biblical theology of liberation was highly dependent on Maya catechists during the weekly Sunday service of the Catholic community. There were variations in the format of the service depending on the preferences and depth of experience of the pastoral worker and the catechists of the parish. In Yibeljoj, a community in the municipality of San Pedro Chenalhó, Chiapas, the catechists used the following format (Moksnes 2003: 130–39). The service opened with an introduction by a senior catechist: "Men and women, we have come to listen to the Word of God and we have lived another week. Now we will begin to pray to God, we will all pray to God, men and women, and we know how it should be. We mustn't stop praying to God since we are living in great suffering; pray that God accompany us and protect us, that God have compassion for us, that there be no more suffering. In the meantime, we must enjoy our lives, men and women."

This was followed by recitation in Spanish of some of the traditional Catholic prayers found in the catechism. Then there were several minutes of simultaneous, semi-chanted, animated spontaneous prayer by each member of the congregation, calling on God's help for whatever was troubling them. Frequently, the prayers were in the form of traditional Maya couplets. There was some crying. Moksnes observed that it was always the most emotional moment of the service.

Following the prayers, the congregation sang, led by several members of the parish's musical group. They had composed lyrics to be sung with music from popular Mexican tunes. "Most of the songs expressed praise to God and happiness for being saved and cared for. Some had more didactic intentions, for example holding out the importance of respecting women's work ... many songs have explicit political content" (Moksnes 2003: 134).

This was followed by the catechist giving some type of moral guidance, from urging his listeners to be attentive during the services to encouraging kind admonitions of those who had been behaving contrary to the group's moral standards.

Next came a reading of and subsequent reflection on a biblical passage. Usually, the reading was a slow monotone with repetition of phrases, as many catechists were only marginally literate and had difficulty with the Tzotzil text. "I found the presentation taking on the form of a reverent reading of a holy and obscure code brought to the group through the diocese, and I knew that many of the creyentes [believers] would blame only themselves if they did not understand" (Moksnes 2003: 136). To help focus the reflection, the catechists asked three questions. The first two concerned the meaning of the text itself. The last one inquired about specific actions suggested by the biblical reflection. Men and women divided into separate groups. In each, a literate person was appointed to take notes about what was said, synthesize them, and when the groups reassembled, read the summaries for further comments by the reassembled congregation. In this way, the community arrived at a consensus. This was an unusual exercise for the women, who were unaccustomed to speaking in public. Part V will examine in more detail the worldview of the Bible and examples of Maya reflections on it.

At this point, an experienced catechist took over the meeting. He explained the meaning of the text, how the questions related to it, and the significance the text could have for the daily life of the community. "This phase of the service . . . is considered highly important by the catechists, since it is here they can influence and guide the church members' understanding of the scripture and Catholic doctrine . . . it requires skills not only in theological reasoning, but also in making a cultural translation into the time and lives of creyentes [members of the Catholic community]" (Moksnes 2003: 138). The meeting closed with announcements of church activities, diocesan events, and political news of interest to the community. A separate service was held for children. It consisted of learning, in Spanish, the catechism prayers, although they had little understanding of the Spanish words. For them, the more enjoyable part of the service was learning songs composed by the parish musical group.

A similar service took place every Sunday in the communities of the Quiché diocese. Families "participated in a 'class' whose purpose was education. They generally used mimeographed materials prepared in the parish center. During a 'discussion,' a period of community reflection, they analyzed the local community and that of the country in light of the faith and Sacred Scriptures, followed by a 'prayer,' a liturgical or informal celebration in which the whole community participated and during which the results of the class and the ensuing discussion were summarized. . . . In these meetings of the community, some time was always given to airing moral problems (drunkenness, fights between neighbors, family quarrels) before the board of directors and the whole community if the case merited it. They were resolved by the guilty party's repentance and the promise to amend. As the community matured in its discussions, there was less discussion of personal moral problems and more of social, economic and political problems" (Diócesis del Quiché 1994: 55).

The Catechists

The catechists prepared for their role by attending courses given by pastoral workers of the dioceses. There are some variations between the different regions, but typically a new catechist attends an intensive course of instruction of up to a week. In the San Cristóbal Diocese, the catechetical schools described in chapter 8 were closed and the training programs moved to regional centers. Originally, the catechists were chosen by the pastoral workers, often from the few who were literate. As the movement matured, the selection was made by the communities themselves. (The introductory photograph shows an Atiteco catechist preaching during a mass.)

Returning to their local communities, the new catechists continued their training under the guidance of experienced catechists. This was reinforced by meetings of all the catechists in a parish or zone; weekly in Quiché; monthly, for two days, in Chenalhó; every six months, for two to three weeks, in Huehuetenango. The purpose was to discuss any difficulties with the biblical texts, community problems, the action programs, and to arrange for coordination with the diocese or other agencies. (Diócesis del Quiché 1994: 55; Santos 2007: 38; Moksnes 2003: 167–71; Kelly 1979).

Annually, catechists attended a refresher course of several days or a week conducted by pastoral workers. In Yibeljoj, "Through the guidance catechists receive from the pastoral workers, their Biblical interpretation evolves in line with the diocese's authority. They also develop forms for how to culturally translate these dogmas for the creyentes of their respective church groups. The courses treat Biblical themes, church history, and traditional Chenalhó religion, the latter as part of the Indian theology effort to integrate Catholic and traditional religious thought [Teología India]" (Moksnes 2003: 171). The Indian theology effort will be discussed in chapter 29. Coordination was arranged with diocesan-wide organizations, and courses given on human rights, women's issues, health, education, and other community services.

Since the political situation of a region impacts Word of God communities, politics were discussed in the refresher courses, but political action was undertaken through independent political organizations. In Guatemala, usually this was the Christian Democratic Party. In San Pedro Chenalhó, it was through Abejas, an independent, ecumenical human rights group modeled on and retaining close ties to the diocese through the more than 110 catechists who were members. Abejas sees its formation as biblically inspired, as will be seen in chapter 21. Abejas is a typical example of a number of action organizations that were initiated or influenced by local parishes or dioceses but that are independent of the Catholic Church.

Other duties of the catechists included giving lessons in preparation for first reception of the sacraments, and visiting the sick and dying to pray and offer consolation. Prayer meetings were held at least two or three times a week in the late afternoons in Magdalenas and Hermosillo (Chojnacki 2004: 530–47; Morales 1985: 182). In the Quiché diocese, once a week a catechist visited each of the five or six families for whom he was a spiritual guide—praying with them, reflecting on the Bible, and giving advice about personal and family problems. This schedule of meetings meant that a catechist attended three meetings in a single week—a considerable sacrifice for men and women trying to provide a meager living for their families.

The older, experienced catechists play a role within the Word of God communities similar to that of the principales for the traditional system (Moksnes 2003: 162–66). They were looked upon as religious, moral, and political leaders. Legitimacy was based on knowledge of and ability to interpret the Bible. As religious leaders, they conducted the Sunday services and prayer gatherings. As moral leaders, they gave advice about family problems and mediated disputes. They were frequently asked to be marriage petitioners, following the traditional Maya custom. Some were political leaders because of their knowledge and experience of the diocese, ladino culture, and the wider world.

Pastoral Workers

These are priests, nuns, and laity who were either assigned to the Maya areas by their religious congregations or who volunteered their services. Many of the pastoral workers in Guatemala were foreigners—Americans,

Canadians, Spaniards, Belgians, Germans, and other Europeans recruited by the foreign bishops of the Maya dioceses. Although a number had been trained in the Tridentine tradition, when they observed the Maya crisis in Guatemala, many quickly absorbed the outlook of Action Catholicism.

When Bishop Ruíz first arrived in Chiapas, the pastoral workers were a few local priests. They were traditional ladinos who conducted catechetical instruction during their infrequent visits to villages in the far-flung parishes. With the reorganization of the diocese, Don Samuel invited mestizo pastoral workers from the urban areas of Mexico. They were attracted by his leadership of a diocese implementing the liberation movement and its option for the poor. This was in sharp contrast to the very traditional outlook of many Mexican dioceses. The Catholic Church in Mexico had been marginalized from community life for almost half a century by liberal and other anticlerical governments (Chojnacki 2010: 51–85). This restricted opportunities for community involvement by church personnel influenced by Action Catholicism. Many Maya in Chiapas were still living in their semi-isolated communities, with little or no governmental presence or assistance. The diocesan policies put the pastoral workers in grassroots contact with Maya communities and offered them an opportunity for many types of community involvement. Some pastoral workers were Christian Marxists who rejected Marxian metaphysics but saw Marxian social analysis as providing the only realistic picture of the Maya subsistence crisis. In the 1990s, there were about two hundred pastoral workers in the diocese.

As Action Catholicism grew in Maya communities and individual catechists matured in their knowledge of the Bible, pastoral workers began to assume the role of advisors rather than teachers. The catechists themselves took on the role of harvesting, synthesizing, and observing how biblical reflections helped to interpret communal realities. The pastoral workers were to accompany the communities in their search for liberation. Exactly what this involved had to be worked out over time by trial and error in the circumstances of each community. As will be seen in the following chapters, it became a particularly acute question as communities became more involved in economic and political organizations as part of their search for liberation.

Women, especially nuns (sisters, *madres*), composed a large proportion of the pastoral workers in Chiapas. Some were from Maya communities. "Don Samuel had noticed that the women did not talk during his visits to their communities. He perceived their suffering. The diocese created schools for the women who came to attend catechetical courses in San Cristóbal for two months. This led to the request to found a group of indigenous women, consecrated by religious vows, who would be at the service of their community since they knew the language and were capable of preaching the Word of God. In 1977 such a group of indigenous nuns was formed" (Coordinator of the Diocesan Women's Commission, in Ruíz García 2003: 95).

An anthropologist describes the work of the nuns: "Madres seem more successful than priests in listening to indigenous people and in breaking down the hierarchy of the Catholic church. Although based in San Cristóbal, Madres spend weeks at a time in indigenous communities meeting with lay workers to discuss community concerns. The times I saw Madres at work in Catechist meetings, remote hospitals, and Guatemalan refugee camps, I saw them listen carefully to Indigenous people's concerns and ideas. They seemed committed to helping them probe the wisdom their people already have. When priests cannot make it to meetings, Madres are there, to talk about whatever problems people might have" (Eber 2000: 226).

The nuns encouraged women's liberation. "The catechists refer to the necessary role of women to construct the Kingdom of God on earth by their active participation. The women are expected to contribute equally with men to a new social order based on the values of justice, equality and dignity for all. Reading the Bible, at times accompanied by Catholic nuns, the men receive continuous messages with the goal of strengthening their attitude of openness and respect toward women, comparing their situations with the treatment Jesus Christ gave those women who accompanied him during his apostolate. For example, the catechists underscore the roles of women in the Bible immediately after the resurrection of Christ, emphasizing that women were the first to witness such an important event in the Catholic doctrine" (Gil Tebar 2003: 150). In Huehuetenango, two Maryknoll nuns were also medical doctors. They initiated a medical program in Jacaltenango that eventually became a small hospital as well as a training center for health assistants who ran clinics in a number of the Maryknoll parishes in the department.

Problems of Accompaniment

Accompaniment involved a wide range of activities. The primary function of pastoral workers was to train and later act as advisors to catechists for their role of helping, encouraging, and focusing the reflections on the Bible during the Sunday service. Since these services followed the "observe-judge-act" format, accompanying the catechists and the community also involved pastoral workers in the actions undertaken as a result of the reflections. Typically, these actions were attempts to alleviate the four major problems afflicting Maya communities, highlighted by the 1974 Indian congress—land, health, education, and justice in the marketplace.

Questions about competency for work in community development arose from the fact that the majority of pastoral workers were priests and nuns (Legorreta Díaz 1998: 62, 77, 88 and passim). In some cases, this was a problem. But in situations where there were no public services, many priests and nuns could play an important role in initiating health, education, and economic programs. Priests and nuns developed a number of skills either before or after entering church service. Some with teaching experience initiated literacy programs. Nuns who were doctors, nurses, or trained health professionals initiated health programs. Some priests were trained in the organization of cooperatives. Many acted as cultural brokers between Maya groups and outside ladino agencies in situations of cultural contact. As programs grew and became more complex, it became time for the pastoral workers to step aside and let better-trained personnel take over, ideally Maya who had been educated for such roles. A number of civil programs in Maya communities owe their origins to the work of pastoral workers.

The problem of political involvement by pastoral workers was never completely resolved. Given the unjust economic and political structures prevalent in the Maya areas, any involvement in their empowerment automatically meant involvement in politics. Empowering Maya in a culture of oppression was, ipso facto, perceived as a political act by the oppressors. How far should pastoral workers become involved in politics? The church sees involvement in partisan politics as compromising its primary role as a teacher of the universal values of the gospels. In the initial stages of Action programs, the Maya needed help in the form of political advice and guidance through the labyrinth of bureaucratic and often corrupt government agencies (although not all individuals in these agencies were venal). As the Maya matured in political matters, pastoral workers needed to withdraw and restrict themselves to their primary function. Frequently, it was difficult to tell when this point was reached. Some pastoral workers became paternalistic or maternalistic, attempting to retain control or to influence civil organizations long after they should have withdrawn into the background.

Floyd (1997: 67–101) examined various statements by Bishop Ruíz and the diocesan assemblies justifying the possible political involvement of pastoral workers. She found five assumptions that gave a theological rationale: "1) poverty and suffering are not consonant with the will of God; 2) the experience of poverty and suffering may lead to conversion [to work for liberation]; 3) change in society is possible through the power of the Resurrection of Jesus Christ [the cycle of birth, death, rebirth]; 4) the church is a catalyst for the construction of the Kingdom of God; 5) the church is called to accompany the poor in their religious and political struggle."

Assumptions 1, 2, and 4 repeat the common themes of the liberation movement mentioned in chapter 10. Assumption 3 springs from faith in a biblical pattern of liberation, to be discussed in chapter 13. In the face of persecution and the temptation to despair, one knew that, ultimately, social change would be possible because of the lesson taught by the Resurrection—Christ's affirmation and promise that life springs from death. Assumption 5 is the conclusion following from the previous four assumption. "Knowing the grave reality of our brothers and sisters, who are the poorest of the poor, we propose to accompany them, as did the good Samaritan, in their effective search for a new society, built on justice and fraternity" (Ruíz García 1993, in Womack 1999: 240; Floyd 1997: 92). The bishop summarized the relationship between faith and politics. "Faith and religion are not to be confused with politics, nor politics confused with faith and religion. But religion and politics have a great relationship between them. Because the political can be illuminated by faith and all practical politics can be one of the diverse ways of professing faith and living religion" (Floyd 1997: 194). The five theological assumptions, the Good Samaritan story, and the bishop's statement lacked the necessary specificity for many of the situations that were to arise.

Given the poverty of Maya communities, invariably, there was social distance between the Maya and the pastoral workers. When the Tzeltal catechists questioned the bishop about the failure of the Tridentine catechesis to meet the needs of the Maya crisis (chapter 9), they posed another question: "You have lived among us and shared our lives. We regard you as our brothers and sisters. Is it your desire to be our brothers and sisters for all times?" (Womack 1999: 29). Given the past history of interaction between Maya communities and Catholic priests, there were some questions about the permanency and depth of the diocese's commitment to its option for the poor. "Ideally the pastoral workers walk *beside* the poor, but this position is difficult given the privilege of pastoral workers in comparison to peasants. The vast majority of the pastoral workers are mestizos, and, although they live modestly (having voluntarily renounced some of their economic privilege in order to walk with the poor), they have the backing of religious orders or other institutions. This is not to deny that pastoral workers sacrificed a great deal; because of their political commitments some received deaths threats and several were jailed in the 1990s. Even in accompanying the poor, a dramatic social and economic distance divided pastoral workers and those they accompanied; pastoral workers are aware of and struggle with this distance. Yet increasing government repression against campesino groups and the diocese strengthened the alliance between the two" (Kovic 2004: 200).

The 1985 pastoral assembly summarized the empowerment activities of the San Cristóbal diocese.

> Initiatives in the economic sphere: production, transport, credit, and mechanical repair cooperatives as well as providing help for collective production, technical assistance, loans and assistance to attend workshops, etc.
>
> In the political sphere: assistance in the demands for land recognition by various cañon groups; political formation by assemblies and workshops; legal assistance for initiatives promoting reflection on and realization of their rights and obligations; critical support to organized groups assisting catechists, refugees, political prisoners, those resisting land invasions, those evicted from their homes, and campesino protest marches. Denunciation in channels of public opinion about repression and similar activities. Encouraging the creation of a union of ejidos; participating in protest marches, assemblies, meetings and actions of public protest, etc.
>
> In the cultural sphere: assisting groups with theological and biblical reflection; liturgies celebrating triumphs and failures in their battles for land; unmasking a traditional sleep inducing religiosity; promoting the integration of faith and life; creating a consciousness that is both critical and just; encouraging the participation of women, etc.; giving information about protestant sects, etc. (Iribarren 1985: 44)

In 1986, the diocese published a forty-six page document, the Plan Diocesano (the Diocesan Plan). It contained an extensive analysis of the civil wars for liberation in Nicaragua and El Salvador. Within this context, it made an extensive analysis of the economic-political struggles in Chiapas. It saw the situation as ominous. "Thus the peasants, finca peons, farm workers, and Indians have remained at the center of the class struggle in Chiapas, and it is this accumulation of laborers, impoverished and taking hard beatings, that points towards a new explosion and social change" (Womack 1999: 203). In spite of criticisms, the diocese pushed ahead with its agenda of liberation.

A Different Trend: Charismaticism

A number of Catholics, both clergy and laity, were not interested in the new direction of the church initiated at Medellín with its option for the poor and Action Catholicism. Some thought that the church should continue to concentrate its efforts on personal reform and salvation in a purely spiritual sense as formulated by the Tridentine paradigm. The charismatic movement was one of the reactions of these dissenters.

The Charismatic Structure

Charismaticism is based on biblical references to the gifts of the Spirit, and on the long history of charismatic spirituality within Christianity (Cartledge 2007: 33–50). The charismatic experience itself consists of a transformation of consciousness during or after a freely structured worship service. Initiation as a charismatic usually would begin by participation in a Life in the Spirit retreat with a communal group under the directorship of persons known for their reception of the gifts of the Holy Spirit. The setting was usually informal. The heart of the charismatic service would involve rhythmic contemporary music for a medley of prayerful songs of praise to God accompanied by full-body swaying, jumping, clapping, and shouting. This would lead to contact with the Holy Spirit, an intense psychological state in which individuals experienced visions or proclaimed prophecies or spoke unintelligible words (glossolalia). An interactive sermon might be preached. The community would answer questions posed by the preacher or interject their own comments, prophecies, and testimonies. The session would conclude with the Baptism of the Spirit. One or more individuals who had previously received the charismatic gift, would lay their hands on those undergoing the renovating experience described as the descent of the Holy Spirit. During this experience, they would recite their sins (Cartledge 2007: 57–68). In Catholic charismaticism, if a priest should be present, he would give a general absolution from sins (Falla 1985: 15).

> Participants in charismatic sessions experience feelings of great joy, bliss, and happiness. "This is followed by dedication to personal renewal and an ongoing growth in a life of holiness. After the Life in the Spirit retreat, participants typically join small prayer groups.

The groups serve to instruct, encourage, and challenge members to greater spiritual growth. The groups are better called communities, and resemble extended families. Members help one another to grow in holiness and to build stronger relationships. Social justice activism is not generally emphasized" (Cleary 2007: 157).

Charismaticism is similar to Tridentine spirituality in that it emphasizes personal reform and ignores social reform. It differs in that the essential worship service does not require the sacraments and, in some instances, sees itself as a replacement for the sacraments. These ambiguities set the stage for controversy with orthodox Catholicism. Charitable works were encouraged, but social reform was not a matter of concern. Social reform was seen as an unwanted mixing of the spiritual with the temporal.

The Charismatic Movement in Guatemala

In 1970–71, small groups of Catholics and Protestants participated in Life in the Spirit retreats. In 1972, two Maryknoll nuns organized charismatic prayer groups in English. In December 1973, at the invitation of the ultraconservative Cardinal Casariego of Guatemala City, a Jesuit priest from the United States directed a charismatic retreat for thirty priests, out of which came five Spanish-speaking prayer groups in Guatemala City. Shortly thereafter, the auxiliary bishop of Guatemala City did the same for a lay group. The same year, the Cardinal formed a diocesan service team led by the auxiliary bishop to supervise the formation of charismatic groups. They drew from middle-class people who had participated in Cursillista weekend retreats. The Catholic branch of the movement grew remarkably in the mid-1970s after Father Francis McNutt and Ruth Stapleton from the United States held a retreat in Guatemala, where thirty-five Catholic leaders were "baptized in the Holy Spirit." By September of 1979, the movement had grown strong enough to fill the National Stadium during a rally led by Father McNutt (Chestnut 2003: 67; Holland 2008: 89).

Charismaticism among the Maya

Led by charismatic pastoral workers, the movement spread slowly among Maya Catholic groups during the 1970s, leading to explosive growth in the turbulent 1980s. Some observations have been made by Maya about charismatic sessions in their communities. Ignacio, in his diary about life in San Juan La Laguna, describes in a negative fashion the charismatic services in the local Assembly of God church. "They have a lot of problems.

They judge one another among themselves and they say they are the best. They pray with wordiness and shouts to be heard by God. I took the opportunity to ask a member of the church why they shout so much and roll about on the floor to pray to God. He told me that they shout a lot and roll about on the floor and bang themselves a lot so that God will pay attention to them. They say they are only heard by God through the shouts and blows since God is disgusted to the point of indifference because of the many sins of humanity. For that reason they use force to get God to listen. But as a consequence of the shouts and the blows, three of the members of this church became mentally ill" (Sexton 1992: 118).

Ignacio also observed Catholic charismatics. Every Sunday, charismatic services were held either in the church or in private homes, with or without a priest. During the week, they were held in private homes. "Who knows where many things are coming from that are appearing in the religions? . . . in the Catholic religion we know the *Santa Misa* [Holy Mass], *Santo Rosario* [the Holy Rosary] and the procession. But now in the Catholic church, those who are called the renovators have appeared. Also, they are called charismatics. They belittle the veneration of the images and do not recite the rosary. They say different prayers. They say that they speak with the Holy Spirit, make miracles and cure sicknesses. They meet in houses and say that they speak new languages. After a while they return to normality and begin to [tell what they saw] when they were concentrating. Some say they saw Jerusalem; others say they saw different celestial things. It seems like a lie, but who knows. The priest of San Martín [pseudonym for San Juan La Laguna] says the charismatics can't be believed, that they are different from Catholics" (Sexton 1992: 87).

In nearby Nahualá, a revered Maya principal in 1996 described the Catholic charismatics in this way. "The Catholic church is just kind of dying. Take the Charismatics, for example—they are not together with the rest of the church. The only reason any of them come to Mass at all is because the Father pleads with them and tries to force them to come: 'Why in the world will you not come to Mass? Here we have this big, beautiful church, why do you not come?'" (Morgan 2005: 90).

In Mayalán, after attending a charismatic retreat, a man who had been quarrelling with his wife describes his charismatic experience. "It came like seeing a lightening bolt. I lost consciousness. . . . I saw various places as if in a dream. The experience left me feeling great joy" (Falla 1985: 16). When he came back to normal consciousness, the man was soaked with sweat. His daughters, who had also participated in the charismatic service, began to shake and receive prophecies about overcoming the family discord. The mother then changed her attitude and forgave her husband, restoring family tranquility. Then they hosted a party for the community to share with them their liberating experience.

The charismatic experience brings about a transformation of consciousness. Such transformations can be achieved by a number of different means, such as strenuous dancing or other intense physical exercises, prolonged meditation, or ingestion of psychedelic substances. Charismaticism sees the transformation brought about by its worship service as a gift of the Holy Spirit, the presence of the third person of the Trinity revealing its abiding presence in the community of worshipers. During the transformation, the individual moves into another world that can unleash powers and intuitions of the human psyche unknown to that part of the psyche ruled by rational processes. Returning from the experience, these intuitions may help individuals renovate their lives.

Summary

This chapter has described how local dioceses in the Maya areas of Chiapas and Guatemala reorganized their structures to implement Action Catholicism. The end result was an upgrading of the role of the catechist. They were to turn the former meetings of memorizing question-and-answer catechisms into sessions of reflection on their communities and on biblical passages for assistance. Leading these discussions gave the catechists greater responsibility. Pastoral workers interacted frequently with them for both initial instruction and follow-up assistance. The reflections were to lead to specific actions in the community to alleviate the subsistence crisis. The final section of the chapter briefly described the charismatic movement, initially a small countertrend to Action Catholicism.

12

Maya Communities Organize for Social Action

This chapter examines some typical examples of community action for Maya empowerment that were undertaken by local parishes. They became centers of reflection about their social conditions interpreted with biblical assistance. They were part of the Catholic Action movement, some in its earlier stages, before Vatican II and Medellín.

San Andrés Semetabaj

As seen in chapter 8, the Tridentine catechesis was introduced in 1951 and the breaking point with the traditionalists occurred in 1958. By emphasizing the primacy of the soul over the body, the catechists' critique of the traditionalists focused on moral reform of individual conduct. A priest visited San Andrés every week to preach and celebrate the mass. In the 1960s, he encouraged participation in the Catholic Action movement (in its restricted sense) that was developing in Guatemala. The movement encouraged reflection on the Bible for the betterment of personal lives, as well as commitment to community action to overcome the subsistence crisis.

The priest encouraged Trixanos to attend church-sponsored courses that discussed the formation of cooperatives. "The proposed solution to uneven land distribution was cooperatives in which Indians could save small amounts of money and take out interest loans so that each individual eventually would have his own land and would not have to sacrifice himself with the *patrones* and live in fatal poverty. The Catholic Church's solution to poverty directs individuals to circumvent the social and economic domination of Ladinos without challenging the foundation of the skewed distribution of economic resources. To this end, the Catholic Church has organized, administered and financially backed 'Indian' cooperatives" (Warren 1978: 140). San Andrés founded a wheat cooperative in the early 1960s, in which members of Catholic Action played a prominent role. One of its goals was to mitigate the amount of seasonal migration to the coastal plantations. The cooperative became affiliated with national and international cooperative organizations, which prevented it from being dominated by local ladinos (140–43).

Implementing the formula of observe-judge-act, members of Catholic Action participated in the activities of the local government school (Warren 1978: 138). Adult members were eager to send their children to school. These students were among the first to obtain scholarships and attend advanced schools away from the town. The motivation was economic advancement. In town, older members attended adult literacy courses for the purpose of defending themselves by being able to read legal documents. In San Andres, one sees the Catholic evolution. The Tridentine catechesis was introduced in 1951; then came the formal break of Maya Tridentine Catholics with the traditionalists in 1958; which was followed by the evolution to Catholic Action (restricted sense) in the 1960s. As seen previously, some first-generation catechists viewed the Tridentine catechesis as a revitalization of the colonial Maya worldview, which experienced a further revitalization by the cyclic evolution to Catholic Action.

Santiago Atitlán

With no paved or graded dirt-road connections during the 1960s, this community was more isolated than San Andrés. Although over 95 percent of the town's population was Maya, its government school of six grades was oriented to the town's small ladino population. The government health clinic had no medicines. Chapter 8 described the emergence of the catechists and their followers as a distinct group from the traditionalists. Following the Maximon saga, there was a ten-year period in which the Tridentine catechesis was the focus of their efforts.

In 1964, there was a further transition to Catholic Action. In Oklahoma, an American Catholic Action group was interested in international relations. It decided to financially aid a poor diocese in the Third World. With the request of Pope John XXIII that all North American religious groups help Latin American parishes, the group persuaded the Oklahoma diocese to send a team of priests and laity to undertake a multifaceted community development project in a Guatemalan parish. The Oklahoma diocese made a contract with the Sololá Diocese to staff the parish in Santiago Atitlán for twenty-five years. A number of other American dioceses staffed Maya parishes in the Western Highlands and developed similar programs.

First Generation of Pastoral Workers

In 1964, the first group of pastoral workers arrived. An initial effort was to try to overcome some of the deep divisions within the community, especially between the traditionalists and the Catholic Action group, given the past history described in chapter 8. Younger members of Catholic Action, not involved in past clashes with Traditionalists, were encouraged to take leadership roles. This was an effort to heal the wounds of the past and to make clear that the Action programs were undertaken for the benefit of the whole community, not just the Catholics.

Given the existing conditions, Catholic Action would have to initiate empowerment programs rather than joining existing ones. The Oklahoma group possessed a number of skills important for such work. The three priests were multiskilled. Father Carlin worked with the catechists. He had training in introductory linguistics, and began work on a written alphabet for the Tz'utujil language. Father Westerman, with experience in construction, supervised and worked alongside wage-earning Atitecos to build a small hospital, an infant-care clinic, and the living quarters for the Oklahoma group. Father Stafford was a grassroots educator, plumber, and electrician. He initiated the agricultural and weaving cooperatives, installed the equipment for the radio station (TGDS) that was used for the literacy program, and also installed the electrical and plumbing systems at the hospital. Among the laity, there was a retired dentist who also worked as an architect, and his wife who worked as supervisor of the living quarters. The

Figure 12.1. Father Stafford (*left*) in traditional Atiteco clothing as a sign of solidarity with the community, including the Traditionalists.

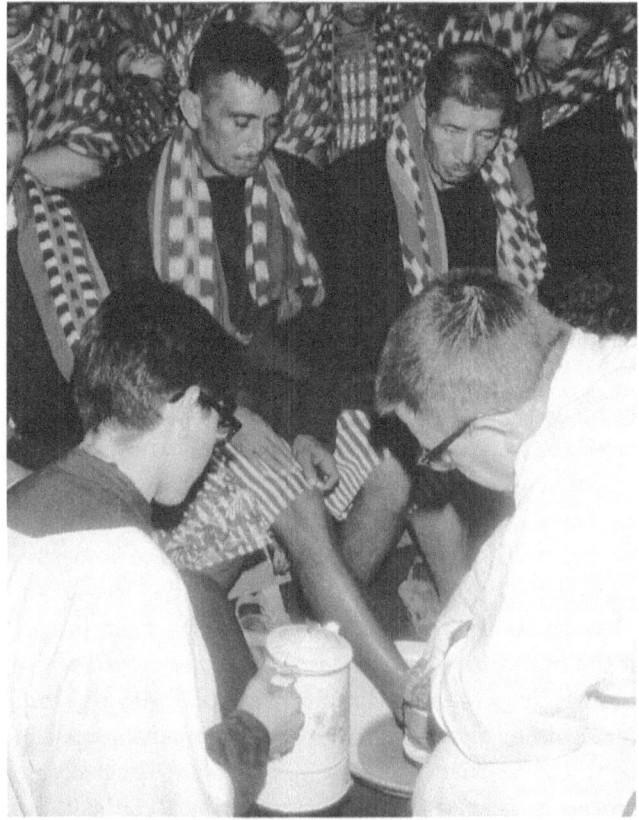

Figure 12.2. The pastor of Atitlán, Father Carlin, washes the feet of cofrades during the Holy Thursday ceremony.

remainder of the original group were papal volunteers (a Catholic peace corps), including an experienced agronomist, a young engineer who initiated the credit cooperative, a former navy nurse with Marine Corps service in Korea who initiated the medical program, and a trained lab technician who initiated the baby-care clinic. A married couple remained in Guatemala City to facilitate logistics and governmental relations. The first-generation staff also included, at one time or another, a resident Guatemalan physician and his wife, two Montessori teachers, a nun as director of the literacy program, another nun who founded an orchestra, and this writer, who was a part-time anthropological consultant. It was a collection of strong-willed individuals whose interactions with each other resembled those in the famous television program M*A*S*H, minus the sex, but all devoted to their work among the Maya.

Programs of Empowerment

Given the health conditions of the community (see the frontispiece and chapter 6), the medical program received priority (Early 1982: 108–30). It operated a clinic, hired a Guatemalan physician, and provided the only functional medical services to the town. It included a strong public health component: instruction about sanitation; an extensive program of inoculations; and a baby-care program that provided prenatal and postnatal information, laundry facilities, and a nutritional supplement to combat the ever present effects of malnutrition. Later, a small hospital was constructed to replace the clinic. The program was effective in dropping the mortality rate of the community. By the 1970s, life expectancy had increased by twenty years, and infant mortality had dropped from 130 per 1,000 population to 90 (98).

A literacy program, an adaptation of Paulo Freire's system (2001 [1970]), was introduced (Early 1973). The few literate Maya in town were recruited as volunteer teachers. Most had already been privately tutoring Atitecos when requested. The program built its own radio station to broadcast literacy lessons to the teachers' homes, which were converted into small schoolhouses in the hours after work (see the introductory photograph to chapter 13). In the Freire system, a keyword is used in each lesson. The words of the first course contained all the letters of the alphabet. The words chosen also represented significant problems in the community. They were selected by the teachers as needing open discussion. Each lesson opens with the teacher leading a discussion of the problem. Then the students proceed to form the letters, syllables, and words. In this

Figure 12.3. An example of the cartelas drawn by Diego Pop with the help of the teachers. The top image was used for the literacy lesson that discussed the importance of education to meet the community needs. Toward the left it shows the radio school, the government school, and the Montessori school. The following lesson discussed the importance of cooperatives. This is visualized by a group of men moving a log that an individual alone could not move. In both images are the words containing the letters to be learned in these lessons. The originals were done in color, with the key words in red.

way, acquiring literacy is not simply working with inert words, but with those that have psychological meaning and significance for the life of the community. Posters (*cartelas*) were made by the teachers showing a scene depicting the selected words, which were written below it.

There were two alphabetization and three mathematics courses taught as a cycle. The preliminary discussions sought to raise self-esteem to replace the internalized stereotype of the poor, ignorant "indio" and to instill an awareness of Guatemala as a country with a national government in which the Maya, theoretically, were citizens. One lesson discussed the national constitution and citizen rights. The focus of the curriculum was the local community and the traditional Atiteco, by far the majority of the population in the 1950s and early 1960s. The *cabecera*, the head of the cofradía system, became a student in a school taught by one of his sons (see the introductory photographs to chapters 14 and 15). It can be extremely difficult for a man of that age, who has worked all his life with a digging stick and hoe in a corn field, to control the muscles of his hand and form the small strokes that make up letters and numbers. To become a student was an act of humility for a man of his status. He appears to have been sending a message to the traditionalists—in this time of crisis, this is something we all have to learn. Eighty-two young girls and women enrolled in the schools, an unusually high number given the traditional view that women's place was in the home and that they had no need of education. The program won high acceptance, with 45 percent of the town's males aged 15–19 in attendance, and 5 percent of the females. There was a strong demand from other lake towns for an extension of the program to their areas. A Guatemalan corporation was formed and the radio station was signed over to it, with the teachers as the board of directors who elected their own officers. This made it independent of the church, although financial assistance continued. As will be seen in chapter 23, the teachers later revised the curriculum to reflect Santiago's greater involvement in the national society and the problems encountered.

The agricultural cooperative borrowed money from the diocese and private sources to acquire a piece of land on the outskirts of town. It was to be used for instruction about production and marketing techniques, as a prelude to an Atiteco enterprise of commercial agriculture. The cooperative acquired a tractor and trained Atiteco drivers (see the introductory photograph, this chapter). Later, the cooperative was unable to carry the debt it had incurred, and turned the land over to the diocese in lieu of paying back the loan. However, the cooperative had opened a consumer products store, selling machetes, flashlights, thread, and other basic goods used in Atiteco work. During their trips to Guatemala City to buy merchandise for the store, the cooperative learned about wholesale buying and the customary 100 percent rural markup over city wholesale prices. This enabled them to charge significantly lower prices. With the increasing integration of Santiago into the national economy, by 1999 the store was no longer needed and it ceased functioning. The cooperative then rented out the store property as a source of income for its members.

The agricultural cooperative also sponsored a Montessori school for the preschool children of its members (see the introductory photograph to chapter 16). Experience in cooperative work impressed on its members

Figure 12.4. Dr. Elizabeth Nick, cofounder and first director of the TGDS literacy program.

Figure 12.5. A graduation ceremony presided over by the president of the station, Diego Coché. He presents a wristwatch to the top student of that semester.

Figure 12.6. A Montessori student studies a didactic puzzle.

the importance of literacy at an early age. Two American Montessori teachers temporarily joined the Oklahoma group and trained two local Atiteco women as Montessori teachers. Under their guidance, the school was very successful. Some time later, an internal conflict developed between the Atiteco teachers and the board of directors of the cooperative, which resulted in the closing of the school. Advisor mediation probably could have saved it, but the second-generation pastoral workers were not interested. A credit cooperative had limited success. Many of the loans were for fertilizer or advances to ambulatory merchants to buy an initial stock of goods.

All of this was Catholic Action in the local community, providing organizations and services to empower Atitecos to confront their subsistence crisis. In conjunction with these programs, Catholic Action sponsored athletic teams, fiesta celebrations, a town orchestra, and other cultural activities. The question might arise as to why the Catholic Church allowed itself to become involved in such a broad spectrum of secular activities? If the community's problems were going to be faced, the church was the only institution with a presence in the community that could initiate them. The goal of the effort was not simply to provide a service, but to train Atitecos so that they could gradually take over the programs and spin off secular organizations. It was hoped that they would retain a moral sense of commitment as the basis of their secular activities. The programs were open to all in the community, and every effort was made to encourage the participation of traditionalists and Protestants. The later history of these efforts will be seen in chapter 23.

The Catholic Action programs in San Andrés Semetabaj and Santiago Atitlán were initiated before Vatican Council II and the Medellín meeting. They were part of the Action Catholic movement at the lower level of Catholicism that contributed to the formulation of the documents from Vatican II and Medellín, not the results of them. The documents helped legitimate the movement's development and its liberation activities against the criticisms of Tridentine Catholics who opposed it. They also brought increased resources of manpower and facilities to the effort.

Diocese of Quiché

The work of the Sacred Heart priests and their Action programs in the Quiché Diocese were summarized in a 1968 self-evaluation (Diócesis del Quiché 1994: 78–80). The diocese trained 36,000 catechists to work with 80,000 members of Catholic Action, more than half of the youth and adult population of the department. In the Nebaj parish there were 400 catechists working in 68 communities (Le Bot 1995: 130). These figures indicate the deep penetration of Action Catholicism's worldview into the lives of the K'iche' Maya. Under the leadership of Father Luis Gurriarán, eighteen cooperatives were formed—ten savings and loan institutions, six consumer co-ops, one industrial co-op, and one agricultural co-op.

In the town of Santa Cruz alone, the departmental capital, the savings and credit cooperative grew from twelve hundred to two thousand members, quadrupling its capital. By 1971, it was the largest in the country (Manz 2004: 55). Father Luis was also instrumental in establishing a center for study of the Bible, and another to raise social consciousness by examining the social problems of the region. Twenty-eight local bars went out of business, a sign that Catholic Action's campaign against the abuse of alcohol was having an impact.

In the municipio of Santa Cruz, a Catholic Action group was formed in each of its thirty-three districts (*cantones*). The cooperatives—savings and loan, agricultural, consumer, and artisan—had three thousand members. Catholic Action built thirty-five chapels, thirty-three schools that also served as community centers, and forty-eight sport fields, as well as paved or dirt roads for vehicles in every district. Throughout the diocese, many programs of social improvement, community development, adult literacy, and study of the Bible were established. All were assisted by broadcasts from the diocesan run Radio Quiché.

These social action programs, especially the cooperatives, resulted in the near disappearance of outstanding examples of ladino domination—the usurious moneylenders, debt-imposing labor contractors, and the ravenous facilitators of written documents (*guisaches*). There was improved agriculture. Some lands taken by illegal means were recovered. Artisanal production was stimulated. All of this reduced the necessity of migratory work to the South Coast plantations. There were some downsides to the cooperative movement. For some, it led to an impulse to buy superfluous and expensive goods, to consumerism and consequent financial indebtedness. For others, it led to an excessive dependence on chemical fertilizer as the sole solution to the land problem without seeking a more extensive and permanent remedy.

Chilón and Sitalá, Chiapas

The Jesuits based in Bachajón undertook an extensive program of Catholic Action in the parishes under their administration (Zatyrka 2003: 171–98). Beginning in 1969, they revamped the catechetical program to help broaden Maya consciousness and to empower them (Bobrow-Strain 2007: 192). In response to the bishop's request to help organize the 1974 Indian Congress, the catechists conducted community discussions as the region's contribution to the final presentations. The Jesuits researched Mexican and Guatemalan archives and found the original grants of land to Maya communities of the area. These lands had been illegally seized by ladinos over the years. "The Jesuits . . . have fought the government to return land to Maya communities, and that ejido status be granted to their residents. They denounced to government authorities the abuses of ladino plantation owners who did not pay a just salary to their workers, but only a half or less. [Padre] Mardonio Morales accompanied Maya on trips to see the state governor and even the Mexican president to explain the land problem" (Maurer 1984: 465). The suits floundered in the local agrarian courts due to bribes and threats from plantation owners (Zatyrka 2003: 173).

Education needs were addressed, especially by literacy programs using Freire's system. Seventy-eight schools were established under the auspices of the mission (Zatyrka 2003: 172–80). A program to train teachers for the local community schools received high priority. The Jesuits taught basic literacy and arithmetic courses, basic agricultural techniques, health, hygiene, etc. Father Morales translated Mexican agrarian and labor laws into Tzeltal and conducted courses on these topics so communities would become aware of their legal rights. Committees were formed to translate into Tzeltal pamphlets on different aspects of community development. This was the transition of the Jesuits from Tridentine to Action Catholicism. "The purpose of the teaching was not simply to make Tzeltales better instructed Christians, but also to give them the opportunity of a coming to consciousness that would help to liberate them from the oppression of western society and to organize themselves" (Maurer 1984: 446).

The Sisters of the Divine Shepherd undertook a medical program that started with them visiting communities to teach public health measures and administer simple emergency medical treatments. The program evolved into developing courses to train local Maya as health promoters and medical assistants. Graduates then held weekly meetings to instruct their communities. They assisted with vaccination campaigns and promoted potable water systems. Later, clinics with basic medicines were established in each community, managed by health promoters. This position was later recognized as a ministry in the parish structure (Zatyrka 2003: 172–80, 267).

To help with the subsistence crises, ninety-six cooperatives were established. There was an internal dispute among the Jesuits about assistance from ladino development advisors that eventually resulted in rejecting them. But for a brief period, some Jesuits collaborated with them. This resulted in the cooperatives being placed under the government agency CONASUPO. All failed (Zatyrka 2003: 180n109). The Tzeltal nuns (chapter 11) established an artisan cooperative for women's weaving that flourished.

In the previous century, plantations in the area were formed from lands seized from the Tzeltales (Bobrow-Strain 2007: 49–104). The Socialist Workers' Party (PST) and other political parties seeking Tzeltal support came to Chiapas and targeted Word of God communities because of their social consciousness. The PST began to encourage invasion of these seized lands. The Jesuits counseled against the invasions (119). In 1980, sixty families from Sitalá were urged by the PST to invade the Wolonchán plantation in an effort to regain possession of their lands. Some were liberation Catholics (Iribarren 1985: 66) and impatient with legal reform. A Tzeltal remembered that they "had been pushing papers, petitioning for land reform, and no one paid us any attention. The PST comes and says, 'now you're going to get land,' well, now that we had the PST, we started to invade the land" (Ovalle Muñoz, in Bobrow-Strain 2007: 122). Father

Morales, knowing the typical response by Chiapas officials, pleaded with the governor that no violence be used against them. But because the state government was controlled by plantation owners, the army was called in, along with local paramilitaries. They forcibly removed the protesting Maya, killing fourteen in the process (Zatyrka 2003: 184). Influenced by the EZLN uprising, in 1994 many other Tzeltales began invading ladino plantations in the Chilón-Sitalá area. Plantation owners blamed the Jesuits for the invasions (Bobrow-Strain 2007: 105–32).

Huehuetenango

Maryknoll is an organization of American Catholic diocesan priests engaged in missionary work. It originally focused its efforts on China, but after the Communist takeover in the 1940s, it was invited to Guatemala to administer the parishes in the department of Huehuetenango and a few in Quetzaltenango. By the 1960s, there were Maryknoll pastors living in most of the municipal centers of Huehuetenango. Upon arrival, they were ardent promoters of Tridentine theology, which led to clashes with traditionalists and the suppression of Mayan rituals (Montejo 2005: 141; and see chapter 22). Copying the American parish model, the Maryknoll priests concentrated on parochial schools and the development of Maya teachers. As they came to realize the subsistence crisis of the Maya, some initiated cooperatives and colonization projects. In 1966, in the town of Huehuetenango, they founded the Center for Integrated Development. Its students, along with some from the capital, gave literacy, agricultural, and cooking courses, as well as religious instruction, in the outlying areas. They also founded the Apostolic Center for the training of pastoral workers (Kobrak 2003: 18). Other Maryknoll efforts will be described in upcoming chapters.

While the numbers affiliated with Catholic Action increased in Huehuetenango, Catholic Action faced a continual problem: Maya involvement for undesirable reasons. "Foreign clergy . . . both Catholic and Protestant often have access to resources from their home countries and have instituted projects that draw followers to their religion. The clergy are asked to loan money, help obtain material goods, or intervene with mayors, governors, army. . . . Many religious workers accept the fact that some members see their Church as means to an end, i.e. to obtain material advantages, to learn Spanish or to gain power within the community" (Manz 1988: 68). Catholic Action encountered this problem in all Maya areas where foreign clergy were involved.

Catholic Action and Maya Culture

As Action Catholicism introduced or encouraged Maya participation in health programs based on science, cooperatives that functioned in a market economy, and especially education that was essentially socialization into the ways of ladino culture, some parents began to worry. In San Andrés, one parent remarked:

> I hope that the educated youths understand and are conscious of their race. But I also fear that there might be a youth who will fail to recognize that he is an Indian, who will disregard his race by wanting to present himself as a Ladino. If so, he regresses, because if he is pure Indian, he belittles the race he belongs to. It is possible that such a person might say, "May those Indians stay in their place. I will be more like a Ladino." And this is only because he changed his dress, while his studies, his culture, his education become worthless because they made him different. I hope that our children's egotism does not increase, because what I desire is the salvation of our race and not that they leave us in the backwardness in which we live because they do not return to their race. (Warren 1978: 163)

As both Maya and Catholic worldviews continued to evolve, this became a problem. While Catholic Action programs started as a way for Maya to defend themselves against ladino oppression and domination, it also raised other questions: Was it also a vehicle for ladinization, or did it have respect for Maya culture? Was conversion to Catholicism intrinsically linked to acceptance of Western forms of cultural expression, to the exclusion of traditional Maya forms, which included the wisdom of the ancestors that had formed the Maya moral sense as it was passed from generation to generation within the family unit?

In San Andrés, some Maya from the first generation of Catholic Action continued to see their efforts as part of the epoch of Jesus Christ, one of many cyclical epochs of the traditional Maya worldview (chapters 2 and 8). Two sons of a first-generation catechist left Catholic Action for the pan-Maya movement. The catechist was not upset, but rather spoke with "equanimity of the cosmic Maya life cycle in which an older generation dies so new generations might be born. He sees the inevitable end of what his generation has been able to achieve through their efforts in the church, schools, commercial crop production and local government. . . . Now the next generation will assume the struggle with different tactics, only to be replaced by still other generations" (Warren 1998: 190). This elder had placed his commitment as a dedicated Catholic Action catechist and community activist within the framework of the Maya metaphysic of epochs and cycles of revitalization.

Summary

This chapter has described the formation of community development organizations influenced by the biblical reflections of Catholic Action. Typically, these included health facilities with a public health emphasis, literacy programs, and cooperatives. The health programs battled the malnutrition and infectious-disease syndrome described in chapter 6. The literacy programs not only taught reading, writing, and arithmetic, but built up Maya self-esteem, ground down by years of injustice. Agricultural cooperatives introduced techniques that increased production to compensate to some extent for the land shortage. They were often associated with ejidos seeking to obtain unused government land. Credit cooperatives were formed to provide necessary funds for chemical fertilizer, to tide members over during the lean period between harvests, and to provide capital for small-scale itinerant merchants. The cooperatives invariably became enmeshed in the region's politics and in the rise of violence to be discussed in upcoming chapters.

V

Liberation Consciousness Assisted by Biblical Reflection

Pastoral workers and catechists spent a considerable portion of their meetings with Maya Catholics reflecting on the Bible. The purpose of these reflections was to reinstill a sense of personal dignity battered by years of domination; to assist in an understanding of the local and national political-economic systems in which the Maya were embedded; and to empower them to undertake specific communal actions that would help liberate them from their crisis of subsistence. The introductory photographs to the chapters of Part V show scenes from the Atiteco literacy programs described in the previous chapter. One of the purposes of these programs was preparation for reading the Bible.

13

The Bible and Its Worldview as a Cultural Document

This chapter examines themes of the Bible and a literary form used to express them. As a document from a specific cultural background, the Bible can be analyzed like any other cultural document. Such analysis abstracts from questions about its validity as divine inspiration, divine revelation, or its truth value. The previous volume and chapter 2 of this volume examined the Maya worldview, showing that it contained a key cultural logic, the Maya covenant. This section attempts to do the same for the biblical worldview.

The Bible, a Moral Worldview

Viewed anthropologically, a theology is a community's religious worldview whose primary function is to help people live their everyday lives. It is a paradigm that uses the categories of a culture to explain the moral dimension of the human condition and its relation to a supreme or some type of superior being(s). The Bible is primarily a book of moral instruction that uses literary forms and cultural symbols of Jewish and Christian cultures with a predominant Hebrew influence given its sources. This includes the final Christian section, since the early Christians initially saw themselves as a reformed Judaism centered on Jesus and often defined him in Hebraic terms.

No document is completely self-evident in itself. The reading of any document requires interpretation of some kind by the reader. Some important considerations are the concepts of the original language in which the document was written and its context, and the intention and cultural environments of both the author/editor and the reader. The original languages, contexts, and compilation of the Bible have been better understood over the past century by the exegesis of its texts, drawing on linguistics, the Semitic languages, archeology, history, comparative literary forms, etc. This work has shown that the present Bible is the end result of theological editors interpreting preexisting traditions. There is much speculation about how many preexisting traditions there were, how many of these traditions were originally oral, when they were committed to writing, who and how many original authors, and later editors, were involved (Kugel 1997: 1-50; Kugel 2007: 1-46, 662-90). In spite of all this uncertainty, there is near consensus that: (1) originally, there were oral and written traditions; (2) the oral materials were later committed to writing; (3) later still, theological editors gathered together these writings, and edited and interpreted them in such a way that they may have changed their original meaning slightly, or radically (the Song of Songs), or somewhere in between; and (4) in so doing there were cultural logics, specifically theological logics, that guided the editing for the purpose of moral instruction. Kugel (2007: xii) calls the editors "interpreters" and estimates that they worked between 300 BCE and 200 CE, although the texts may have already contained the work of earlier interpreters.

Their work was guided by four underlying assumptions (Kugel 1997: 14-15):

1. "They assumed that the Bible was fundamentally a cryptic text: that is when it said A, it often meant B."

2. "Interpreters also assumed that the Bible was a book of lessons directed to readers in their own day. It may seem to talk of the past, but it is not fundamentally history. It is instruction, telling us what to do."
3. "Interpreters also assumed that the Bible contained no contradictions or mistakes. It is perfectly harmonious, despite its being an anthology."
4. "They believed that the entire Bible is essentially a divinely given text, a book in which God speaks directly or indirectly through His prophets."

The Theological Pattern

A cultural logic can join together individual themes to form a pattern. An important cultural logic used by the interpreters and later editors of the Bible was that of the Deuteronomistic History (Noth 1981 [1967]; von Rad 1962; Cross 1973: 274-77). The theology of the Deuteronomistic pattern links together six major themes that form a cycle (some have more than one descriptive term): election/covenant; sin; judgment/punishment; the faithful few; liberation/redemption; and salvation, the last being a return to the first theme, election/covenant. The distinction between liberation/redemption and salvation is not always clear-cut. Sometimes they are blended together. The moral logic unifying the themes can be expressed in this abstract manner: God, out of pure love, initiates a covenant with the cosmos and human communities within it. This is election. Love, justice, resulting in harmony, should mark the life of these elect communities. However, communities rebel by sin; they turn in on themselves by some form of greed that fails justice and love toward others and God's creation. God refuses to accept this rebellion. By the terms of the covenant, the sinful community will suffer punishment as judgment for their sins. But because some members of the community remain faithful to the covenant, God will ultimately liberate and forgive some of the community for its sins. The community then enters into a period of salvation, a restoration to the condition of election. This is a logic of hope based on God's love. This moral pattern tells people how they should live their lives as God's elect, and what are the consequences if they do not. This is a highly simplified, abstract statement of the pattern, which has many shadings, complications, and different expressions in the biblical books themselves. Figure 13.1 is a simplified chart of this biblical logic and its expression. Across the top of the chart are listed the abstract categories of the pattern.

Election Covenant	Sin	Judgement	Just Remnant	Liberation	Return to Election
Historical Format of Phases					
1. Adam & Eve & Descendants	Pride Many	Banishment - Pain - Death Flood	Noah	Ark on Waters	Abraham Election of Israel
2A. EXODUS Israel in Egypt	Egyptian Oppression	Plagues in Egypt	Moses	Crossing Red Sea	Journey to Canaan
2B. Israel in Sinai	Idolatry, Complaints	Wander in Desert	Moses	Crossing Jordan River	Canaan
3. Tribal Israel	Idolatry	Defeat by Neighbors	Judges	Victories of Judges	Independence of Tribes
4. Kingdom of Israel	Oppressed by Kings	Defeat by Assyrians Exile by Babylonians	Prophets	Return from Exile	Rebuild Temple & Jerusalum
5. Ezra & Law	Disobey Law	Seleucid Persecution	Law Faithful	Maccabean Victory	Hasmonean Dynasty
Origins of Christianity					
6. Israel	Oppression by Scribes & Pharisees	Fall of Jerusalem to Romans	Jesus and Followers	JESUS Death & Resurrection	Descent of Spirit Early Christian Communities
7. Christian Communities	Systemic Oppression	Wars General Judgement	People of the Golden Rule	Eschatological Kingdom of God	
7A. Contemporary Segment Within the Seventh Phase					
Maya	Oppression	Fall of System	Just Community	Social Justice = Kingdom of God	

Figure 13.1. The biblical pattern and its expression in a historical format of phases.

Phases of the Pattern and the Historical Format

To express the moral instruction of the pattern, the Bible uses a literary device, a format of "historical events" to present its instructional material. These "events" are drawn from stories, myths, legends, actual events, poems, etc., that accurately express the moral dimension of the human condition. The Bible divides Jewish and early Christian "history" into a number of phases. Seven of them are sketched in figure 13.1, because of their clarity, or frequent usage to express the pattern. More could have been used. The division of the Bible into books sometimes obscures the pattern.

The "events" of this "historical" format are used in a symbolic manner, as metaphors illustrating the moral instruction of the pattern. Given this purpose, the phases should be taken seriously as an expression of a group's morality and its problematic nature. There is no necessity that they be taken literally. Some of the "events" actually took place. Others took place, but not as depicted in the Bible, and for others, there is no known evidence that they ever took place. But this is unimportant, as the Bible is trying to teach the reality of moral values and their existential impact. It is not trying to prove the events occurred, nor is it trying to narrate history in the modern sense. Consequently, the validity of the moral values does not depend on the historical accuracy of the various phases. All are narrated as if they were actual historical events. Such literary devices help to emphasize the reality implications of the moral instruction. A similar technique is used by much of the world's great literature. Readers, unaccustomed to analogical or metaphorical modes of expression, tend to take all the narratives as actual events, as historical in the contemporary sense.

In spite of the linear depiction in figure 13.1, a phase is cyclical, because the last stage, liberation, is a return to the characteristics of the first stage of the cycle, election/covenant. If all the cycles had been portrayed, the last stage of each cycle would have been the first stage of the following cycle. Perhaps the format should be called a "spiral of cycles." The phases are an example of parallelism, which Mary Douglas (2007: 10) sees as the Bible's dominant literary technique. Some books of the Bible, such as the Psalms and the Wisdom literature, can be seen as commentaries on the themes of the pattern.

First Phase

This is the cycle of Adam and Eve and their descendants as narrated in the book of Genesis. Election is symbolized by the idyllic harmony of the

Garden of Eden scene. Adam and Eve sin, and this is repeated by their descendants—the murder of Abel by Cain being one example. God's wrath is aroused by these sins. As their punishment, Adam and Eve are banished from the garden and become subject to the pains of the human condition, the examples being childbirth and working the soil. Their descendants are destroyed by the flood. But because of Noah's faithfulness as a just man, God tells him to build an ark that liberates his family from the punishing flood. After the flood, they begin over again as a redeemed community partially restored to a condition of election. God makes or renews a covenant with the descendants of Noah—Abraham and his progeny, who are the Israelites—a community called to justice and observance of the Jewish law. Literally interpreted, this is a covenant between God and only the Jewish people. Symbolically interpreted, the Jewish people stand for all human communities and the following phases are a moral theology of human community. Genesis places Abraham in the context of world history (chapters 1–11, and especially 12:1–3); what God is doing, through Abraham, is a transforming process of liberation for all humanity. Israelites passed on and eventually wrote down these stories to express and shape their self-understanding as a moral community.

Second Phase

The Israelites are in Egypt because Joseph sold his brothers to Egyptian slavers. There they suffer oppression under the Egyptians. As a sign of his covenanted love for this community, God punishes the Egyptians by the ten plagues. Abraham's descendant, Moses, frees the Israelites, as they pass through the liberating waters of the Reed Sea (Red Sea) to their freedom, expressed in the Sinai covenant with the promise of land in Canaan for his elect. On the way, Israel itself falls into sin, by complaining and by the worshipping of idols, and is punished by a prolonged wandering in the desert. But finally the Israelites cross the Jordan River, a redemptive moment. Salvation is symbolized by their taking possession of Canaan. The purpose of the narrative is the moral instruction contained in the categories of the pattern, not historical narration in the contemporary sense. If the text is interpreted in a literal sense, the validity of the values exemplified depends on the historicity of the events. From the viewpoint of symbolic analysis, the question of exactly what happened historically and when, or did it happen at all, is beside the point. The "events" are being used symbolically to instruct, not to prove. (Secular history dates the crossing of the

Jordan River into Canaan around 1200 BC. There is no known event in the contemporary historical sense that corresponds to the Exodus.)

Third Phase

Once the Israelites have taken possession of the land, they fall into sin again, by the adoration of Canaanite gods. God punishes them by allowing the Canaanite tribes to rise up and conquer the Israelite tribes. But because of a group of just men, called "judges," God later liberates the Israelite tribes from their neighbors and saves them by restoring their independence, a redemptive and salvific moment. This phase is from the Book of Judges and a clear example of the logical pattern used by the theological editors. (This phase uses events known from history. They took place from approximately 1200 BC to 1020 BC.)

Fourth Phase

The Israelite tribes, united under King David, build the Great Temple under Solomon, a period of election. But the kings sinned by their disobedience, their idolatry, and especially their oppression of their own subjects. The prophets denounce these sins against the covenant. As a result of forsaking the covenant, punishment follows. Israel is defeated by the Assyrians and Judah by the Babylonians, the latter forcing the Israelites of Judah into exile in Babylonia. (The kingdom had divided into a northern kingdom, Israel, and the southern kingdom, Judah.) But because of the faithfulness of the prophetic group, God liberates the Israelites by allowing their return from exile, leading to a salvific period in which the temple at Jerusalem is restored. (This phase uses events from secular history: around 721 BC for the defeat of Israel by the Assyrians, 587 BC for the defeat of Judah by the Babylonians, and 539 BC for the return from the Babylonian exile.)

Fifth Phase

This cycle repeats the previous ones. After their liberation, the Jews once again sin against the covenant, and as punishment, they are persecuted by the Seleucids, who desecrate the temple. But because of the continued observance of the law by a just remnant, with God's help the Maccabees revolt and are victorious in liberating their people. Thus initiates the period of Maccabean descendants, the Hasmonian dynasty, and the rededication of the temple in Jerusalem. The rededication is celebrated by the feast of Hanukkah. First and Second Maccabees and the Book of Daniel, important for the apocalyptic literature, relate much of this cycle. (This phase uses the rededication of the temple in 165 BC and the Maccabean revolt in 167 BC.)

Sixth Phase

This cycle initiates the Christian section of the Bible. The earliest Christian church was composed of Jews who initially considered their group as practicing a reformed Judaism centered on Jesus. Therefore, the early Christian oral traditions, and many later, edited forms, state the meaning of Jesus's life in light of the same theological pattern and its logic. Jesus was a prophet who condemned Israel's sins against the covenant, especially the injustice and hypocrisy of its priesthood and scribes. For doing so, Jesus himself was rejected, brought liberation by his blood, and initiated the new covenant of Christianity. (Saint Paul and others saw the fall of Jerusalem to the Romans, who desecrated the temple in 70 CE as the punishment.)

Seventh Phase

A new elective period, a new covenant, and a new cycle begin with the Descent of the Spirit at Pentecost and the formation of the early Christian church under the leadership of Saints Peter and Paul. Soon there is a strong disagreement between them about the implication of Jesus's message. Was the new covenant restricted to the circumcised—the sign of the Hebraic covenant as Peter first thought—or was it a covenant for all communities without the necessity of circumcision? Paul's position of universalism prevails. Christianity then separates from Judaism. The Catholic Church considers itself the direct descendant of the first Christian community.

Many see this phase as the contemporary period. It includes the sinfulness of the current epoch, which will end with a period of liberation—the reign of the Kingdom of God brought about by the remnant fighting for social justice. The "Kingdom of God" can be interpreted as referring to both liberation in this present phase, as well as some kind of final transformation, symbolized in the apocalyptic literature. Liberation in the present phase may be followed by another phase of the cycle that includes sin. The Kingdom of God on earth is not a utopia. In terms of Maya metaphysics, the present cycle of existence will end with cataclysmic destruction, out of which will arise the cycle of another epoch.

The biblical description of the last judgment emphasizes moral action in the present period.

> When the Son of man comes in his glory, then he will take his seat on his throne of glory. All nations will be assembled before him, and he will separate people one from another as the shepherd separates sheep from goats. He will place the sheep on his right hand, and the goats on his left. Then the king will say to those on his right hand, "Come you whom my Father has blessed, take as your heritage the kingdom prepared for you since the foundation of the world. For I was hungry and you gave me food, I was thirsty and you gave me drink, I was a stranger and you made me welcome, lacking clothes and you clothed me, sick and you visited me, in prison and you came to see me." Then the upright will say to him in reply, "Lord when did we see you a stranger and make you welcome, lacking clothes and clothe you? When did we find you sick or in prison and go to see you?" And the king will answer, "In truth I tell you, in so far *as you did this to one of the least of these brothers of mine, you did it to me*." (Matthew 25: 31-40; emphasis added)

This biblical passage emphasizes the importance of justice and love in everyday community life. To the damned, the same catalog of their refusals to do these deeds in this life is repeated with the reminder, "In truth I tell you, in so far as you neglected to do this to one of the least of these, you neglected to do it to me." Biblical theology is deeply concerned about the lives of people and their communities in present time.

Who Sinned?

There is a variant in the pattern depending on who sins and who is sinned against. Usually, the elect community itself, or part of it, sins, bringing God's judgment on it. God ultimately redeems a portion of the elect and restores the community (phases 1, 2B, 3, 4, 5, 6). But the oft-cited Exodus account describes a variant—the elect community is sinned against by the oppression of others, the Egyptians, who undergo the punishment (phases 2A, 7). There is no mention of sin by the elect community itself, nor is it denied. The emphasis is on liberation from oppression by an out-group.

The Book of Judith uses both versions of the pattern. The Jewish elders, in their panic before an attack of the Assyrian army, interpret the threat in terms of their own misdeeds, although they had not worshipped any false gods and could not assign a reason for their impending punishment. Judith rebukes them for their lack of faith. Invoking the God of Abraham, who kept the covenant with Israel through Moses in the Exodus story, she liberates the Jews from the out-group, the Assyrians. Her methods are quite different than those in the Exodus event. She seduces the Assyrian general and cuts off his head. This is a didactic story using striking imagery to make its moral point—God helps communities that stay faithful to the elect life.

The Pattern as a Covenant

The categories of the biblical pattern were used by the editors to express an underlying long-range, moral truth of all the day-to-day and year-to-year activities of human communities. Keeping to the theme of a covenant, the pattern can be stated as covenant protection (election), failure to live up to the covenant (sin), community crisis (punishment), covenant revitalization (redemption). The Maya metaphysic in chapter 2 describes the process as the cycles of birth, death, and rebirth/regeneration.

How Does Biblical Reflection Empower?

The book *Pedagogy of the Oppressed*, by Paulo Freire, was very influential among pastoral workers in the Maya dioceses. Freire (2001 [1970]: 30-33) points out that a characteristic of the oppressed's worldview is self-depreciation leading to extreme lethargy and a fatalistic outlook. Before those outside this culture, its members fall silent, bow their heads, keep their eyes cast down. *Caído* ("fallen") is the Spanish word frequently used to describe this picture. Ladino culture frequently applied it to traditional Maya, where it was seen as stupidity or laziness. The Word of God communities, with the assistance of the Bible, sought to break through this culture of self-depreciation and silence. It involved helping them to come to a consciousness of their dignity as a person. Biblical reflection was telling them that God cared for them, that the Maya people and each individual Maya were objects of his love and concern, that if they would take cognizance of their situation, his presence would help them remove the oppression. The biblical messages helped the Maya realize that their contemporary situation was not a punishment for their sins, but was due in part to the sinful oppression on the part of the social system in which their communities were, and had been, embedded for centuries. This was a message of hope and empowerment to do something about it.

Chapter 6 summarized the Maya crisis of subsistence—land loss resulting in lack of food, sickness, and early death. This was accompanied by confusion about the Maya covenant of the traditional worldview, and its psychological correlate, self-identity. The Maya worldview was contained in oral traditions scattered through numerous myths, stories, and rituals that made little attempt to harmonize its elements. As long as the worldview was not seriously challenged, diffuseness did not weaken its acceptance by Maya communities. But in the confusion occasioned by their cultural crisis, the Maya saw the biblical message as a divine revelation, the direct Word of God that broke through the confusion and questioning of the God-humanity relationship of the traditional religion. Through the written word, perceived by illiterate Maya to be the form for all important matters, God was speaking to the them. The Exodus story and the moral teachings of Jesus were relatively simple, straightforward messages that they readily understood, and were seen as concise, coherent expressions of the "wisdom of the ancestors": the Maya moral code.

In the village of Yibeljoj, the bishop of San Cristóbal is seen as a sainted disciple of Christ who has arrived among the Maya to carry out the mandate of the verses in Matthew's gospel 28: 19-20: "Go, therefore make disciples of all nations; baptize them in the name of the Father and of the Son and of the Holy Spirit, and teach them to observe all the commands I gave you. And look, I am with you always; yes, until the end of time." Since the bishop cannot personally be among so many, he brings the Bible with him, to teach the word of God. The catechists learn the Bible and help their communities understand it. The discussions make frequent use of the vine-and-branch metaphor (John 15: 1-8). God the father is the owner and keeper of the vineyard. The main stem of the vine is the bishop. The smaller stems, or branches, are the Maya communities. The fruits are the diocesan organizations and services in which the communities participate. Those that remain faithful to the teachings from the main stem, remain attached to it and bear much fruit. Those who are not faithful separate themselves from the main stem, bear no fruit, are cut off and burned (Moksnes 1995-96).

Summary

Taking an anthropological point of view, this chapter has examined the Bible as a cultural document that employs theological categories of Hebrew culture to give moral instruction. It uses a literary technique of parallel "historical" phases to symbolize and teach the moral instruction and its metaphysical import, not to prove the occurrence of events nor to narrate history.[1] To fully understand the pattern, the Bible needs to be seen in its entirety. There is a resemblance between the phases of the biblical pattern and the structure of Maya cycles. Many other liberation movements have used the biblical themes as a message of hope and empowerment (Walzer 1985). In the nineteenth and twentieth centuries, the Black Liberation Movement in the United States frequently employed it. The speeches of Martin Luther King Jr. were peppered with its passages. Few use all of its many phases. The following chapters will examine what sections of the Bible were more frequently used by the Maya for their reflections and resulting empowerment.

14

Methods of Reflecting on the Bible

The preceding chapter has described the structure of the Bible as a book of moral guidance. Figure 10.1 indicated the key role of biblical reflection for the liberation movement. This reflection is done from the viewpoint of the poor seeking liberation from social injustice and has certain characteristic marks.

> It is a hermeneutic that favors *application* rather than explanation. . . . a pursuit that was neglected for a long time in favor of a rationalistic exegesis concerned with dragging out the meaning-in-itself.
>
> Liberative hermeneutics reads the Bible as a book of life, not a book of strange stories. The textual meaning is indeed sought, but only as a function of the *practical* meaning . . . as interpreting life "according to the scriptures. . . .
>
> Liberative hermeneutics seeks to discover and activate the *transforming energy* of biblical texts. In the end, this is a question of finding an interpretation that will lead to individual change (conversion) and change in history (revolution). (Boff and Boff 1987: 33)

As noted earlier, the Bible stresses moral action. "In accordance with its Hebraic origins, the truth of Christianity was not to be seen or theorized on; rather, it was to be done, practiced.... The Christian concept of truth was originally not contemplative and theoretical like the Greek concept, but operative and practical" (Küng 2001: 26). The liberation movement uses biblical reflection for guidance in finding the practical meaning and empowerment in contemporary situations of oppression.

Problem: A Common Usage of the Bible

As seen in previous chapters, there have been a number of ways the Bible has been included in Catholic catechesis. For years, the Bible itself was seldom used, but interpretations of it were included in the question-and-answer catechisms that were simply memorized. Later, when the Bible was included in the catechesis, it was read as the definitive divine revelation that took place in past historical periods, a revelation that was to continue from generation to generation by reading the Bible and its ecclesiastical interpretations. At times, the meaning of a biblical text appeared evident in terms of the reader's own cultural categories. If not, a search was made for a literal meaning that would fit these same categories. When this has been determined, the text is then directly applied to the life of the reader. People have been helped by using the Bible in this manner.

But this method has also created many problems. The cultural categories of many readers of the Bible are usually different from those of the original Hebrew or Koine Greek text, and from the cultural context of the times talked about in the Bible. Biblical applications made in this manner are often forced or trivial or meaningless, because the situation of a reader or community is seldom an exact replica of a past historical situation talked about in the Bible. This usage makes past time definitive and normative, a template to be applied to present and future time. In its extreme form, this method is biblical fundamentalism. When taught in this manner, the catechist, as teacher, was active, and the rest of the group, as students, passive. The liberation movement found this traditional method of using the Bible a hindrance to understanding its message of liberation. It distracted from the Bible's ability to bring into consciousness present conditions of oppression and to give assistance and empowerment to confront them.

Bishop Ruíz's Recommended Method of Reflection

In reaction to this literal usage of the Bible, a treatise, *Biblical Theology of Liberation*, was first published in 1974 and reprinted in book form a year later (Ruíz García 1975a). Bishop Samuel Ruíz was listed as its author, although it was reportedly written by two young theologians who were afraid to attach their own names for fear of ecclesiastical censure (Meyer 2000: 125). The bishop's willingness to allow the use of his name indicates his agreement with its contents.

Seeds of the Word of God as Already Present in Maya Communities

The diocese had seen the virtues long practiced in Maya communities. "Gradually we were discovering the fidelity of the poor, their openness, their affirmation of faith, and then we felt a strong need to ask for an explanation and we gave thanks to our Lord for having allowed us to discover these things" (Ruíz García 2003: 49). "Nuns and priests find in native practices values and practices that they hold dear, for example patience, faith and sacrifice" (Eber 2000: 225). These observations confirmed a principle established by Vatican II, that the seeds of revelation already existed in many cultures (Ad Gentes, #9, #11).

These insights resulted in a different use of the Bible for purposes of religious instruction. The seeds of revelation in Maya culture needed to be harvested. Bishop Ruíz observed: "We are living in a time of transformation with regard to biblical reflection. Because of the life style of indigenous communities, they do not reflect about the meaning of a biblical text in order to later apply it to present reality, but the text throws a light on reality and in reality itself they discover the meaning of the text in a very natural way. Sometimes they discover surprising interpretations that would not be approved by exegetes, but they do not do violence to the text nor do they betray its meaning" (Ruíz García 2003: 153; also, Ruíz García 1999a: 6). "Faith is the reading of the signs of the times. It is looking at events and discovering there what God is saying to us so that we can go forward with our lives. This is a fundamental point."(Ruíz García 1999a: 28).

Biblical Reflection

Given reflection on present reality—the signs of the times—as the starting point, this reading is helped by analogies between the present and biblical situations. The Bible's usage has been compared to shining a light on an object in order to better see it. The light assists the comprehension, but does not impose it. For the purpose of understanding liberation, the treatise uses two of the phases in figure 13.1, the life of Jesus and the oppression of the Jewish people by their own rulers. Jesus was engaged in frequent conflict, with his denunciations of the scribes, the Pharisees, the money changers, and all those who oppressed the community by their sins (Ruíz García 1975a: 33–39). The longest section of the treatise draws on the phase of the Israelite kingdoms and the thundering denunciations of the prophets against the sins of unjust oppression by their rulers. It lists the many specific sins condemned by the prophets, especially Jeremiah and Isaiah, the prophet of social justice (48–70). These denunciations gave theological insight to the Israelites about the meaning of their sufferings and exile, followed by their liberation.

The treatise insists that biblical reflection shows that all human beings ought to be involved as liberators of the oppressed. This includes the oppressed themselves as actors in their own liberation, as well theologians, pastoral workers, and all concerned Christians. The treatise points out that working for liberation involves conflict. It negates a pietistic theology of passive, withdrawn people who say that all depends on God and that human liberating action is fruitless; who say that the goals of liberation will only be accomplished in a life to come. Biblical theology does not see liberation as utopian, as solving all present problems. Liberation is improvement of the present, but also begins the next biblical cycle, which includes sin.

Biblical liberation demands action in present time. The treatise talks about the seventh phase of figure 13.1, that a living Christian faith demands here-and-now action or else it is a false faith. "The essential for a Christian who wishes to live their faith, is to know what God wishes. For the Christian all the commandments can be reduced to only one that includes all the others. 'You shall love your neighbor as yourself'" (Ruíz García 1975a: 73). This is the golden rule. To illustrate its meaning, the treatise cites the parable of the Good Samaritan in Luke 10: 25–37. It begins: "And now a lawyer stood up and to test him [Jesus], asked, 'Master, what must I do to inherit eternal life?' Jesus asked him, 'What is written in the law? What is your reading of it?' He replied, 'You must love the Lord God with all your heart, with all your soul, with all your strength, and with all your mind, and your neighbor as yourself.' Jesus said to him, 'You have answered right, do this and life is yours.' But the man was anxious to justify himself, and asked Jesus, 'And who is my neighbor?'" Jesus answers by telling the story of a man who was attacked on a highway by robbers who stripped him naked, beat him, and left him by the side of road. A priest and Levite, both specialists in ritual ceremonies, came by, saw the man at a distance, went to the opposite side of the road to avoid him, and continued on their way. A merchant came by, saw the victim, went over to him, dressed his wounds and transported him to an inn, where he cared for him.

Leaving the next day, the merchant paid the victim's expenses and told the innkeeper to take care of the man, keeping a tab of all further expenses, for which he would be reimbursed at the next visit of the merchant. The story is an example of the norm of the final judgment: "In truth I tell you, in so far as you did this to one of the least of these brothers of mine, you did it to me." This was the golden rule in action, and answered two questions: Who is my neighbor? And when should liberation take place?

In brief, the method of biblical reflection recommended by the bishop compares contemporary events to a phase of the pattern outlined in figure 13.1. Placing them in this framework gives insight into the abstract theological meaning of the event, but because of its abstractness, it does not impose itself as a template for action. It helps define the context of the situation, leaving the specifics of action to be worked out in terms of the specifics of the event itself.

Nowhere in the treatise is there mention of the poverty and oppression of the Maya, or of those responsible for it. But every step of the argument can be applied to the Maya situation (phase seven). The bishop probably agreed on the use of his name because he found it useful at that time for clarifying his own evolving theological vision, which will be discussed in chapter 28. Probably, it was also intended to give a theological framework and method for pastoral workers of the diocese. They in turn were to communicate its methodology to the catechists for the reflections in the Word of God communities.

Revised Role of the Catechist

To implement these insights, the pastoral workers met with catechists and began by asking: What does it mean to be a catechist? The answer of each catechist was written on a blackboard. Some of the answers were: one who lives the Word of God; one who teaches; one who gives good example to the community. Then they were asked: Should people who are not catechists live the Word of God, etc.? All agreed they should. All the previous answers were erased from the blackboard because they pertained to all Christians, not just catechists. Then the question was again asked: "What does it mean to be a catechist?" The catechists talked, but gradually a silence settled over the group as they arrived at a consensus. "The agreement was that a catechist was one who collected a harvest consisting of the community's reflections. He is not the one who teaches because he has learned something during their catechetical courses, but one who goes with pen and paper to gather together the community's reflection. This idea changes not only the concept of what it is to be a catechist, but the attitude of the catechists themselves. They become harvesters of the community's reflections" (Ruíz García 2003: 34). Later, they began to use tape recorders.

The Questions: An Adaptation of the Bishop's Method

Many found the method described in the treatise was too broad an approach in itself, too vague. There was need to give the reflections some focus, some structure. Consequently, a compromise method was developed for the Sunday meetings. It was called *tijwane* from the Tzeltal verb *tijel*, meaning "to move, to stir." It describes the role of the catechist as one who stimulates the community to reflect.

When the Jesuits first arrived in the Bachajón area, they found the catechists using the traditional question-and-answer catechism by memorizing it. Later they adopted the method suggested by Bishop Ruíz. But they soon found that the biblical reflection needed focus. The bishop's method presumed a general knowledge of the Bible. This was lacking in many communities. Consequently, specific passages would be selected for reflection and some questions asked about it, to focus the ensuing reflection. Community services for reflection would follow a step-by-step procedure (Zatyrka 2003: 143–45):

1. Presentation of the theme (*Yochibal k'op* [first word]). The catechist explains the reason and theme of the meeting. He then asks provoking questions to stimulate people's interest in the subject.
2. Narration (*Scholel k'op* [announcing word]). The catechist expresses the theme of the text (biblical passage, community proposal, etc.) in his/her own words, transposing the theme into ideas proper to the local culture.
3. Reading of the reflection materials (*Yilel Hun* [seeing the book]). The biblical text is read to the assembled community. If needed, it can be repeated several times.
4. Activation (*Stijel k'op* [animating word]). The catechist clarifies concepts, explains the reasons behind and the importance of the text being presented, and makes clear references to the life of the community. This helps people realize the importance of their reflections, and that they have the ability to reflect in a meaningful way.
5. Conversation (*Snopel k'op* [knowing the word]). To start the discussion, the catechist asks some questions. The group is then divided into smaller groups of four to six people, so that each individual

will have the opportunity to express him- or herself during the discussion of the text. Eventually, each group arrives at conclusions. They name a representative to report them to the catechist or the assembled community.
6. "Gathering" the word (*Stzobel k'op*). The various groups reassemble and the catechists encourage the groups to share their conclusions with the other groups.
7. "Returning" the word (*Sutel k'op*). After hearing the opinions and conclusions from the different groups, a catechist synthesizes what he has heard in his own words. He comments on what he considers to be important points and tries to bring the reflections to a conclusion. The catechist may ask for, or the people can add, clarifications or elaboration of some ideas.
8. The reflected word (*Te nopbil k'op* [the "apprehended" word]). At the end of the meeting, a secretary writes down the conclusions of the meeting, which are read back to the people to make sure everybody agrees with the summary.

This is an adaptation of the bishop's method, since it is focused by questions prior to biblical readings. In liturgical settings in Bachajón, the community always favored starting with Biblical reflection, and then they would reflect on their situation. But in meetings to analyze a community problem, communities always started with a detailed description of their "reality" and only then did they search for assistance from their knowledge of the Bible, the catechism, or the wisdom of the ancestors.

The questions of the catechists are of great importance.

> They are asked to reflect upon parallels between the events or relations described in the Biblical text and their own lives today, and what this should imply for their own behavior. With the Questions [capitalized to stress importance], the church members began to evaluate their position within the broader Mexican society. Comparing this with conditions that Jesus had condemned during his time on earth, they have found that their own living conditions were against the will of God. Converts told me repeatedly that it was with the advent of the Questions that the political reflection began. A senior catechist described the change: "It seems that it wasn't until then that we began to reflect on what God wanted. There it's asked what we should do in different cases, what we want, what God wants, what he asks from us. That's the Questions: they ask us and we reflect." (Moksnes 2003: 143–46)

The reflection of this catechist's community about their position within Mexican society resulted in its participation in the political organization Abejas, to be described in chapter 21.

Selection of the Biblical Passages for Reflection

The biblical texts influenced the reflection on reality and any consequent action taken as a result. Consequently, the selection of the texts to be used in the Sunday services was of particular importance. A Vatican congregation, guided by the themes of the liturgical year, compiles a list of texts for use on each Sunday of the year. However, some dioceses prefer to draw up their own lists, as obligatory or suggestive, for their parishes. In the Maya dioceses, the pastoral workers appear to have selected most of the texts, although at times selection may have been left to the senior catechists. The pastoral workers could select texts that contained themes they wanted emphasized in the communities with whom they were working. In Yibeljoj, a pastoral worker wanted to emphasize the place of women in Word of God communities. She selected Romans 16: 1-6, where Paul praises the work of women in the early Christian communities. In an attempt to make the community pay strict attention to the importance of reflection, the pastoral worker told the community that the transcripts summarizing their reflections were read by the bishop himself, to see how the community was progressing (Moksnes 2003: 137). It is highly unlikely that the bishop had time to read the transcripts, although pastoral workers and catechists may have discussed them in meetings with the bishop.

The Sunday Sermon for a Large Gathering

The above methods assume a group small enough to be easily divided into several subgroups for the reflection period. In Santiago Atitlán there was also the Sunday sermon before a large congregation. How was the Bible to be used in this context? As described in chapter 12, in the early 1960s a Catholic Action group from the diocese of Oklahoma assumed administration of the predominantly Maya parish of Santiago Atitlán, Guatemala. The American pastor worked with the catechists. He understood the problems of the cross-cultural situation in which he was involved, and expressed the theological dilemma of the Oklahoma effort in this way:

> Initially we had the idea of helping the people to fulfill their religious life by learning the Good News of God in his invitation to all to be members of His community. This was to be attempted by means of a

twofold approach, namely through the liturgy and through a somewhat formally structured catechesis. Both of these areas were to be approached with great caution and no little degree of trepidation, since we were approaching an almost completely unknown set of two distinct cultures [ladino and Maya], the one of which seems to be more oriental than western. Thus we would have liked to avoid sweeping them into not only an American influenced liturgy, but also that of a Roman type which was necessarily quite foreign to them. But this seemed to be beyond our control due to the rigidity of the Church in insisting on the use of the Roman liturgy plus our complete ignorance of just what a suitable Indian liturgy would be like. We knew nothing of their signs and symbols, and were even highly skeptical as to just how much we conceivably knew about the signs and symbols of the ladino culture.

As a result of this, and because we could find absolutely nothing relevant to this dilemma during our six months of so-called cultural and language training, plus less from our local bishop and other priests of this diocese, we were necessarily tremulous as to just what we should do about the catechetical area, other than the idea of training native catechists to work among their own people. Hence we resolved to begin by way of introducing them to Christ and His Word as they have been presented to us in the Bible. The reasoning was that they had at least the right to know Christ and His Word, and then leave the interpretation to their own cultural resources. (Early 1969)

This emphasis on Christ as a teacher distinguished the catechesis of Catholic Action from the Tridentine emphasis on finding Christ in sacramental rituals and concerns about belief. Father Carlin took a realistic view of the problems confronting him. But acknowledging the problems did not provide a solution. Therefore, he simply went over a biblical passage selected for reading at the Sunday service, explained any necessary details, and let the catechists reflect and free-associate on it and deliver their own interpretations at the Sunday mass. This was in place of the customary sermon in Spanish by the priest, that was then translated into the local language by a catechist. The Maya moral sense found that it could easily express itself through the parables and sayings of Jesus. As an indigenous Catholic priest has noted, "If our grandparents have placed Jesus Christ in the center of our communities, it is because He speaks our language and gives life to our communities" (Teología India Mayense 1993: 49). "The deepest desires of our people are the same as those of Christ, the differences are of appearance, not content" (Flores Reyes 1996: 260).

Summary

This chapter has described various methods of biblical reflection used to find meaning in contemporary Maya reality. A method approved by Bishop Ruíz began with reflection on the current situation of the community, the signs of the times, and then proceeded on to the Bible to help understand the theological significance of the current reality. It was not to be seen simply as a book dealing with past time that should be imposed on present time. Some found this method was too vague in its starting point. They began by asking questions related to the text to help focus the reading and discussion. These usages of the Bible helped overcome group passivity stemming from a lack of self-confidence and from previous catechetical methods.

15

Biblical Reflections in Maya Communities

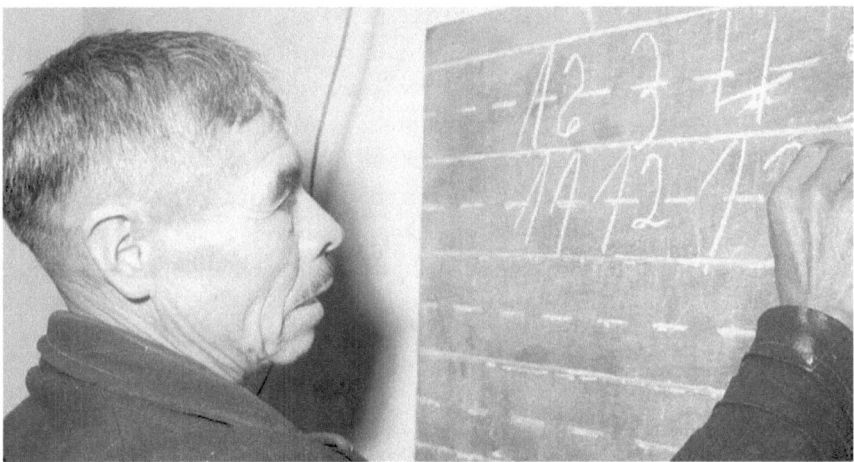

This chapter examines some examples of Maya biblical reflections and their influence on communal activities.

Magdalenas

This is the community of Rudolfo, whose early life was described in chapter 6. It is a highland village on the outer edge of the San Pedro Chenalhó municipio. Because of the mountains, communication by trail is easier with the town of San Andrés Larraínzar in the neighboring municipio of the same name. Consequently, the parish is administered by the priest in San Andrés, while in civil matters it is dependent on San Pedro. Into the 1990s, the traditional worldview was still well entrenched in the cargo system and shamanic practices. Initially, there was strong opposition to the efforts of the Catholic catechists.

Rudolfo had married, begun his family, and struggled to obtain subsistence on his small inherited milpa and by working on coffee plantations (Chojnacki 2004: 29–33). He started his ascent in the civil-religious hierarchy of Magdalenas. While serving as *mayordomo* of the church, he listened to the daily services conducted by the catechists for their small group of followers. He liked the praying and singing. He approached the community's first catechist about singing with them and started attending Catholic services. He also accompanied the catechists on their visits of consolation to the sick and dying. He later assumed the traditional cargo positions of *major* and *regidor*, while continuing to attend the Catholic services and accompanying the catechists.

In 1980, the head catechist invited Rudolfo to attend a preparatory course for new catechists in San Andrés. It was given by Padre Diego, who will be seen in chapter 23. Rudolfo agreed, but found the reading, praying, and singing difficult. Later, when the head catechist asked what he had learned, Rudolfo said he had understood very little. The head catechist told him not to be discouraged but to keep trying. He related his own first experiences preaching as a catechist, how blood flushed his face, but that he gradually got used to leading the services.

The Breaking Point

At this point, Rudolfo had not completely broken with the traditional system even though he was now a catechist. Incidents involving its rituals led to a complete break (Chojnacki 2004: 34–36). In 1984, Rudolfo was serving as *capitán*, a cargo service for the community's festivals. The Christmas Festival of the Niño, a child-saint not identified as the infant Jesus, required ritual dancing and drinking all day and night. Rudolfo recoiled at the thought of doing this, as well as at the expense involved. Another ritual requirement was riding horses at racing speeds around the town square. It was dangerous, as drunk riders often fell and sustained serious injuries. Rudolfo refused, and initiated an effort to end the obligatory taking of cargo positions. As a capitán, he was also responsible for recruiting two men to be *maxpat* on Good Friday during Holy Week (Chojnacki 2010: 127). They performed the *cavilto*, a ritual in which they reenacted Christ's Way of the Cross through the community while beating themselves with whips to whose tips were attached wax balls embedded with shards of glass. With blood flowing, they usually collapsed at the finish. Infection frequently followed.

Rudolfo refused to recruit anyone for the bloody ritual. This angered the principales. At a raucous meeting, the community discussed the matter. With some pushing, shoving, and threats of violence against Action Catholics, the meeting ended without resolving the problem posed by Rudolfo's refusal to carry out a traditional cargo duty. Local authorities dropped the matter when civil lawyers in Chenalhó were consulted and said that taking ritual cargo positions was a voluntary matter under Mexican law.

With these events, the Word of God community definitively split with the traditionalists. In 1980, Rudolfo finally convinced his father to join the Word of God community and stop drinking. Rudolfo's younger brother also joined. In the brother's view, "You don't have to pay anything to practice the Word of God. People are poor here and it's really hard to practice costumbre. It's a better life now, with people working for the community" (Chojnacki 2004: 48). Rudolfo went on to become a lay deacon, a role to be described in chapter 29.

Biblical Reflection

In a Word of God service, the community reflected on Luke 1:39–45, the Visitation of Mary to her cousin Elizabeth.

> Mary set out at that time [following the Annunciation] and . . . went into Zechariah's house and greeted Elizabeth. . . . as soon as Elizabeth heard Mary's greeting, the child leapt in her womb and Elizabeth was filled with the Holy Spirit. She gave a loud cry and said, 'Of all women you are the most blessed, and blessed is fruit of your womb. Why should I be honored with a visit from the mother of my Lord? Look, the moment your greeting reached my ears, the child in my womb [John the Baptist] lept for joy. Yes, blessed is she who believed that the promise made her by the Lord would be fulfilled.

The first catechist reflected,

> It wasn't just that they loved each other. No, they weren't just wasting time in vain, no. Their work received power from God. So it was that the child grew with the help of the Holy Spirit. They [Mary and Elizabeth] raised and cared for the little child . . . women who came together. . . . But she [Mary] became the mother of the savior. . . . They carried out the Word of our Lord Jesus Christ. They carried out the Word of our Father God. . . . She did everything with greatness of effort. . . . She obeyed with great effort in [her] work, our mother Mary. . . . But for us now, for all men and women, for all youth, this is our work . . . such as we learn from her we must carry it out. We are like her. . . . she also exhorted all men and women. She exhorted women of that time to take good models. She explained everything

that is right before God. This is what our mother Mary did. . . . We must quit everything bad, even in the small things here on earth . . . drinking alcohol, scolding each other, mistreatment between men and women . . . jealousy and hatred . . . slander. . . . at dawn, at noon, in the evening, let us ask our holy mother Mary to give us more of her power . . . to be able to do good works, just as she did when she was in the world. . . . we can't know how great Mary is . . . we can't compare it to what we are doing now, it's not the same . . . she did everything that was important . . . she obeyed with all her power in work. . . . But for us now, for all women and men, this is our work . . . God takes into account all the efforts of human beings. . . . We must work with all our effort, as we can do. . . . we have to . . . work while we live. . . . there is no rest in all of life. . . . Mary is at the side[of], near to God now. (Chojnacki 2010: 181–84)

The catechist is focusing on Mary as an example of good works in this life, rather than on her metaphysical status as interpreted by Catholic theology. The theme of action is stressed by the repeated use of the words "effort," "power," "carry out," "work," "hard work." The particular acts of work involve overcoming negative interpersonal relationships and habits stemming from the traditional costumbre. Mary will empower those who strive to imitate her life of faith and devotion to the hard work that her calling involved. The interpretation of faith as action rather than assent to the theological propositions about her is more in accordance with the Hebraic mode of biblical thought than the style of Tridentine formulations.

A second catechist then stood and answered questions about the biblical text. Most were about details of the story, such as: the fact that Mary's visit was to her aunt's house; that Elizabeth was in her sixth month of pregnancy; that John Baptist was born three months later; that he did not die of illness but was killed, and that a soldier carried his head on a plate; that Joseph was a carpenter who worked with a saw and lived in Bethlehem, where Jesus was born; that Mary nursed him. In some places, the catechists went beyond the biblical text. Nowhere, for example, does the Bible mention Elizabeth helping to raise Jesus. During the reflection, a catechist remarked that Mary had sinned in the past. This is contrary to Catholic theology. But these are insignificant matters for the liberation movement oriented toward action. Liberation theologians use the expression "emphasis on orthopraxis rather than orthodoxy"—correct action rather than correct belief (Gutiérrez 1973: 10).

The reflections concern the main problem for this Word of God community—perseverance in the break from the degenerate aspects of traditional customs. Community pressure to revert to traditional ways remained strong, and the reflection emphasizes the effort needed to persevere. In this example, the catechist himself does the reflection on the text, an earlier method of biblical usage retained in this community.

Another Reflection

At another gathering, the reflection was on Matthew 11:2–15, about John the Baptist.

> Now John had heard in prison what Christ was doing [teaching and preaching] and he sent his disciples to ask him, "Are you the one who is to come, or are we to expect someone else?" Jesus answered, "Go back and tell John what you hear and see, the blind see again, and the lame walk, those suffering from virulent skin diseases are cleansed; and the deaf hear; the dead are raised to life and the good news is proclaimed to the poor, and blessed is anyone who does not find me a cause of failing."
> As the men were leaving, Jesus began to talk to the people about John, "What did you go out into the desert to see? A man wearing fine clothes? Look, those who wear fine clothes are to be found in palaces. Then, what did you go out for? To see a prophet? Yes, I tell you, and much more than a prophet: he is the one of whom scripture says:
> Look, I am going to send my messenger in front of you to prepare your way before you [Malachi 3:1].
> In truth I tell you, of all the children born to woman, there has never been anyone greater than John the Baptist; yet the least in the kingdom of heaven is greater than he. Since John the Baptist came, up to this present time, the kingdom of heaven has been subjected to violence and the violent are taking it by storm. Because it was towards John that all the prophecies of the prophets and of the Law were leading; and he, if you will believe me, is the Elijah who was to return. Anyone who has ears should listen!"

The catechist comments,

> Nobody has received more strength for work than John had. . . . he who wants to believe, to learn what God in heaven wants . . . can be very great. He who confronts suffering, is hungry, is cold . . . finds the glory of God. Nobody looks well on him. But this person will possibly find salvation. . . . we have suffered so much poverty, cold, hunger,

> thirst; we don't have clothes, food, energy, we live who knows how, we seem ... like sheep without good pasture.... The reign of God, how to find it? Jesus said it is by making an effort. It is by the strength of human beings. But only when I make an effort. Only in this way can I enter the reign of God. He came to open the way to those who make an effort. (Chojnacki 2010: 187)

A second catechist, commenting on the last sentence of the text, says,

> We know that we have ears to hear. We have noses, we have mouths. It's necessary that we understand the commandment because the Lord gave us our head, or ears so that we understand well the commandment of God, dear brothers. Christians, true Christians, prophets of Jesus, we must make an effort to do the good. We must be strong like John the Baptist who was concerned about changing the world.... If we go to the *cantina* [local bar], it's horrible. We must open our understanding, just as our Lord said. (Chojnacki 2010: 188)

Here again, the emphasis is on action, empowerment, encouragement to persevere in the converted life in spite of discouragement, community pressure, and poverty. The contrast is with the traditional life, as exemplified by the customary drinking, and its self-defeating role in community life. For empowerment, they can count on the Jesus's presence and assistance, as related in another reflection: "Our Lord Jesus Christ goes before us. So we shouldn't think we are alone.... Even though we don't see him physically, he is here, men and women. If it weren't so, we wouldn't understand anything, maybe we would throw ourselves into some abyss" (Chojnacki 2010: 189).

A Summary Version of the Magdalenas Catechesis

For a synthetic view of the catechesis, an elderly deacon of the community explained that he initiates reflection by asking three questions: Where is God? Who is God? What is God? (Chojnacki 2010: 124–27). The questions assume that there are strains within the community, as described in chapter 5.

In response to the first question, Where is God?: "God is in the love of each person ... God wants us to have love. We need to act like brothers to all. If you do this, you are going to have more friends because all now share the same right attitudes, their friendship is no longer based on drinking. It's better for everybody. Now it's another way, not the way of drinking, but carrying on with work, good counsels and examples" (Chojnacki 2010:

125). Here the catechesis focuses on one of the major problems disturbing the equilibrium of the traditional community, the rampant ritual and secular drinking and its web of destruction. Again, it stresses that the converted life requires effort.

To the second question, Who is God?: "He is a God who always has his way of acting in a people or a family. . . . A people's way of acting requires good conditions. To burn candles, incense, drink—without thinking how is God—though one loves God, it's in vain. God wants the truth; they [traditionalists] act without thinking, through *costumbre*. God is always in all places, permanent. If God is permanent, people need to move ahead in life" (Chojnacki 2010: 126). Here the catechesis rejects the robot-like offerings of the traditional cargo and shamanic rituals, while ignoring moral action. This highly ritualistic interpretation of the Maya covenant had become commonplace in many Maya communities.

As to the third question, What is God?: "He is not a God who hides, he is a God who comes near. He is with you during all your work. If you don't know what God is, don't have a line [guidance]—do whatever you feel like doing—if you don't think carefully about what you're doing, what good is it?" The catechesis concludes, "If you grasp these three questions, now you will no longer have difficulty, no one can dominate you because you understand" (Chojnacki 2010: 127). The main focus of the catechesis in these traditional Highland communities was a commitment to a life liberated from traditional ritual performance and its worldview, especially the abuse of alcohol, with its web of destructive interpersonal relationships. The biblical figures provide the role models for the effort to reform.

Community Action Resulting from Biblical Reflection

Reflecting on the early Christian communities described in the Acts of the Apostles, the Action Catholics of Magdalenas formed cooperative groups to increase production and spread the risks involved in their subsistence agriculture. Rudolfo describes these groups.

> In Genesis it says God intended the land for everyone, not just some few people. In Acts four we heard how the apostles worked. When Jesus died they made an agreement. They decided to have all their goods in common. We talked about it for a long time. Some people [in the community] had more land and money than others. All the land we had originally came from our fathers, and it's all in separate plots spread over a huge area. Sometimes a person has to walk two hours from home to one parcel he holds. Then he has to walk another two

hours to a different plot on the other side of Magdalenas. We decided to form a cooperative and work all our land together.

We started little by little. First it was eight of us catechists. . . . Those who . . . [had] land lent it the others. But no one pays rent as almost everyone did before. And we all share the harvest together, each getting the amount of corn or beans according to how many people there are in the family. We also take turns getting firewood for each other, especially those who have horses [to haul the firewood]. The women help too. (Chojnacki 2004: 23)

Rudolfo continues the theme of cooperation, relating how some occasionally go to work on neighboring ladino ranches for short periods of time. But they first ask permission of the group. With the money earned, the group buys fertilizer and tools needed for their work. They also bought a small piece of land at a lower altitude to grow coffee and share the profits from its sale.

The example of this group encouraged others (Chojnacki 2004: 408, 411–15). Many Magdalenero families lived near their small holdings, isolated from other families. A catechist encouraged these scattered families to move into small cooperative hamlets and work some nearby unused communal lands. This led to a dispute with the traditional faction. A committee historically controlled by the principales oversaw the assignment of communal lands. When the Catholic community had been exempted from traditional cargo service, the enraged principales no longer considered them members of the community, and consequently without rights to use communal lands. An argument arose over the composition of the overseeing committee. Civil authorities decreed that the three members of the land committee had to be elected by the entire community rather than appointed by the principales, who no longer represented the whole community. At this point, the Catholic group became involved in political action, and they were able to elect three catechists as the committee. The committee, in turn, recognized the right of the Catholic groups to cultivate unused communal lands.

The communal lands were inadequate for the community's needs. Their area had been greatly reduced by twenty-eight ladino ranches that had fraudulently claimed or bought land from the Magdaleneros. Between declaring cargo service as voluntary in 1986 to the time of the study in 1996, twenty-six of these ranches were bought by Maya groups. Fourteen were acquired by groups led by catechists, three by traditionalists, and three by

Maya from the neighboring municipio of Chamula. Ladinos retained two. A catechist remarked: "The ranchers sold because they saw we were organized. . . . The indigenous grabbed everything through the word of God, working and giving up drinking" (Chojnacki 2004: 408). Also, the Mexican economic depression at that time, and the growing unrest in Chiapas about land claims and land invasions, may have influenced the sales.

In brief, the biblical catechesis of Magdalenas and its implementation in the cooperative work activities of the Word of God community are examples of the results achieved by Bishop Ruíz and his diocesan organization encouraging Action Catholicism in Maya communities.

Additional Biblical Reflections

Numerous other communities use biblical reflection to focus on their problems and to take specific actions to resolve them.

Hermosillo

In 1973, families originally from the area around Margaritas settled in a cañada, a small river valley, on the Rio Jataté in the Lacandón (Morales Jimenéz 1985: 179–85). For five months, they lived under the trees while they built dwellings and started their milpas. In 1974, they organized as a community and invited two catechists to give instructions and help them form a Word of God community. Soon they appointed two of their own members as catechists. In 1978, they constructed the community church, joined the Dominican-encouraged cooperative, Quiptic, and elected two village representatives to it. That same year, they attended Quiptic-sponsored workshops on politics and economics given by ladino organizers (see chapter 20). The community met every evening in the chapel for biblical reflection and discussion of needed actions. They split into small groups for discussion, and then reassembled to formulate a community consensus. "With the Word of God we were learning to talk and resolve our problems" (Morales Jimenéz 1985: 182).

When the first settlers arrived, they had claimed the better lands for cultivation. When additional families arrived later, only less-desirable lands were available. In 1980, the community decided that "this was not a Christian life," and citing Romans 12:9–15 on brotherly love, decided to combine and work the individual milpas as a communal holding. Each family retained about two-and-a-half acres for personal use. Citing the parable of the vine and branches in John 15:1–11, they decided to do the same with their pasture land. This was reverting to the traditional Maya custom of communal landholding. "The guide for the future path of our community was helped by the themes talked about in the lives of the first Christians." Cited are Acts 2:43–47, 4:32–37, and 5:1–11, which relate how the first Christian communities held all possessions in common; James 2:14–20, whose theme is faith without good deeds is dead; I Corinthians 13:1–10 on the necessity of love that is patient and kind; Matthew 7:21–28 on the necessity of realistic action, like the man who built a house on rock instead of sand; Matthew 12:33–37 about sound fruit only from a sound tree; and Mark 6:34–44, which relates the loaves-and-fishes miracle. In 1979, the community built a school, a basketball court, and a canoe for community use on the river. In 1980, they sent representatives to attend information and skill workshops. When they returned, the representatives would give similar workshops for community members. In this way, they learned Spanish; tailoring, using a purchased sewing machine; carpentry; how to build an oven for baking bread; and how to cultivate vegetable gardens. They built latrines and a small airfield. In 1981, they built a *trapiche*, which squeezes the sweet liquid out of sugar cane. They finally were able to arrange for a schoolteacher. Starting out with nothing but raw jungle, this was an accomplishment after eight years of work. The communal spirit nourished by the Word of God was an important factor in uniting and motivating the community.

Biblical Reflection in a Children's Group

Bishop Ruíz attended a service in the town of Margaritas (Ruíz García 2003: 153–54). The community divided into sections for the biblical reflection, including a separate one for children. The Bishop sat in with the children expecting that he could help them with the text, John 15:1–6.

> I am the true vine and My Father is the vinedresser. Every branch in me that bears no fruit he cuts away, and every branch that does bear fruit he prunes to make it bear even more. You are clean already by means of the word that I have spoken to you. Remain in me, as I in you. As a branch cannot bear fruit all by itself, unless it remains part of the vine, neither can you unless you remain in me. I am the vine, you are the branches. Whoever remains in me, with me in him, bears fruit in plenty, for cut off from me you can do nothing. Anyone who does not remain in me is thrown away like a branch—and withers; these branches are collected and thrown on the fire and are burnt.

A young boy commented on the passage: "The gospel is right, we have not understood it and we have not lived it since we have not understood that Jesus Christ is present in our brothers and sisters. Two women [in the town] are sick and who knows how their day goes, whether they eat or not, whether anyone cleans their house or washes their clothes. We don't pay attention to them since we are not branches united to the vine, we do not live the life of Jesus Christ" (Ruíz García 2003: 154). As a result, each day one of the group carried water to one of the houses, while another cooked for the second woman. The bishop was surprised at their ability to understand and use the text.

Biblical Reflections in San Pedro Chenalhó

Chapter 21 will describe the massacre of forty-five Liberation Catholics in Acteal by government-supported paramilitaries. After the government returned the bodies, in December 1997, the catechists approached the pastor of San Pedro, Father Miguel (see chapter 24), about a commemoration mass. "Miguel, tomorrow is the first of January. Tomorrow begins a new year, a new cycle. We do not want the blood of our brothers and sisters shed in vain. So early tomorrow you will celebrate for us a Mass of the Resurrection" (Chanteau 1999: 104). The catechists did not want the customary mass for the dead, but the one for the resurrection, regeneration. The deaths of their fellow Catholics would help revitalize the effort for human rights or, in terms of the Maya cycle, their deaths were not simply deaths but would help to bring about regeneration. At the mass, Father Miguel took as his theme the Exodus account, "that had been used so many times in our catechesis" (105). Here, the Maya and biblical cycles coincide.

Eber (2000: 225) observed reflections in another community in the San Pedro area. "In classes and meetings Catholic Action [pastoral] workers lead followers to see Jesus Christ as another powerful invisible being to whom they can pray for guidance and whose life they can study for wisdom about their struggle for justice. Catechists relate events in Jesus' life to their people's history, as they do their ancestors' stories to their memories. They worry less about chronology and more about understanding. Treating Jesus' parables and stories like signs and symbols from their ancestors, they learn to see Jesus' love and commitment as they do the enduring support they receive from *totilme'il* [divinized ancestors]." This catechesis can be easily related to the traditional worldview. Jesus is seen as a "powerful" being, perhaps emanated, whose teachings are comparable to and sum up the wisdom of the divinized ancestors, bringing harmonious balance to the community.

There were degrees of understanding the biblical catechesis. In Yibeljoj,

> There are ... differences among church members in how they understand the new teachings. It varies between those converting late in life and those who are born into the church; between active church goers and those seldom attending; and between ordinary followers and experienced catechists. The followers also differ in what attracts them to the new religion. It is often young, literate men who most easily grasp the Biblical discourse and who gain confidence when they notice their agility in learning. However, those who have incorporated the new religion most deeply in their lives are commonly women who have become mothers, whether married or divorced. Generally, they have little or no schooling to facilitate their Bible study and many complain that they are ignorant and slow in understanding the texts. However, having the primary responsibility for the health and care of their children, many women have God as their steady companion and source of encouragement. Thus, it is they who most faithfully attend church service. (Mosknes 2003: 126)

Fray Bartolomé de Las Casas Center for Human Rights

The San Cristóbal Diocese decided to take action against the repeated violations of human rights of Mayas and pastoral workers involved in many conflicts throughout the region. In 1989 it established the Center for Human Rights. The Center was staffed by priests, nuns, and laity. Its main functions were to provide legal assistance to victims of human rights violations and to educate people about their human rights, usually by means of workshops (Kovic 2005: 102–13). In these workshops, the demand for equality was expressed in religious terms drawn from biblical theology. As in meetings and ceremonies of Word of God communities, biblical texts were read and reflected on. The most common texts were drawn from the prophetic tradition (110):

> You will not ill-treat widows or orphans. If you ill-treat them in any way and they make an appeal to me for help, I shall certainly hear their appeal, my anger will be aroused and I shall put you to the sword; then your own wives will be widows and your own children orphans. (Exodus 22:21–27)

> Woe to those who enact unjust decrees, who compose oppressive legislation to deny justice to the weak and to cheat the humblest of my

people of fair judgment, to make widows their prey and to rob the orphan. What will you do on the day of punishment, when disaster comes from far away? To whom will you run for help and where will you leave your riches to avoid squatting among the captives or falling among the slain? After all this, his anger is not spent. No, his hand is still raised. (Isaiah 10:1–4)

They hate the man who teaches justice at the city gate and detest anyone who declares the truth. For trampling on the poor man and for extorting levies on his wheat; although you have built houses of dressed stone, you will not live in them, although you have planted pleasant vineyards, you will not drink wine from them: for I know how many your crimes are and how outrageous your sins, you oppressors of the upright, who hold people to ransom, and thrust the poor aside at the gates. That is why anyone prudent keeps silent now, since the time is evil. Seek good and not evil so that you may survive, and Yahweh, God Sabaoth, be with you as you claim he is. Hate evil, love good, let justice reign at the city gate: it may be that Yahweh, God Sabaoth, will take pity on the remnant of Joseph. (Amos 5:10–15)

Well now, you rich! Lament, weep for the miseries that are coming to you. Your wealth is rotting, your cloths are all moth-eaten. All your gold and your silver are corroding away, and the same corrosion will be a witness against you and eat into your body. It is like a fire which you have stored up for the final days. Can you hear crying out against you the wages which you have kept back from labourers mowing your fields? The cries of the reapers have reached the ears of the Lord Sabaoth. On earth you have had a life of comfort and luxury; in the time of slaughter you went on eating to your heart's content. It was you who condemned the upright and killed them; they offered you no resistance. (James 5:1–6)

You must see what great love the Father has lavished on us by letting us be called God's children—which we are! The reason why the world does not acknowledge us is that it did not acknowledge him. My dear friends, we are already God's children, but what we shall be in the future has not yet been revealed. We are well aware that when he appears, we shall be like him because we shall see him as he really is. (I John 3:1–2)

The first four texts are prophetic in tone, confirming that God knows about the oppression and injustices occurring in Chiapas and that punishment will follow sins against the covenant. This is the biblical pattern seen in chapter 13. The fifth text stresses God's love for the remnant fighting for justice. In addition to these most commonly used texts, other biblical reflections were drawn from Deuteronomy 24:12–15; Proverbs 6:3; Isaiah 25:6; Matthew 25:34–45; and Luke 3:1–14; John 3:37, 15:1–6, and 19:11 (Kovic 2005: 200n9).

The final sections of the workshops are reflections on the biblical texts. Here are two examples.

> For us in front of our oppressors, we didn't value our dignity and accepted what was done to us and we walked not for our own good, not to free ourselves of that which our ancestors worked for. The people didn't realize what rights they have and continue to accept things as they are. In knowing our rights we have the obligation to demand that others respect them so that we can change society. (Kovic 2005: 111)

> We have gotten rid of the customs that we used to be obligated to carry out. Before the fiestas were obligatory, it was the custom to drink alcohol, and those who had to pay for the cargo of the fiesta were poor. Thank God that now we continue with fiestas, but not as before, because that caused many people to be drunk and even die. When people drink they begin to fight in front of church with yelling. (Kovic 2005: 111)

From the records of thirteen workshops given by the organization from 1989 to 1991, and observing five more between 1993 to 1995, Kovic (112) summarizes their impact. "The courses stress that rights come from God and from laws. The idea that rights are given to all as 'children of God' is powerful. It illustrates that the current political and economic system of Chiapas goes against God's will. Following this reflection is another powerful idea—that people must organize to defend their rights. At the end of each course, agreements are reached and commitments are made. The most common is that the participants of the course will share their knowledge with other members of their communities. They also agree to put into practice what was learned by 'following the Word of God' and working in their families and communities to see that rights are respected."

Diocesan Women's Organization

In 1992, the diocese created an organization, Diocesan Coordination of Women (CODIMUJ), to discuss the problems of liberation specific to women (Kovic 2003: 132). It was composed of nuns who were pastoral workers and Maya women with whom they were working in local communities. They met periodically to discuss women's issues. As described by a member from Oxchuc,

> Around 1991, Sister Doiro came to work with the women and to look for the coordinators for a women's group in each community. They pulled out my name and said that we're going elsewhere to see how the groups are formed in other communities. Sister Miriam said that we'd organize the women to bake bread and plant a vegetable garden. . . . The bread and garden projects help the women in the group a little bit because it gives them a little bit of money and we can eat a little bread and vegetables and not only corn and beans.
>
> We meet every Wednesday in the chapel to reflect on the Word of God. We look at the actions of the women in the Bible and how we are in comparison. Sister Laly came to accompany us and to speak with the women so that we don't get tired, so that we work together.
>
> Now we are more than 40 women in the group. Beginning in 1995, Sisters Nuria, Justa and Salud came and we had a workshop to learn to sew and knit. We pray in the chapel. There the sisters speak to us about the readings in the Bible and we share our thoughts on the problems that exist in our community, Chiapas and in all Mexico, and on what we have to do as women. It is good to pray [reflect] because it helps to wake us up. (Gil Tebar 2003: 153)

A coordinator selected from each local group attended regional meetings and workshops conducted by CODIMUJ.

In the workshops, a text frequently used for reflection by the women's groups was Matthew 28:1–8,

> After the Sabbath, and toward dawn on the first day of the week, Mary of Magdala and the other Mary went to the sepulcher. And suddenly there was a violent earthquake, for an angel of the Lord, descending from heaven, came and rolled away the stone and sat on it. His face was like lightning, his robe white as snow. The guards were so shaken by fear of him that they were like dead men. But the angel spoke, and said to the women, 'There is no need for you to be afraid. I know you are looking for Jesus who was crucified. He is not here for he has risen, as he said he would. Come and see the place where he lay, then go quickly and tell his disciples, "He has risen from the dead, and now he is going ahead of you to Galilee"; that is where you will see him.' Look! I have told you. Filled with awe and great joy the

women came quickly away from the tomb and ran to tell the disciples. (Kovic 2003: 134)

At a CODIMUJ workshop in April 1997, the women summed up their reflections on this passage.

> It is our turn to announce, Jesus gives important work to us women. Today we must have courage like the women who went to the tomb. The work of women is important. Some have strength, but those who stay in their homes do not have sufficient strength. The women of earlier times had strength. We can also have this strength. The women went to visit the tomb and the angel asked them, "Whom do you look for?" Jesus brings peace to the women, because of this. The reading is important for us because through the women we are given notice to take the message to our brothers and sisters. It teaches us not to be afraid, to take the message to our communities: it tells us to be without fear, like Mary Magdalene and other women. If we go out [of our homes], if we come to the meetings, we wake up, we learn, and we gain courage. If we learn from this text, we have hope and the clarity to walk. We are the first messengers for the rest of the *compañeras* (women friends) in order to encourage them. We need to have more faith, courage, and confidence; we must be steady and determined in our work so that the announcement of the Kingdom [of God] advances. (Memoria from CODIMUJ Workshop, April 1997, in Kovic 2003: 135)

Another summary reflection on this biblical passage was, "We women, like the women in the gospel, are called to announce the Word [of God], but also to denounce that which is unjust, the exploitation in which we live" (Memoria from CODIMUJ Workshop, April 1997, in Kovic 2003: 131). The women also wrote songs in Spanish to express their hopes of liberation. One called *Today, the Women*, is sung to the tune of La Cucaracha. It mentions the negative effects of the stereotype of Eve bringing all evil into world because of her transgression; and how all through the Old Testament women struggled against oppressors to bring about liberation (Kovic and Paz y Puente 2003: 147).

Matthew 9:20–22 was also used in the meetings. "Then suddenly from behind him came a woman, who had been suffering from a hemorrhage for twelve years, and she touched the fringe of his cloak, for she was thinking, 'If only I shall touch his cloak, I shall be saved.' Jesus turned around and saw her; and he said to her, 'Courage, my daughter, your faith has saved you.' And from that moment the woman was saved." The pastoral workers pointed out that the law at that time prohibited a woman from touching a man's cloak. The women attending the meeting reflected that maybe she "lost her fear, was filled with courage, and pushed away any obstacle to be near Jesus. . . . she was filled with happiness and felt free to go to Jesus, even though the law did not permit it." In their reflections, women considered the possibility of breaking social norms and laws to fulfill their commitment to Jesus (Kovic 2003: 136).

The reflections in the workshops influenced similar reflections by women in their home communities. A Sunday meeting in the village of Yibeljoj reflected on Paul's letter to the Romans 16:1–6. In it he introduces a woman from another Christian community and asks the Romans to help her. He sends his greetings to three women of the Roman community who had previously worked with him. The text was chosen by a nun pastoral worker to initiate discussion about the role of women in the community and the work of the diocese. "Several [women] spoke out and some gave lengthy arguments on how women were needed in the church, to attend the meetings, pray, and reflect, and that they should support one another" (Moksnes 2003: 135).

The biblical reflections gave women the self-confidence and courage to actively participate in the discussions of community affairs. Traditionally, in community meetings, women were not expected to express their opinions, but to listen to the men. This was part of the subordination of women in Maya communities described in chapter 5. There were over 400 local CODIMUJ groups involving thousands of women throughout the diocese.

Summary

This chapter has described how reflections on biblical passages helped various Maya groups to persevere in their break with traditionalism, to form sharing communities, to demand their rights, and for women to realize their self-dignity and take part in struggle for human rights. The biblical figures are seen as models of behavior. There is no concern about their metaphysical status, as discussed in dogmas of the Trinity, Incarnation, Immaculate Conception, Virgin Birth, and the condition of sainthood, so prominent in the Tridentine catechesis concerned with theological formulations that became Catholic orthodoxy.

16

Biblical Reflections in Lacandón Migrant Communities

The Lacandón is a large area of eastern Chiapas that stretches to the Usumacinta River, the boundary with Guatemala. Around 1900 it was a jungle populated by only a few hundred Lacandón Maya and several timber companies. The unused lands belonged to the government and were open to settlement. By the 1990s, heavy immigration by Tzotzil, Tzeltal, and Tojolabal Maya, other indigenous peoples, and some ladinos had swelled the population to an estimated 200,000 to 300,000.

Social Conditions in Lacandón Regions

There was an arc of large plantations stretching from Palenque south to the area around Margaritas. Beginning in the 1920s, workers from these

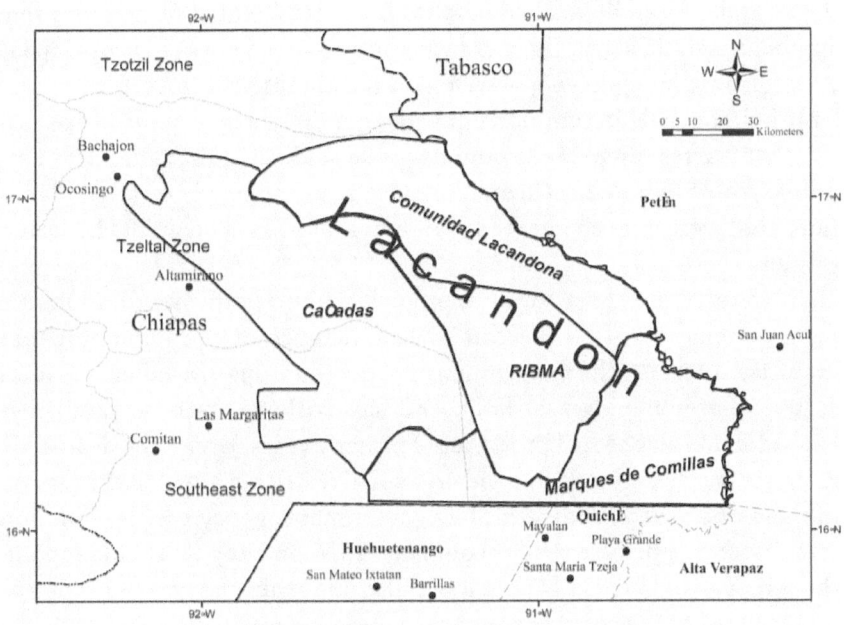

Map 4. The Lacandón in eastern Chiapas, with Its Four Subdivisions

plantations and other land-short Maya moved east to the Lacandón jungle searching for unused land. With the towns of Ocosingo, Altamirano, and Margaritas as jumping-off points, they penetrated the high ridges and narrow river valleys of a district called Las Cañadas (the Canyons). Settlements occupied the narrow river valleys where land and water were available, with the surrounding ridges supplying firewood, building timber, and an area for hunting game. A large part of the area was part of the Ocosingo parish administered by a pastoral team of Dominican priests, nuns, and laity. Among them was Javier Vargas, a Catholic layman, who, before its closing, had been a Marist brother at the Marist catechetical school in San Cristóbal. In the Cañadas, the pastoral team developed a network of catechists who produced their own catechism, the focus of this chapter. Understanding the problems faced by the Lacandón communities throws light on some of the catechism's contents, as well as on the later role of some of these communities in the Chiapas insurgency.

The immigrants formed ejidos to lay claim to unused government lands (*tierras baldias*). Land for peasants was the driving force behind the 1910 Mexican Revolution. By Article 27 of the Mexican constitution, poor and landless peasants could petition the government for titled ownership of these lands. An individual could obtain a small holding of no more than approximately 25 acres. The more common occurrence was a group of peasants forming an ejido cooperative and petitioning the government for a large holding, held in common but worked individually in plots of approximately 50 acres. Ejido land could never be sold. The northern part of the municipio of San Pedro Chenalhó had a similar history of large plantations that were later abandoned and then sought as ejido grants by former workers.

In spite of the constitutional provision, the government became the most serious threat to the continued existence of the Lacandón communities. In 1973, the government created La Zona Lacandóna, a huge, 2,400-square-mile tract of land, and gave title to sixty-six families of Lacandón Maya who had lived there for many years. Under the decree, the sixty-six families were to live within a small restricted area. All the remaining land was to be administered by a government agency, which would contract it out to a private timber company. This company would monopolize the removal and export of valuable cedar and mahogany trees to the economic profit of the company and some government officials (Leyva Solano and Ascensio Franco 1996: 167).

A portion of this grant included land in the Cañadas that had already been settled and worked by four thousand families in thirty-seven Chol and Tzeltal communities, many for ten or more years while they made their ejido applications. A presidential decree threatened to evict all of them. Twenty did relocate, but the remainder were not about to give up the land they had worked and developed from nothing but raw jungle (Womack 1999: 16–19). Violence flared as villagers blocked construction crews attempting to build a road into the area. This incident, know as La Brecha (the Breach), became a rallying call for the defense of repeated incidents throughout the Lacandón where communities were threatened with eviction from lands they had settled and worked.

In the southeastern part of the Lacandón, at Marquéz de Comilas, Tojolabal Maya had immigrated, encouraged by the national government under its policy of developing commercial agriculture (Vos 2002: 157–59). The government promised to build an access road, to survey and fix settlement boundaries in preparation for obtaining land titles, and to make a study of the social and economic needs of the area as a prelude to a program of public services. The colonists arrived and built settlements from nothing except what was provided by the jungle itself. But none of the promises of the government were carried out, except for the building of the road, which was used for military purposes. In 1979, the government, in cooperation with UNESCO and the international environmental movement, set aside a large land tract of slightly more than 1,200 square miles as the Reserva Internacional Biosfera de Monte Azul (RIBMA in map 4). Twenty-seven migrant settlements were dispossessed and resettled (Womack 1999: 19). These evictions fueled more resentments.

Many communities had their land titles constantly delayed by government inaction or corruption. In some cases, lands already settled by Maya were titled to Chiapas elites for commercial timber or cattle enterprises. Settlers lived under the constant threat of eviction. These relatively isolated communities became the objects of predatory practices. They needed to sell some agricultural products with which to buy necessary goods. This put them at the mercy both of the few tradesmen who came to sell goods, and of those with pack animals who could haul out their agricultural products to market. In this monopolistic situation, both providers charged exorbitant prices. The settlers were also set upon by corrupt government employees—forestry, treasury and land survey personnel—for bribes or charges for services they were not providing (Vos 2002: 252).

In addition to the continual fear of eviction and the failure of the government to provide the promised assistance for opening up this last unsettled area of Mexico, the most devastating blow came with the proposed repeal of Article 27 of the Mexican constitution. This had been the legal basis for obtaining ejido land from the government, the last hope of many Maya to overcome their crisis. The proposed repeal was the result of the impending trade agreement with the United States and Canada: NAFTA. One of its provisions called for the abolition of small, subsidized agricultural holdings, such as ejidos, that were considered relatively unproductive by economic experts. Article 27 was repealed in 1992, with the repeal to go into effect on January 1, 1994. At that time, pending ejido applications from Chiapas comprised 25 percent of all applications in the entire country. All settlements that had pending applications became potentially subject to eviction. With the abolition of ejidos, land could now be sold. The Maya in difficult circumstances who had obtained ejido land could now be forced to sell it, a repetition of one of the original causes of the Maya crisis (chapter 4).

Composing the Tzeltal Catechism

The Tzeltal catechism originated at the beginning of these conflicts and evolved as the conflicts became more intense. In 1971, at a meeting with the catechists and the Tzeltal pastoral team, community members expressed objections about the lack of participation in formulating the content of the catechesis. They said that the catechesis was ignoring the local culture, and that the pastoral team and catechists themselves were acting too much as "instructors," contrary to local custom of community discussion. This resulted in four decisions: "(1) to respect the cultural identity and social reality of those receiving the catechesis; (2) to recognize the presence and value of the preexisting Christian tradition [the colonial Maya worldview]; (3) to encourage the participation of the whole community in reflecting on the Word of God; and (4) to allow the catechists and pastoral workers to become 'facilitators' of this process" (Vos 2002: 223). These objections were rooted in the Maya custom of open discussion and reflection in order to reach a consensus position, as discussed in chapter 14. The introduction to the catechism notes: "Christ is in our lives, we will not find him if we look for him outside of our lives. And what is life? It is our history, our work, our joys. Life is the customs, the prayers, the hopes and all the works of the community. Here in the life of the community is where we will find Christ who is the way of liberty" (Los Tzeltales 1972-74: 3).

The catechists were instructed to ask each of their two hundred communities three questions in conjunction with a biblical text assigned for a particular Sunday: What was the plan of God when he created the world? How have human beings lived this plan? How are faith, hope, and charity to be lived as part of this plan? The reflections produced interpretations and stories that were collected by the catechists, who taped and transcribed them. The pastoral workers then synthesized them as a series of topics (Ruíz García 2003: 38). Copies were distributed to the catechists for further discussion in their Word of God communities. Later, the catechists themselves took over the task of editing and synthesizing the reflections. Over time, a considerable number had been accumulated, which were then duplicated and privately distributed as a catechism titled *Estamos Buscando La Libertad: Los Tzeltales de la Selva Anuncian la Buena Nueva* (We are Seeking Liberty: The Tzeltales of the Jungle Announce the Good News). The remainder of this chapter examines the catechism, with its synthesis of community reflections about their current reality and how the Bible was used to help understand this reality.

Format of the Catechism

The catechism contains fifty lessons divided into the three main themes of the original questions. "The Plan of God from the Beginning" explains the kind of life God intended humans to live when they were created. "How We Live in Oppression" details how the oppressed community is living in conditions contrary to this plan. "Faith, Hope and Charity" shows how these foundational virtues are necessary in the struggle to overcome the oppression that thwarts God's plan. Each lesson is a guide for the catechist to conduct a meeting. A lesson opens with a prayer for enlightenment or thanksgiving, along with an original song, both of which are focused on the topic of the lesson. To help focus the reflection, a lesson contains fifteen to forty statements for discussion. The statements are derived from previous reflections about the significance of community conditions, elicited in meetings of this or other communities. Sometimes these statements will be interspersed with a biblical verse or an illustrative story or questions about the statements. The final part of the meeting is silent reflection and prayers of personal petitions, followed by a repetition of the opening song and a closing prayer. Again, the songs and the prayers relate to the topic of the meeting. (Womack 1999: 133–47 has translated ten of the fifty lessons into English.)

As the bishop noted, some usages of the Bible are not in accordance with biblical exegesis, but that is beside the point for the purposes of these communities. The catechism mainly served to help the catechist lead the discussion. Many members of the communities were illiterate, or their literacy was weak. Communication was primarily oral. As an aid to memory, there is considerable repetition in the statements, hymns, and prayers, not only within a lesson but also between the various lessons. There is certain amount of free association in the statements, so that topics can appear and be repeated almost anywhere in the document, rather than in their logical place within a theme. Given its purpose and method of compilation, the catechism was not intended to be a well-chiseled theological treatise.

Observed Reality

What were the external and internal problems of these communities that furnished the basis for the first step of reflection? As seen above, the national and state governments were the source of the external problems. The catechism was composed in the early 1970s. Ejido communities were promised titles as well as government public services—potable water, schools, electricity, roads. But titles were delayed, services were not provided, and there was the constant threat of eviction for many communities. Using bribes and threats, local economic elites who coveted land for its timber or for cattle pastures, were obtaining titles. In some cases, private armies of the elite violently evicted Maya settlers from the lands they had worked. The violence resulted in some deaths. As the situation worsened in the late 1970s and subsequent decades, the reflections on the external problems probably became more pointed than in the early version of the catechism cited here.

As settlements in a newly occupied area, there was not a division between entrenched traditionalists and orthodox Catholics. The main internal problem was to turn these new settlements into communities with which people identified and for which they felt a strong obligation to perform community service. This was a difficult task. The settlers were from different Maya areas and spoke different Maya languages. Among the Maya, the primary unit of social identification was the community where one was born, socialized in a kin group, and where one worshipped the community gods in the church and nearby mountains. This sense of identification was not lost with out-migration. People returned to their original communities to perform rituals and celebrate the saints' festivals, to keep in contact with kin, and to fulfill reciprocity obligations (Pitarch 2010 [1996]: 218n8). Being Maya, that is, sharing a similarity of languages and customs, did not and never had unified the people described by this generic designation. All of this militated against forming a strong sense of unity and obligation to one another in the newly formed settlements. Yet only with such a unifying force could they hope to confront the external problems they faced.

There were additional problems. There was the ladino stereotype of the "indio" as a poor, stupid creature incapable of anything except working the soil. Some Maya, internalizing this stereotype, would develop an inferiority complex, become passive, and fail to take the initiative in confronting internal and external problems.

First Theme: The Plan of God from the Beginning

Trying to understand the significance of their contemporary situation, the Word of God communities sought insight from biblical passages. The first part of Genesis gave them a framework of finality, the way things should be. God created all things. His force continues in them, especially in the land, where one sees this force in its fertility, and in the sun, where the force is seen in its light and the energy it radiates. "God wishes us to enjoy all the things he has given us. He wants us to make use of the things he made for us, making them grow and be improved by the force that is within us. We save ourselves if we work with the work of God which is to make the world grow and seek a new life that is constantly improving. With God making all these things and giving them to us as a gift, we human beings are the only owners and administrators of all creation. Therefore let us say to God that we accept his gift. Let us say that we accept the responsibility to work for the betterment of the world. Let us say that we promise that all these things and our work will be to serve the whole community. Because only in this way will each one have what he requires for his life, that all of us may be free" (Los Tzeltales 1972–74: 17).

The only cited, but not quoted, biblical passage in this section of the catechism is Genesis 1:28–31. "God blessed them [Adam and Eve], saying to them, 'Be fruitful, multiply, fill the earth and subdue it. Be masters of the fish of the sea, the birds of heaven and of all the living creatures that move on earth.' God also said, 'Look, to you I give all the seed bearing plants everywhere on the surface of the earth, and all the trees with seed bearing fruit; this will be your food. And to all the wild animals, all the birds of heaven and all the living creatures that creep along the ground, I give all the foliage of the plants as their food.' And so it was. God saw all that he had made, and indeed it was very good. Evening came and morning came: the sixth day."

In the context of creation, the third lesson mentions God sending Christ as a human being. It quotes Matthew 1:18–23, which describes an angel's appearance to Joseph, instructing him to name the infant in Mary's womb Jesus, meaning the one who "will save his people from oppression." Most versions of the Bible translate this phrase as "save his people from their sins." Here the Tzeltal catechism specifies the contemporary role of Jesus in restoring the goal of creation in Chiapas.

Second Theme: Oppression as Thwarting the Plan of God

Given God's plan in the first theme, this theme reflects on what is happening to God's plan in the contemporary conditions of Maya communities.

What Oppression Means

This section describes how Maya families have lived and still live oppressed by others contrary to God's plan. The equation, lack of land equals lack of food equals sickness and early death, is constantly repeated in one form or another in many places in the catechism. Referring again to Genesis, the catechism reflects: "In the beginning God made the earth. Then God said to humans: I give you the earth and everything I have created. I give you the mental ability in your heart so that you may govern the earth and be its owners.... All humans have the right to own land in order to work it: this is the will of God.... Our fathers and mothers were peons on the plantations of the rich. They and many other Mexicans suffered slavery and they began to fight for liberty. They fought so that each man should have land because God has given us all that he made as our inheritance. A person who works under physical threat does not consider himself an owner and master of the earth, nor does a person who works for a plantation owner out of fear. There is no justice when one works from sunup to sundown and suffers hunger along with his family. One is a slave who does not receive a fair wage for his work" (Los Tzeltales 1972-74: 19).

The final lesson is a suppliant prayer about Maya needs in contemporary conditions: "Look, Señor [Christ], on our sufferings: we are sad and lacking the necessities of life because many of us do not have land. And when we seed our milpas, they yield poorly because the land is worn out. This is the reason we lack food and suffer hunger and sickness. We know, Señor, that we must improve our land, but we do not know how. We cannot continue living like this, Señor. We need both land of which we are the owners and [assistance about] how to take care of it. We need to work without fear, planting other fruits that are still unknown to us. We need to understand how to have more money to cure our sicknesses, to buy clothes and all the things we are lacking. Today we have come together to speak with you about our necessities. We hope, Señor, that you will give us the strength to work together in community so that we may escape from our poverty" (Los Tzeltales 1972-74: 27).

In their struggle to overcome oppression, the Maya see themselves mirrored in the struggle of the Israelites to free themselves from the oppression of the Egyptians. The Exodus theme helps them interpret their present reality. The catechism quotes Exodus 3:7-8: "Yahweh then said, 'I have indeed seen the misery of my people in Egypt. I have heard them crying for help on account of their taskmaster. Yes, I am well aware of their sufferings. And I have come down to rescue them from their oppressors and bring them up out of that country, to a country rich and broad, flowing with good fruits.'" Instead of "oppressors" and "good fruits," most versions of the Bible use the more literal translations, "Egyptians" and "milk and honey." Again, the catechism's translation uses terms appropriate for the Maya situation. God's saving presence in the midst of the Israelites' struggle empowers the Maya in theirs.

God's Law and Overcoming Oppression

This section contains twelve biblical quotes regarding the law of God's love written in the hearts of humans. To overcome oppression, the Maya themselves must live this law in their communities. This law is the love of neighbor, of justice, of deeds not words, of helping the poor, of pardoning and helping one's enemies. The biblical passages cited are: John 31:33; Matthew 5:17-18, 20; John 13:34; I John 2:7; Hosea 4:1-2; I John 3:17-18; Matthew 25:31-46 about the Last Judgment; Proverbs 17:5; Matthew 7:7-12; Luke 6:20-35, 36-49; and a verse poorly referenced. The section also paraphrases the Beatitudes without citation.

The reflections continue about the traditional Maya culture and its customs that sought to carry out God's plan. Jeremiah 31:33 is quoted: "I will put my law in all humans, writing it in your hearts and I will be your God and humanity will be my community." The reflection on it, "This law . . . tells us that we ought to take care of the life we have; ought to live as brothers not doing harm to anyone; that each one should have the land, food and all things necessary for life" (29). Citing Luke 6:20-49, the biblical reflections encourage helping the poor, pardoning your enemies, listening to Christ and struggling for that which is good (Los Tzeltales 1972-74: 37).

A further section continues reflecting on the necessity to preserve Maya culture even as adaptations are made to improve their impoverished condition. Oppression is seen in ladino disrespect for Maya culture. "Culture is everything that distinguishes us, all that we will leave to our children: history, language, customs, stories, traditions. All of this is a richness built up over many years of life. We must take care of, respect and improve our culture. Our culture is like a well of water and from it we have drunk all that we have, all that we know, and all that we are" (Los Tzeltales 1972-74: 39). A story is told of a ladinized Maya who looked down on his own people and ridiculed them for their customary ways (44-45). This is a danger to be avoided. The Maya must learn to speak out and protest when they experience disrespect from the ladino world, when teachers do not show up in the schools, when they are cheated in the market for their coffee, pigs, and many other things; when they cannot obtain medicines to cure illnesses (49).

The Value of Our Culture

This section emphasizes that there is value in Maya culture. It needs improvement, but this should not be interpreted as ladinization, completely disrespecting and rejecting the traditional culture, as urged by the Tridentine catechesis. The biblical reflection again cites Exodus 3:7–12, the mission God gave Moses. It helps the Maya to understand that in spite of their oppression, God is present in their culture and will help with doing away with unjust domination.

A final section of Part II reflects on oppression against the traditional religious worldview. It describes how Maya ancestors constantly sought God in their thoughts and rituals. Seeing the force of God in the sun and moon, they saw them as gods in themselves rather than leading them to God. Many of their ways of seeing God can be improved, but they are not to be done away with. It was the Maya way of attempting to fulfill God's plan described in Genesis 1:28–31 that is again cited.

Such considerations about the traditional culture marked the new type of catechesis represented by the Tzeltal catechism. In the introductory letter to the catechism, the pastoral workers remarked, "The dynamic of the previous catechesis prescinded from and even destroyed the culture of [Maya] communities, seeing many of its elements as incompatible with the Gospel because of confusing the content of the Good News with the western cultural patterns" (Los Tzeltales 1972-74: ii). This raises the question of a Teología Indígena (Indigenous Theology), a movement to be discussed in chapter 29.

Third Theme: Faith, Hope, and Charity in Seeking Liberation

These are the theological virtues in Christian asceticism that bring about union with God. The catechism places them after the theme of oppression because they empower the faithful in their quest for liberation, the restoration of God's plan for creation in this life.

> Our union with God the Creator and Savior is shown by our faith. Our faith seeks salvation, liberty for all. A person who seeks salvation must fight against hunger, sickness and pain. The person who seeks salvation cannot remain tranquil where there are injustices. The person who loves God also loves that which God has made. To love God is to work that we may be a community where no one disrespects another. . . . Let us not believe that salvation will arrive when we

pass from this life of suffering. Salvation will arrive when we decide to put an end to the bad things and sufferings we endure. If we have faith in God, let us fight that there be no more suffering. When the Israelites were living as slaves, they had to leave and fight to obtain their liberty. When our ancestors were living as peons, they also had to fight together to retake their lands. These were men of great faith and they showed it by their work. . . . Salvation will not arrive after we die. Salvation is taking place each time we exchange the bad for the good. Salvation began when God made the world—But by our work we must finish the task. (Los Tzeltales 1972-74: 78)

On the biblical chart (figure 13.1), liberation/salvation is the return to a state of election, of balance, of harmony in the Maya community. But the spiral of cycles continues taking them into the next epoch.

Hope is the inner force empowering the Maya to seek liberation and salvation (Los Tzeltales 1972-74: 115). In working with the earth, Maya live in hope for their families, for their communities relying on the promises of God's help. After living in oppression, they hope for a new earth shared by all humans, with new hearts of love as exemplified by Christ. The section about hope has twenty-one biblical quotes or citations, 45 percent of all such quotes/citations contained in the catechism. They repeat topics seen previously, this time with the added note that the virtue of hope based on faith empowers a community in its struggle to overcome oppression and realize God's plan. Five texts from Genesis stress the role of marriage in God's plan. Texts from Matthew, John, and Paul emphasize unity in a community, comparing it to a body working with many parts but one spirit. There are no reflections about charity, because all the communities replied that a definition could not be given, it simply must be lived (Ruíz García 2003: 35).

Some Characteristics of the Catechism

Given its purpose and the circumstances of its compilation, the catechism is somewhat unique. Jesus is presented as a teacher of how people should conduct their lives, as providing a moral code that liberates individuals and communities from psychological and social oppression. By his abiding presence he empowers communities to fulfill God's plan. "Christ came to live among humans to show us the way of salvation. He continues helping us so that all may attain salvation. Human beings are the only ones responsible for bringing to fulfillment the salvation God offers through

Christ. If we do not work united in this task from God, never will we have land, sufficient food, liberty, never will we save ourselves" (Los Tzeltales 1972-74: 15). This theme is repeated throughout the catechism. It is the theme of harmonious balance expressed in scriptural terms. Texts of Jesus's frequent exhortations about the need for love and unity are interspersed among many of the recorded reflections. There is no mention of Christ's death on the cross. The resurrection is mentioned only once, as a sign that salvation, the overcoming of oppression in this life, can be accomplished. Other than a brief mention of baptism, there is no mention of seeking Christ through the sacraments. All this is very different from Tridentine Christology.

Some concepts from the traditional Maya worldview are used in the catechism. The description of God in the first lesson of the entire catechism is:

As He is life, He gave us life
As He is light, He gave us light,
As He is force, He gave us force,
As He is the word, He gave us His word. (6, #4)

All these qualities can be seen as emanations of the one into the many. God as force or power is basic to the traditional pantheistic conception of God. Words can be seen as having an intrinsic power to actualize what is uttered (chapter 2). The double couplet is reminiscent of Maya prayers. Throughout the catechism, salvation is defined exclusively as a goal to be attained in this world, the same as the Maya covenant. A life after death is not denied, simply not discussed. As a corollary, the primary role of Jesus as Savior is seen to be that of a liberator from oppression in this life.

The reflections do not discuss the oppressive economic and political structures of Chiapas. The catechism mentions in a general way the oppressors: plantation owners, government officials, merchants. But it does not dwell on them or go into detailed descriptions. Rather, it continually repeats the results of the Maya's oppression—lack of land, food, and health—but with few details. There are no discussions of specific strategies to be used against their oppressors. Such topics, which could involve legitimate differences of opinion, are omitted. These matters are left to the civil organizations dedicated to specific economic and political purposes. The catechism sought to put a moral foundation, a theological worldview of love and justice, as community unifiers, upon which these specific organizations could build.

Chapter 10 discussed biblical theology as a basis of the liberation movement. Chapter 13 showed how the logical structure of the Bible was built around an abstract pattern of liberation, with its themes of election, sin, punishment/judgment, just remnant, and liberation/salvation. The themes of the Tzeltal catechism took guidance from this pattern and provided the framework of the Catechism—God's plan is election, sin is oppression by local economic and political structures and a contradiction of God's plan, liberation is salvation from oppression in this life. The catechism's use of the Bible is a typical application of its use by the liberation movement.

The catechism may have influenced Bishop Ruíz to allow his name to be used as the author of the book detailing a method of biblical reflection. The bishop saw the catechism as useful for transmitting his vision of liberation theology. It was composed in the Ocosingo parish. Its introduction clearly states that it is not to be used in other places, but can serve as a methodological model for other communities to compose their own catechisms based on their own community reflections (Los Tzeltales 1972-74: v). The bishop regarded it as implementing the documents of Vatican II. "The council represents a phase of maturity within the Church, and a very clear transition from a church that announces the Word of God as something already given in order to apply it to history: rather from history itself, to see the illumination that it gives to the message. . . . The African bishops demanded during the Council, not a summary of the traditional missiology, but at least some clues, some orientations, some light to begin to find a solution to the forceful critique of the missionary activity of the church from sociologists and anthropologists" (Ruíz García 1999a: 6). Ruíz cites what he considers to be the key documents and key sections (the equivalent of the "two sentences" to be looked for in ecclesiastical documents): Lumen Gentium (Dogmatic Constitution on the Church), section 23; Gaudium et Spes (on the Church in the Modern World); Ad Gentes (Decree on the Church's Missionary Activity), and that "marvelous and dynamic" paragraph eight of Dei Verbum (Dogmatic Constitution on Divine Revelation) (Ruíz García 1975: 13, 16; Ruíz García 2003: 37, 153; Ruíz García 1999: 6).

Summary: Selective Use of the Bible

Figure 13.1 charted how editors of the Bible used a historical format to illustrate a pattern of morality centered on reform that liberates and restores human communities to an intended divine order. This chapter, and the preceding one, have given examples of Maya reflections on the biblical themes. Certain sections were given greater importance in helping the

Maya understand their crisis and what needed to be done. What sections of the Bible were of importance and for what purposes?

The creation story from Genesis was cited by the Tzeltal catechism and the community in Magdalenas. Given their land shortage and the consequent sickness and death from inadequate nutrition, the community used Genesis to establish the fact that land and the results of its fertility were for the use of all men, not a select few, as embodied in the ladino elites of Chiapas and Guatemala. This gave the Maya a sense of rights to land, a sense that had been lost over the years of privation of sufficient land and the many things it produced. Their closeness to nature made them aware long ago that every item of food, fiber, and mineral came from the earth and its waters. Genesis told them that as human beings, they had a right to share in the earth's natural resources. For those Maya who had passively accepted their oppressive situation, this gave them the realization that oppression was not God's intention.

The Exodus account was cited by Magdalenas, the Tzeltal catechism, the human rights group Abejas, and frequently in the San Pedro Chenalhó area (Moksnes 2003: 224–25). It extended the Genesis account to a consideration of oppression, that it was not in accord with God's intention as stated in Genesis, that God had a concern for oppressed groups, and that his presence would be with them in their efforts to throw off the oppression by punishing their oppressors.

The prophetic tradition of the Bible, the ringing denunciations of social injustice, of elites oppressing their own community, runs all through the Bible, but especially it is expressed by the prophets of the Kingdom phase. The diocesan human rights organization frequently used passages from this phase, and it is the most detailed section of Bishop Ruíz's book on the method of reflection. These usages are by non-Maya engaged in helping Maya empowerment. While one would expect frequent use of these passages by the Maya themselves, there are no examples in the limited data available. In the Maya view, the political rulers of Guatemala and Chiapas were not their rulers, merely those with the ability to impose an oppressive system. In earlier times, many Maya had little or no sense of being Mexican or Guatemalan, or of a national government. Consequently, the prophetic tradition, which primarily deals with oppression by the rulers of one' own group, was not of great importance. Their rulers were the principales of their own communities, who in most cases were not seen as oppressors.

The central role of Jesus Christ in the reflections was described above. All the uses of the Bible were consistent with seeing contemporary reality as the epoch of Jesus Christ and his key role in the revitalization of Maya culture. The Maya identified with the early Christian communities that embodied the morality of the elect and are cited in the reflections of almost every community. This involved doing away with the destructive aspects of costumbre. In its place, the scriptural quotations and citations stress that there should be a uniting in love and justice and the giving of an example of sharing and working together among those who have little.

Vos (2002: 251) summarizes the Bible's influence on the Maya. "They found in the 'Word of God' . . . the only place where in addition to a religious formation, they became conscious of the social, economic and political implications of their lives. For those who had the good fortune to attend this 'multidisciplinary school' the Word of God was much more than the usual evangelization. It included individual and communal development on all levels, although the introduction of the Good News was, without doubt, the foundation of an integral liberation." Reflection on the biblical worldview was the driving force behind Action Catholicism. It helped the Maya understand the causes of their crisis, to be empowered to take action to overcome it, and in so doing, to revitalize Maya culture. Analytically and methodologically, Part V has shown the importance of theological worldviews in helping to explain social action.

VI

The Worldviews of Insurgency and Counterinsurgency

While Action Catholicism was developing in Maya communities, both Guatemala and Mexico were going through national crises due to an inability to meet their growing problems, especially those of the rural sectors. Governmental corruption was an important factor. In reaction, national leftist parties organized and began building bases of power and armed resistance in rural areas from which they hoped to eventually march on the capitals and bring down the dictatorial governments. The radical left targeted Maya communities as offering the ideal conditions to initiate their various strategies of revolution. As a result, three competing external worldviews were at work in Maya communities: that of the national government, that of the Marxian radical left; and that of the Catholic Church. The history of the economic, political, and military aspects of the Maya involvement has been detailed by other writers. The following chapters will refer in a summary fashion to this history while concentrating on questions of worldview. They lead up to the question of why did some Word of God communities and pastoral workers join armed insurrections of the radical left and others did not.

17

Guatemala

The Role of the Maya in the Worldview of Marxist Insurgency

For many years, Guatemala was ruled by military dictators whose main purpose was to maintain their own interests and those of the economic elite. The government's role in fostering the Maya crisis was an oppressive part of a larger national picture of systemic injustice and consequent unrest. The chain of Guatemalan dictatorships was broken for a brief ten year period, from 1945 to 1954, under the presidencies of Juan José Arévalo and Jacobo Arbenz. Democratic institutions were installed and a land-reform program inaugurated. The United States, infected by a paranoia that saw all demands for social justice as Communist conspiracies, labeled the Arbenz government as such. The CIA financed and equipped a Guatemalan army that overthrew the Arbenz government and designated a military general, Castillo Armas, as the presidential dictator. For a thirty-one year period, until 1985, military dictatorships ruled Guatemala. (There was an exception from 1966 to 1970, with a civilian president, but he was usually controlled by the military.)

An Early Armed Effort for Social Justice

The first phase of the Guatemalan revolution took place from 1960 to 1967 in the predominantly ladino Eastern Highlands. On November 13, 1960, disgusted by the political corruption within the army itself and the affront to their nationalism when the United States was allowed to train mercenaries on Guatemalan soil for the 1962 Cuban invasion, four hundred military officers planned a revolt. It was quickly put down, as many of its original supporters backed out at the last moment. Some of the young officers fled to the low mountains of eastern Guatemala, where they made plans for a national revolution. Originally, there were three groups: the Rebellion Movement 13th of November (MR-13N) led by Marco Antonio Yon Sosa; the Armed Forces in Rebellion (FAR), which later became the Guerrilla Army of the Poor (EGP), led by Luis Turcios Lima; and the Partido Guatemalteco de Trabajo (PGT), led by the Guatemalan Communist Party. The latter group launched a brief uprising in 1962 in the Concuá region of Baja Verapaz, but was quickly defeated. In 1962, these three groups formed a loose alliance under the leadership of Turcios and, after his death, César Montes. (The connections of these revolutionary leaders with church personnel will be seen in later chapters.) They developed a two-prong strategy. Relying on guerrilla tactics, they would form a peasant army in the Eastern Highlands and an urban combat group within Guatemala City. With success in the rural sector, the army would march on Guatemala City, which would be in the midst of an urban uprising. With the military victory, a socialist government committed to social justice would then be installed.

The FAR initially gained control of portions of the Eastern Highlands. Alarmed at the prospect of a revolution leading to a reform government, the United States again saw events only in terms of an expanding Communism. It came to the aid of the Guatemalan army, equipped it with the latest weaponry, and formed it into a counterinsurgency organization. The model was the Phoenix program then being developed in Vietnam, which was based on the prior experiences of counterinsurgencies in Malaysia and Indonesia (Thompson 1966). Trained in this manner, the Guatemalan army reacted to the insurgency with a terrorist program of kidnappings, tortures, and murders of many civilians over a two year period from 1966 to 1968. Using the reversal of Mao's metaphor for the rural sector, the army's strategy was one of draining the water (the civilian population) to catch the fish (the insurgents). After a similar reign of terror within Guatemala City itself, the insurgents were defeated in 1968. The estimated loss of life was 6,000 to 8,000, and possibly 15,000, Guatemalans (Centro de

Investigación y Documentación Centroamericana 1980: 83). Many were innocent civilians. In retaliation for the military aid and training in terrorist tactics from the United States, the insurgents assassinated the U.S. ambassador and two of his military attachés. For the most part, this was an uprising of poor rural ladinos and urban university students led by a middle-class military and urban intellectuals. The surviving insurgent leaders fled to sparsely inhabited jungles or went into exile in neighboring countries. There they began plotting their return, this time to the Maya sector of the country.

The Revised Worldview of the EGP Insurgents

The Guatemalan Left split into four main groups. They agreed on social justice as the overall goal of the revolution, but disagreed about the strategy to attain it. Given their differences, they concentrated their activities in different areas of the country. In 1971–72, the MR-13N attempted to organize revolutionary groups among the scattered ladino and Maya populations of the Petén, but were overcome by the rigors of jungle living and military pursuit. Yon Sosa was killed by Mexican police while attempting to cross the border into Chiapas. The PTG (the Communist Party) worked in Guatemala City and on the South Coast. ORPA (Organization of People in Arms) worked on the flanks of the chain of volcanoes in the Sierra Madre between San Marcos and Guatemala City. The EGP (Guerrilla Army of the Poor), first led by Turcios and later Montes, was the largest of the four groups. It worked primarily in El Quiché, Huehuetenango, and the Verapaces. The following chapters concentrate on the EGP, because of its larger size, greater importance among the Maya, and because it displayed an important Catholic influence that was lacking in the other three groups.

A few Q'eqchi' Maya from Rabinal in Baja Verapaz were involved in the conflict in the Eastern Highlands in the early 1960s as members of the PTG (Payeras 1997 [1984]: 48). This led the ladino theoreticians of the Left to discover the Maya as potential revolutionaries for their renewed efforts to bring down the Guatemalan government. They began to understand the subsistence crisis of the Maya, as well as their developing social consciousness, owing to the impact of Action Catholicism (49). While the Maya were about 50 percent of the national population, they comprised 90 percent or more of the populations in many Western Highland departments (Early 1982: 31), an area with minimal military presence. More importantly, the Maya had serious grievances against local governments. The EGP saw these as ideal conditions on which to build a popular peasant army. A 1967 document by the leader of the EGP, Rolando Morán (whose wartime alias was Ricardo Ramírez), laid out the new strategy (Payeras 1987: 37). Based on a neo-Marxian model, the Maya should come to see themselves, primarily in terms of their position in the class struggle, as a large segment of the rural exploited class who must unite with rural ladinos to overthrow the exploiting class. They must learn to subordinate their consciousness of being from one of the many Maya groups to a class consciousness. Otherwise they would not unite for the national cause and their hopes for their culture within it would fail. The analysis pointed out that the Maya failure to defeat the Spaniards at the time of the Conquest was due to each Maya group retaining its own primacy rather than uniting as a single group (Payeras 1997 [1984]: 71–89). Of paramount importance was the education of each combatant and support member of the EPG to understand the class structure of the country and the primacy of the political goal. All military actions were to be guided by long-term political considerations. For some Maya intellectuals, giving priority to class consideration over Maya cultural concerns caused them to stand off from the EPG (Konefal 2010: 133–50).

Even though the military action was to be concentrated in the rural areas, initially, the urban sector played an important but secondary role (Payeras 1987: 18). It maintained a logistical support network to supply the rural insurgency with arms, additional manpower, and any goods unobtainable in the rural areas. Recruits, mainly ladino university students and workers from the South Coast, were trained and sent to the Western Highlands. Since the university students were educated and better understood the political philosophy of the class struggle, they conducted indoctrination sessions for new combat recruits and were the spokesmen during the propaganda phases of the insurgency. People "thought of the guerrillas as long-haired, unshaven ladinos who were adept at avoiding danger. . . . Summoned to listen to guerrillas' speeches, the people realized that the guerrillas forces were made up of ladino leaders who spoke in Spanish and Mayas who translated their message into the local language" (Montejo 1999: 45). The insurgent's worldview was formulated by a small group of urbanized, educated ladinos, with no input from the Maya.

To implement the worldview, a small armed force was to assemble in Mexico, enter Guatemala, initially recruit Maya in the sparsely populated sectors of El Quiché, Huehuetenango, and Alta Verapaz, and build a popular army. At the same time, the ORPA was to recruit in Chimaltenango, Sololá, Quetzaltenango, and San Marcos to build another popular army.

The two armies would meet at the Pan American Highway and from there sweep into Guatemala City, which would be in turmoil from urban guerrilla activity. Catholic Action communities would be especially targeted for EGP recruitment (Stoll 1993: 66; Payeras 1997 [1984]: 49). They provided a large number of already organized people. Assisted by biblical reflection, their worldview included a social consciousness of the sources of their crisis. They had come to understand governmental corruption and injustices, and they were actively working for peaceful, systemic reform. Economically, they had developed an extensive system of cooperatives. Politically, most had joined the Christian Democratic Party and embraced its reform philosophy. For the Marxists, such reform was an illusion. The ladino Marxists needed to take Catholic Action Maya one step further—they needed to dispel the illusion of peaceful reform and supply them with arms to fight for social justice by the only realistic means, an armed uprising.

Maya Evangelicals were not of great interest to the insurgents. They were not as numerous at that time, and were considered to have a less-mature social consciousness. Most Maya Evangelical groups stressed obedience to governmental authority while working to improve themselves within the system (Annis 1987, notable exceptions in Samson 2007). Some Catholics had similar attitudes (Goldin and Metz 1991). With some important exceptions, many rural ladinos were cultural Catholics with little or no connection to the church's theology.

There were to be four stages of the insurgency (Colom 1998: 219). The initial step would be the clandestine establishment of revolutionary Maya communities from which armed combatants would be recruited. In the second stage, armed contingents would temporarily occupy other Maya communities and conduct propaganda campaigns to win their support (Colom 1998: 219–29). They would kill any known army collaborators, Maya or ladino. At the same time, armed combatants would conduct hit-and-run attacks against the army, to protect the propaganda effort and to show that the army was not invincible. The third stage was to be an insurrection leading to a liberated rural area and the formation of a popular army. In the fourth stage there would be an insurrection in the capital aided by the rural forces advancing from their liberated zones. This would result in the final military victory and the takeover of the government by the rebel's political party. Details were always sketchy, but at that point, the revolutionaries envisioned the future as a utopian golden age in which all wrongs would be righted. The plan was a secular version of the biblical battles of the final days and the establishment of the messianic kingdom.

The Plan in Action

The initial entry of the EGP from Mexico was in El Quiché, a department with three distinct geographical regions: the then sparsely inhabited Ixcán in the northern jungle lowlands bordering the Lacandón jungle in Chiapas; the Ixil region to the south of the Ixcán where the lowlands give way to the Kuchumatán mountains; and the mountains and intermontane valleys of the Sierra Madre in the southern sector of the department (see map 1). To the west was the department of Huehuetenango, which became closely linked to the conflict in El Quiché.

On January 19, 1972, sixteen Guatemalan guerrillas crossed from the Lacandón of Chiapas into the Ixcán jungle. One drowned shortly after crossing the border. Of those remaining, eleven were ladinos—four from Guatemala City, five from the South Coast, and two from eastern Guatemala. Four were Maya from the Kaqchikel and Achí areas. None spoke the indigenous languages used in the Ixcán. Among the ladinos was the Guatemalan poet and ecologist and later member of the EGP National Directorate, Mario Payeras, who chronicled the EGP effort. For several months the group got lost in the jungle before they finally stumbled onto communities that had been established by early Ixcán colonists. One was Santa María Tzejá, a Catholic Action colonization project of the cooperatives of southern Quiché. The insurgents bought some supplies at the community store, talked with Padre Luis Gurriarán, and departed (Payeras 1983: 36; Manz 2004: 75). Visits such as these were the beginning of the clandestine recruitment process that was carried out for three years in the Ixcán prior to the first public action of the EGP. In 1978, clandestine recruitment began in Huehuetenango (Kobrak 2003: 29).

This recruitment took place amid mounting frustrations. In the 1970s, economic depression came to the Western Highlands. The price of oil skyrocketed, and with it, the price of chemical fertilizer and other agricultural inputs, causing a slowdown in all aspects of agriculture. As profit margins shrank, members of cooperatives were unable to repay their loans and the cooperatives began to fail (Samandú, Siebers, and Sierra 1990: 81–84). Presidential elections were held in 1974. The Christian Democrats won the presidency and local offices in El Quiché. But the military dictatorship intervened, annulled the elections for all offices, and set up its own general in the presidency (Stoll 1993: 87). All of this resulted in a growing discontent within Catholic Action. One member remarked: "We looked and looked. The coops didn't work; the schools were no good; none of our projects had helped. What were we going to do? We thought the Christian

Democratic Party would be our salvation. We would win the mayor's seat. We would take over the local government and run things our way" (Carmack 1988: 50).

Around 1974, the EGP moved south into the Ixil area and the towns of Chajul, Cotzal, and Nebaj. By 1975, the EGP had about fifty combatants (Manz 2004: 77). Their first public military action came in 1975 when they killed the military commissioner in Xalbal, Ixcán. One month later, they occupied a plantation and killed its owner, as a symbol of their strength and purpose. In 1976 they killed army collaborators in Cotzal, and in 1978 did the same in Nebaj

In 1976, an earthquake of 7.5 magnitude shook Guatemala. The death toll reached 23,000, the injured 76,000, and many more were left homeless. Maya homes, with their adobe walls, tumbled, as well as those in the predominantly ladino shanty towns surrounding Guatemala City. For this reason, the quake was called the *terremoto de clase*, the earthquake of the underclass. Reconstruction committees from communities throughout the country rushed to the aid of the victims, and private international aid was distributed through them. Assistance from foreign governments had to be funneled through the military, which sidetracked substantial portions for their own benefit, creating more antagonism and frustration.

In 1977, there was a bitter strike of the Maya miners in San Miguel Ixtahuacán, in Huehuetenango. They marched to Guatemala City to protest the condition of the mines, the dangers they encountered, and their low wages. Along the way, solidarity groups of the Committee for Campesino Unity (CUC), both ladino and Mayas, greeted them with food, encouragement, and lodging. The event generated much publicity, which made many middle- and upper-class ladinos aware of and uneasy about the growing rural unrest and the collaboration of the Maya and rural ladinos. They began to understand that the Maya were a political force to be reckoned with. It was against this background that the EGP was clandestinely recruiting in El Quiché.

In 1979, the EGP inaugurated the second phase of its plan: armed propaganda. It temporally seized Nebaj and many of its aldeas. It summoned the residents to the town squares, invited them to join the revolution, and threatened death for anyone collaborating with the army. Guerrilla squads began ambushing army patrols with success. The same tactics were initiated in Huehuetenango and in southern Quiché. As Maya protests (see the introductory photograph[1]) about injustices were ignored, sizable numbers of Catholic Action Maya, especially young members, willingly joined the revolutionaries because of their resentments and frustrations with the national government—for its procrastination in providing land titles, for its refusal to provide public services, for the corruption of local governments by landowning elites, and for low wages and harsh treatment by plantation owners and their agents (Colom 1998: 219). As the presence of the guerrilla organizers became known, the EGP received handwritten invitations to visit Maya communities.

The community of Santa María Tzejá is an example of how clandestine recruitment was carried out. Its inhabitants were highly motivated by the biblically influenced training they received in their highland communities. They had formed a colonization cooperative, and had founded Santa María Tzejá in 1970. They worked closely with Padre Luis, who lived with them until 1975, when he left due to death threats from the army. The catechists continued their biblical reflection sessions, with its liberation theme, giving the community a moral foundation for its strong cohesion (Manz 2004: 107, 111).

The guerrillas recruited by returning a number of times to the community, getting to know the people, gradually revealing the reason for their presence, and involving the Maya in logistical support for the armed combatants (Manz 2004: 97–103; Kobrak 2003: 40–48). They would begin by talking about their grievances against the national and local governments. Exaggerated promises were made for a new government that would address their grievances. As the first step of actual involvement in the revolution, they would ask the community to collaborate by supplying them with food and information. Then the level of involvement was increased by the community agreeing to receive political education. This education included learning about the overall goals of the revolution, whose implementation required patience and a long-range view; about their own place in the class structure; and about the class they were fighting against. Armed with this understanding, the community was then asked to form a part-time civilian armed force, a militia, to protect themselves against the army. They were given some basic military training. They were also asked to perform auxiliary services for the guerrilla combat force, such as informing about army activities, collecting food, acting as couriers or guides, transporting goods, and recruiting relatives and neighbors. When a community committed itself to the EGP, it was ruled by a local, five-person committee, whose members, in turn, took orders from an EGP civilian regional commander. In other communities, the EPG recruited but was unable to take control of the entire community. This network of communities comprised the revolutionaries support base, the civilian collaborators of the armed combatants.

Recruits from these communities would also undergo intense military training and be formed into combat units. These units were to be integrated into regional guerrilla armies (fronts) named after historical revolutionary figures. Within the combat units, the physically strongest did most of the fighting. Those with speaking skills were in charge of political education, both within the unit and in the communities they entered for purposes of open recruitment. Others were primarily concerned with medical or other logistical functions. All were armed.

A Case History of Rural Recruitment: Committee for Campesino Unity (CUC)

The CUC was a broad coalition of rural ladino and Maya groups demanding social justice. They were independent groups who worked together, but with each retaining its own autonomy. The EGP was strengthened when some groups from the previously nonviolent CUC joined its ranks in 1980. A Quiché Maya group of the CUC had its beginnings in 1972, when an idealistic group of Jesuit seminarians and university students from Guatemala City visited San Antonio Ilotenango during their vacation period. They wanted to learn about and assist the Maya. They built a suspension bridge for the schoolchildren over a local river. They were shocked by the poverty they encountered. Wishing to become more involved, they moved to Santa Cruz, the departmental administrative center, where they made contact with the younger members of Catholic Action. These were second-generation, young, educated, politically knowledgeable Maya. They thought that the leadership of the older men of Catholic Action had become complacent, that their vision was not growing to accommodate contemporary events, and that they were too passive in the face of governmental corruption (Diócesis del Quiché 1994: 104; Samandú, Siebers, and Sierra 1990: 81, 107; Konefal 2010: 37).

Developing a Network

Two Jesuits who had recently arrived from Spain, Fernando Hoyos and Enrique Corral, visited the communities on weekends. They instituted a program of consciousness-raising through a literacy program based on Freire's system. While studying literacy, students reflected on the problems of their communities. But the consciousness-raising discussions went beyond the local community to include discussions of systemic problems: unjust land distribution and its protection by a corrupt government and the dysfunctional political system that sustained it. These discussions employed biblical reflection, especially Exodus, to help understand the current situation.

Hoyos became the informal leader of the literacy work, which was developing into a liberation network, drawing many of its adherents from Catholic Action. With the help of sympathizers at the diocesan radio station, their influence spread throughout the whole department and beyond. The Quiché network joined other rural networks providing earthquake relief. "In conjunction with this assistance, Kaqchikel were delighted to have volunteers who offered building materials, foodstuffs, water and clothes, but they point out that in some cases aid was conditional. Some say they had to attend meetings to gain access to these donations. They note that during these meetings, organizers espoused Marxian ideology and made lists of names of people interested in obtaining materials" (Carey 2001: 145).

These activities brought together the poorer sections of Guatemalan society—the Maya, the poor rural ladino, the urban poor. They were forming a coalition of like-minded groups. The coalition took part in demonstrations to demand social justice for plantation workers and other oppressed groups. They were helped by a pastoral letter from the Guatemalan bishops, "Unidos en la Esparanza," in which the bishops clearly denounced the oppression and exploitation that had occurred in the reconstruction effort after the earthquake. As a Maya leader noted, *"Unidos en la Esperanza* helped us to clarify many things, since there is a large sector of the church that has been accompanying the journey of the people, now not only with prayers and the mass, but with denunciations of injustice" (Fernández Fernández 1988: 9). After Cardinal Casariego refused to sign the letter, it was issued by the Guatemalan bishops' conference during his absence from the country (Samandú, Siebers, and Sierra 1990: 51).

In April of 1978, the rural networks formed a loose coalition, the Committee for Campesino Unity (CUC). They defined themselves as being "not a federation nor a new governing body, but a committee whose only demand is that its members conduct themselves with honesty, firmness, sacrifice and perseverance in collective undertakings, in the struggle on behalf of rural agricultural workers as well as being prepared to fight on behalf of Guatemalan victims of exploitation" (Fernández Fernández 1988: 15). The group brought together rural ladinos and Mayas—and among the Maya, it brought together those from the highlands with those who lived on the South Coast plantations. On the plantations, the CUC organized a strike of more than 70,000 sugar-cane cutters and 40,000 cotton harvesters that was successful in raising their minimum daily wage from Q1.20 to Q3.20 (González 2002: 328).

Biblical Influence

The CUC was a political coalition that drew assistance from the Bible. Hoyos, who was ordained a priest in late-1973, visited Maya communities to celebrate the mass and to organize reflection groups using the Book of Exodus (Hoyos de Asig 1997: 120). A Maya CUC leader remembered the role of Hoyos in laying the foundation of the CUC in the Quiché area. Hoyos always went to the poorest outlying communities. In such work, Toj saw the "revolution within the church and the new way of reading the Bible. . . . He [Hoyos] never failed to talk about the reality in which we live as poor people, as indigenous people. This was developing a consciousness in many people of these communities so that when the CUC was organized, this seed fell on fertile soil because it had been prepared. Fernando repeated this message at every opportunity he had to talk to groups, he always talked about reality and the necessity of an organization that in time became the CUC" (Toj Medrano 2008: 145). Although drawing many of its members from Catholic Action, the CUC was independent of both the Catholic Action organization and the diocese. Bishop Gerardi did not trust the EPG and opposed what he considered to be a Communist CUC, although he never publicly denounced them (Carmack 1988: 51). (The trajectory of Hoyos's life will be described in chapter 25.)

Another founder of the CUC in the Quiché area was Domingo Hernández Ixcoy. He bought a Bible and started to read it. He found that "it was always speaking about liberation of the community, of equality, of respect for others, but I was alone, I could not do anything." His Catholic family had the custom of having a thanksgiving mass celebrated at harvest time, for which they paid the customary stipend. In 1972, Domingo decided not to ask for the mass, but to give part of the stipend money to the poor. He and his wife went to the market in Santa Cruz, sought out those who appeared to be the poorest, and gave each one or two quetzales. "The scripture says it is necessary to do something for others. That was the first step we took" (Hernández Ixcoy 1984: 296).

In talking with his in-laws he discovered there were others who wanted to do something for the community. The small group began recruiting, and was gradually assumed into the CUC network. Domingo went to people's houses with Bible in hand. He began telling them about the life of Christ.

To persuade them to organize and improve local community conditions, he compared passages from the scriptures with life in the community. When specifically asked if they were forming a religious group, Domingo replied negatively, citing the recruitment of both traditionalists and Protestants as members of the CUC. He also cited the past use of religion by the rich and the government to keep the people submissive, to accept their repressive condition as part of "whatever God wants." The CUC steered away from sectarian religion, although it drew many of its members and leadership from Catholic Action. One of its organizers stated, "When CUC was formed, we no longer talked about religion, but about exploitation, about the struggle for equality, freedom for workers, better wages. People from all the groups could relate to that" (Carmack 1988: 51).

From Nonviolence to Violence

As part of the army's counterinsurgency plan, it kidnapped people, routinely torturing them for information, then killing them or returning them to their communities with their horror tales and the physical evidence of torture. In 1980, such kidnappings took place in the Uspantán and Ixil areas of Quiché. Not knowing anything about their relatives whereabouts, family members went to Guatemala City to protest and demand information from the government. The method of protest was to occupy important sites, thereby gaining media coverage for their cause. One such site was the Spanish Embassy. The police retaliated by storming the embassy, which became a burning inferno in which thirty-six people died. Among them were five members of the CUC (Stoll 1999: 71–88). "After the Spanish Embassy massacre on January 31, 1980, the nature of the CUC changed, and it turned to armed struggle. . . . The massacre showed that open and peaceful movements would not bring about any change. This act of repression was blatant proof, and its effect in Quiché was electrifying. From that point, the CUC gradually began to change into a guerrilla-linked organization, then a seedbed for guerrilla members, and later a guerrilla force" (Falla 2001 [1978]: 234–35). Other sectors of the CUC's broad coalition also joined, although some Maya groups dissented, as they did not wish to lose their autonomy. Some Maya intellectuals held back in a dispute over the priority of social justice for all versus Maya cultural recognition (Konefal 2010: 68–80 and passim). The testimony of Juan, a K'iche' in his early twenties, appears to tell a typical story of the EPG recruitment of combatants from the CUC.

In those days no one wanted to leave his land. We always had a thousand reasons for not leaving our homes, but after awhile there was no freedom, we were always in hiding, so we decided to see who would rise up once and for all. In the beginning the army would come to your house, and if you weren't there, they would still respect your family. Later though, if you weren't there, they would grab whoever was around. By January 1981 all you had to do was complain about one little thing . . . so many people 'disappeared' . . .

I'll tell you how I got involved in the EGP. The heads of the guerrillas had their contacts in the CUC, and the first thing they did was to recruit leaders, and from there, the leaders began to move things along. But it was a slow process.

It was August 12, 1980. I remember that they asked me, "Are we just going to stay in CUC, more politics and more politics? [The army] have guns—what are we going to do?" "Shit" I said, "I'll join. I'll fight them with machetes if I have to." "Really?" they asked. "If you're interested, we're going north of Chajul and Cotzal." . . .

I was just sixteen. . . . There were 150 of us who went up. . . . With us it was enthusiasm, pure and simple. Everybody was under twenty five years old, men and women. Almost all of us had been in Catholic Action. There were a couple of atheists, too. . . . When we had political orientation, they talked to us about social and economic injustice, and how to be politically disciplined with the people. . . . They did talk about Marxism, but it was like something out of a textbook. (Simon 1987: 128)

Strategic Revision: The Urban Sector

The decision to relegate Guatemala City to a secondary logistical role became controversial within the EGP. In response to the growing strength of EGP, the army initiated a military buildup in the Quiché area. Payeras (1987: 18, 33) and others thought the EGP needed to engage in urban guerrilla warfare to draw army forces away from Quiché. They also thought that they needed to recruit urban working-class people who would more easily understand the political goals and aims of the EGP than Maya recruits from isolated highland areas. Payeras returned to the city and helped organize urban units. The combat unit went into action in December 1975 with the assassination of a member of Congress followed by the robbery of an armory (42). The EGP urban efforts continued through to 1981.

During these years the Guatemalan military sought the help of Israeli and Argentine intelligence agents, who taught the Guatemalans methods developed in their own urban guerrilla wars with Palestinian and Argentinean insurgents (Payeras 1987: 76). They were highly effective, as Guatemalan intelligence slowly assembled a complete picture of the EGP's small urban group. They continually frustrated many of the EGP's efforts that, in turn, limited the support the urban EGP could give to the rural sector. In 1983, the army destroyed all the EGP's safe houses, capturing most of the urban guerrillas along with all of their weapons (104).

Summary

This chapter has sketched the background that led to Guatemalan Maya involvement in armed insurrection. There was a national crisis fed by governmental corruption. In reaction, national leftist parties began to plan and organize an armed revolution. The Maya became a key element in these plans. Among the Maya, Catholic Action communities were targeted because they were well organized, had been brought to social consciousness, and were actively seeking to reform the system by nonviolent means. The EPG needed to bring them a step further, dispel the illusion of peaceful reform, provide them with arms, and make them see that armed insurgency was the only possible means for reform. Driven by frustration at the failure of their reform efforts, and by the brutality of army tactics, many young Catholic Action Maya readily joined the revolutionaries. In 1979, the insurgents' morale was high, encouraged by their successes in the Ixcán, Ixil, and northern highlands of Huehuetenango. The victory of the Nicaraguan insurgency with the fall of Somoza, and the successes of the insurgency in El Salvador further raised their morale. It encouraged them to move up their timetable of military confrontation, to the detriment of political education.

18

Guatemala

The Maya in the Military's Worldview of Counterinsurgency

The Guatemalan army had taken over the Guatemalan government, with a series of generals holding the presidency (Schirmer 1998). The officer corps saw itself as entrusted with national security against the threat of Communist subversion. To keep the country secure, they said that it was necessary that the government be run by the military. By taking over the presidency, the country became a military dictatorship that was more concerned with preserving a privileged status quo for the officer corps and the economic elite. To legitimate this state of affairs, the military allowed elections and other trappings of democracy. If they disagreed with the results, they annulled them. Their unchallenged position also afforded the military many business opportunities, as they became an important sector of the national economy. This allowed them to enrich themselves by various legal and illegal means (9–34).

The army knew the revolutionaries were planning to keep the defeated insurgency of the late 1960s alive, and was aware of the arrival in the Ixcán of the small EGP force in 1972. Looking at the experiences of Malaysia, Indonesia, and Vietnam, Thompson (1966: 50) emphasizes that "any sensible government should attempt to defeat an insurgent movement during the subversive build-up phase before it enters the guerrilla phase, and if that is not possible owing to circumstances outside the government's control, then the movement must be defeated as early as possible during the guerrilla phase."

Initial Strategy of the Rural Counterinsurgency

The army strengthened its presence in the Western Highlands and built a new installation in the Ixcán at Playa Grande. It undertook a counterinsurgency plan known as the "100% Solution"—kill every individual who was a member of or collaborator with a subversive organization. To discover the EGP's clandestine network, the army recruited spies in Maya communities to report all known or suspected subversives and any pertinent information. Usually, these were Maya who had served in the army, some of whom were now in charge of the local military draft (*comisionados militares*). Others were paid spies (*orejas*, meaning "ears"). Based on their information, as well as certain a priori assumptions described below, the army would enter a community, kidnap suspected supporters of the revolution, and torture them using any method that yielded information. As an intimidation measure, some would be returned alive to their communities with their bruised and battered bodies for all to see. More often, the army would kill them. Sometimes, the mutilated bodies of the dead would be left in the town square or some other prominent place for all to see, again as an intimidation measure. More often, the murdered were simply buried in unmarked graves. They became known as the "disappeareds" (*desaparecidos*), as their families were never informed of their whereabouts or final disposition. From the information it obtained, the army would continually compile death lists and the killings would continue. The 100% Solution sought to kill all subversives without really knowing who comprised the "all." Overkill was the order of the day (Schirmer 1998: 35–38). It was a strategy that involved repeated incursions into Maya communities without permanently occupying them.

Some Assumptions of the Counterinsurgency Worldview

The army's view of the situation included assumptions about who were insurgents, either as combatants or collaborators. Any person who fit one or more of these assumptions immediately became suspect and a potential victim for torture and death. There was no attempt to establish innocence or guilt.

Catholicism among the Maya embraced differing positions regarding their attitudes toward the government. There were Tridentine Catholics who abstracted themselves from political problems by concentrating on orthodox Catholic belief and sacramental observance. The Catholic Action programs had sought empowerment of the Maya in local communities and worked for reform that would enable Maya to fairly participate in the national system. As Catholic Action evolved into the liberation movement, some Catholics rejected it, not wishing to become involved in attacking the government. Adherents of the liberation movement splintered into two camps. Some advocated a restructuring of the total system by nonviolent means. Politically, they were usually aligned with the Christian Democrats. But others, especially younger members of the movement constantly frustrated by trying to work within the corrupt system, quickly agreed with the Marxists that armed violence was the only solution and readily joined the insurgency. Given the continuing crisis conditions in which the Maya found themselves and the government enforcement of the crisis, it was not a difficult leap for some, as seen in the case of the CUC in the preceding chapter.

But as noted previously, in Guatemala the term "Catholic Action" was used generally to describe most of these political positions without distinction, and included the majority of Maya Catholics as well as pastoral workers serving Maya parishes. "Army repression . . . involved persecution of the Catholic and Evangelical churches alike. The persecution demonstrated the army's deep mistrust of the Church, which was no longer playing the traditional role of legitimating the ladino state. The offending image of the Church was that of an institution that, out of faith, promoted the liberation of the poor ('incited them to subversion' the army would say). Mistrust was an attitude, like racism and discrimination, that caused the army to misinterpret the behavior of those with the Church and gave another slant to Guatemalan counterinsurgency" (Falla 1994: 187).

Another assumption of the counterinsurgency was a subconscious equation of race with class. The ladino officer corps looked down on the ordinary rural Maya agriculturist, not only because he was Maya, but also because he was poor and uneducated. It was both racial hatred and class hatred. In carrying out the tactics of the insurgency, the officers would conduct a mental experiment: What would I do if I were a typical rural Maya? The result was the fateful equation: Maya = campesino = poor = guerrilla (Montejo 1987: 56). This equation dominated the officers' subconscious. If any Maya gave the least sign of what the army interpreted as subversive behavior, some type of ruthless torture, usually ending in death, would often follow.

For an extended period of time, people did not believe the army was acting as blindly and ruthlessly as it actually was in carrying out its counterinsurgency plan. People initially believed the army was considering each individual case, was trying to make reasonable judgments about the innocence or guilt of individuals, and had arrived at mistaken conclusions perhaps based on false information from orejas. Some community elders or parish priests would try to reason with the soldiers carrying out abductions. The army's automatic assumption was that their prisoners were subversives, and that anyone who attempted to intervene on their behalf must also be aligned with the guerrillas in some fashion. Consequently, the interceders were also often marked for torture and death (Montejo 1987: 13). A variant of this assumption concerned any attempt to inform the outside world of army atrocities in Maya communities. It was assumed that priests, schoolteachers, cooperative officials, or anyone with some education, would be revolted by the atrocities and would inform the outside world, possibly bringing international disrepute and sanctions against the government. These people, too, were to be silenced.

Social Class within the Army

The officer corps had watched the insurgencies during the 1970s in Nicaragua and El Salvador. The Nicaraguan insurgency was successful in 1979, leading to the installation of the Sandinista government that included three Catholic priests in its original cabinet. In the same year, the Salvadorian insurgency fought the government forces to a standstill, leading to a coalition government. The officer corps vowed that what had happened in Nicaragua and El Salvador was not going to happen in Guatemala, especially by an EPG insurgent army largely composed of Maya Indians. The United States backed this determination with training and arms.

The low-ranking government soldiers were usually from rural areas and included many Maya. They were drafted into the army, or had joined to earn the meager pay. They were not assigned to service in their home areas.

Since the Maya self-definition was that of their home community, a sense of unity among all Maya was weak (Early 2006: 209). A Maya ex-soldier related: "They brainwash and indoctrinate us in such a way that we could torture our own parents, if we were ordered to. I spent three years in the barracks, and what did I learn? Fucking zero. The only thing you are taught is to kill and kill, again and again. Just because you're fed well and warm blood runs in your veins, you soon want to make the bullets fly, as if to say, it makes you feel real macho and no one can stand in your way. In my case, for example, during my time in the barracks I was always spoiling for a brawl, and if it didn't happen, I would go out looking for it. Just imagine" (Montejo 1987: 86).

Special units, Kaibiles, were trained for counterinsurgency work. Each solider, at the beginning of his training, was given a young dog to take care of. The soldier would develop an affection for his pet, who remained close to him all through the training period. On his graduation day as a Kaibil, the soldier would then be required to kill the dog and eat it (Porras Castejón 2009: 68). This was part of the dehumanizing process so that the soldier would readily carry out the atrocities of the counterinsurgency.

As Montejo, a Maya schoolmaster, watched the slaughter, he reflected: "foot soldiers . . . all of dark complexion and ill-educated—alienated Indians who like rabid dogs, had been sent by their ranking officers to the villages to finish off their own people" (Montejo 1987: 54). As long as it did not involve his home area, a Maya soldier followed his orders to carry out the atrocities called for in the army's counterinsurgency plan (Montejo 1999: 83–103; Montejo and Akab' 1992).

Extensive looting also took place when the army entered a Maya community. A soldier explained, "You should know that we have our own wives and children to feed, and so we were deceived into joining the army. It was against our wills that they armed us and sent us to the mountains to combat the guerrillas. The fifty dollars they pay us a month is not enough to support our families with. That's why they allow us to loot and steal, so we can supplement our meager wages. We were tricked into coming here. The officers tell us we Indians must do away with our own people because the Indians are enabling the guerrillas to carry out the revolution" (Montejo 1987: 68).

Santiago Atitlán under Army Occupation

The infiltration of Santiago was part of ORPA's plan to eventually link up with the EGP. ORPA began successfully recruiting in Atitlán in the

1970s. The town was temporally occupied at least once by ORPA to carry out armed propaganda. In response, the army established a garrison on the outskirts of the town and initiated the disappearances and murders of many suspected persons (Tarn 1997: 339–52; Carlsen 1997: 143–69). The town reacted by becoming an ORPA support community. Between five hundred and two thousand Atitecos lost their lives during this period (Tarn 1997: 353–63) A priest from Oklahoma, Father Stanley Rother, was among the victims (see chapter 23). In December 1990, after the army killed protesting civilians, the president of the country ordered the army garrison to vacate Atitlán (Carlsen 1997: 157–59).

Failure of the 100% Plan

In the 1970s, the strategy of kidnappings, tortures, disappearances, and murders resulted in deep resentment against the army, as they killed many innocent civilians, causing their relatives and friends to join the EGP or ORPA. The use of spies led to pervasive fear and mistrust within communities.

The army's counterinsurgency plan was unsuccessful in stopping clandestine recruiting or in preventing armed guerrilla bands from entering communities to carry out their strategy of armed propaganda. The army would then appear, threaten the community, kidnap some suspects, and then withdraw. The army was reluctant to spread out its forces by permanently occupying many small communities that would be difficult to defend against guerrilla attacks. Likewise, the guerrilla combatants were never able to build a force of sufficient size to permanently occupy Maya communities. This left the communities vulnerable to both sides.

A New Strategy: Ashes

As the guerrilla threat grew, the army was suffering increasing casualties among its junior officers and soldiers. Its strategy of killing targeted individuals was not achieving its goal. The guerrillas were expanding their clandestine networks of collaborating communities and continued to have some success in their hit-and-run attacks. The army needed to counterattack. A conventional counterattack was impossible against small, highly mobile guerrilla bands that hid in the jungles or highland forests. Additional problems confronted the military. The situation in the country was deteriorating. There was massive corruption within both the military itself and in the government under its control. The uncertain conditions caused a flight of capital from the country. There was growing isolation by the international community for Guatemala's human rights violations stemming from the 100% Solution. Some generals thought the country was headed for a crisis and that the guerrillas would triumph in a few years (Schirmer 1998: 18–20).

Pondering the ineffectiveness of the "100%" strategy, a group of Guatemalan senior officers drew up a new strategic plan—drain all the water, thereby killing the fish. The strategy was to completely destroy the support bases of the guerrilla units: the collaborating Maya communities. The plan was carried out by entering designated communities, torturing and killing in the most barbaric fashion every possible man, woman, child, and animal; looting every dwelling; and, finally, burning every single edifice and dead body. There would be complete devastation; nothing would remain but ashes. No questions would be asked about innocence or guilt. The fish and their families could be caught by this highly successful military tactic. The slaughter of countless innocents would be tolerated. This was the 100% Solution applied to entire communities, not just to individuals. The end, national security, would justify the means.

Implementation in the Western Highlands

Beginning in late 1981, in Chimaltenango, and proceeding to the Ixcán, the army carried out its program of human slaughter and total destruction of communities. In the Ixcán, the strategy resulted in 773 "disappeared" or confirmed dead (Falla 1994). Many from Santa Maria Tzejá were massacred, and the settlement itself torched. Also torched were the colonies of the Catholic Action cooperatives that had been initiated by Maryknoll in the western Ixcán. Gory descriptions and lists of the murdered have been compiled elsewhere (Falla 1983, 1994; Iglesia Guatemalteca en el Exilio 1992). The introductory photograph is a painting by a child who escaped from a village outside Chichicastenango. She retained vivid memories of the tragedy. It shows the army entering the area by helicopters and armored vehicles. A villager has been tied to a cross as a soldier readies his machine gun to riddle him with bullets. In the background can be seen the burning village—ashes. The painting was done after the war as a therapeutic exercise of the peace and reconciliation program.

The same pattern of insurgency and counterinsurgency was repeated in Huehuetenango. The worst incident took place in July 1982 at the San

Francisco plantation. The army slaughtered the entire Maya community of 350 men, women, and children. All the houses were burned. Nothing was left (Falla 1983; Kobrak 2003: 69–98; Melville 2005: 34–46).

The Maya anthropologist Victor Montejo has described the situation in his home area, Jacaltenango. The insurgents, after clandestine organizing, appeared in Jakalteko communities to carry out their armed propaganda and skirmishing with the army (Montejo 1999: 45, 52–59, 64–65): "The people were afraid of them. . . . They were cautioned not to denounce the guerrillas and were asked instead to collaborate with them to overthrow the repressive government of Romeo Lucas García. After the guerrillas' visits, many people were confused and fearful, as it was known that the army arrived soon after the guerrillas did. Army spies . . . regularly reported on guerrilla activity and, indeed, were the source of information about army violence in other communities. This was the beginning of the killings, robberies and disappearances of Mayas in Kuchumatan communities as the army started its counterinsurgency."

Another tactic was for army personnel to dress as guerrillas with masks over their faces, enter a Maya community, summon the people for a talk as the guerrillas did, offer money to any Maya who would join them, then take them away and kill them. In one community, during the fiesta of its patron, the people were told to shout "Viva el EGP." Several drunken men did so. The disguised soldiers approached them and said they would give them ten quetzales for joining and a hundred quetzales for their families for the time they were away from the community. The drunken men were then put in a truck. An ex-soldier approached and asked for the release of his brother-in-law. Instead, the soldiers also put him in the truck. All were taken to a river just outside the town where they were murdered. This was a constant pattern. Anyone who pleaded for the release or for the life of a prisoner must be a guerrilla collaborator of some kind, according to the army's logic (Montejo 1999: 49–51). To escape the counterinsurgency, many Jakaltekos fled to Mexico and attempted to reconstruct their lives and culture in refugee camps (105–21).

By 1984, in the northern region of the department, forty-six communities had been abandoned (Manz 1988: 66–95). The autobiographical account of the Maryknoll pastor of the San Mateo Ixtatán parish from 1981 to 1985 gives many similar details of the brutal dialectic of insurgency-counterinsurgency in that area (Melville 2005: 29–270, 463–607).

Implementation in Alta Verapaz

Insurgent groups did not become significant in Alta Verapaz until the 1980s. The EGP was the largest, with a presence in Cobán and some communities to the north and west. Clandestine insertion proceeded at a slow pace. Compared to the Western Highlands, Alta Verapaz was an isolated area with little experience of development projects and political parties. The most politically conscious members of the community were those with outside experience, the merchants. Since they were literate and spoke Spanish, many were also catechists. Given this combination, "informants repeatedly claimed that the catechists were the first to join the rebel forces" (Wilson 1995: 212). Gradually, the EGP built a network. Many of the later recruits were permanent residents on plantations who wanted redress for the low wages and poor living conditions. Additional recruits were colonists who had settled on lands of the northern region taken from Germans during World War II. The Germans, in turn, had seized them from the Maya. As now-unclaimed lands, they were owned by the government, who wanted reimbursement. Many Maya rejected this claim. By 1982, the guerrillas were in the armed propaganda phase. They were killing army collaborators, ambushing army patrols, and cutting the pipeline to the new oil fields in the Petén.

The army reacted with its counterinsurgency plan. It carried out selective repressions against catechists and anyone suspected of collaboration (Conferencia Episcopal de Guatemala 2007: 165–76). When these tactics were unsuccessful, the army reacted with its ashes strategy. "Massacres of Q'eqchi' communities began in 1981 in Chisec and to the west of Cobán. . . . seven thousand refugees were held at the airstrip outside Cobán between September 1981 and March 1982. Operations then intensified to the west of Cobán. The majority of the communities between Cobán and Chamá were destroyed. In some cases, their inhabitants were massacred; in others the people had already fled to the jungle" (Wilson 1995: 218–19).

The United Nations Investigation

Stemming from the reports of massive human rights violations in Guatemala, the United Nations conducted an investigation of the alleged atrocities.

> The CEH [the United Nations Commission] has noted particularly serious cruelty in many acts committed by agents of the State, especially members of the army, in their operations against Maya

> communities. The counterinsurgency strategy not only led to violations of basic human rights, but also to the fact that these crimes were committed with particular cruelty—with massacres representing their archetypal form. In the majority of the massacres, there is evidence of multiple acts of savagery that preceded, accompanied, or occurred after the deaths of the victims. Such acts as the killing of defenseless children, often by beating them against walls or throwing them alive into pits, where the corpses of adults were later thrown; the amputation of limbs; the impaling of victims; the killing of persons by covering them in petrol and burning them alive; the extraction, in the presence of others, of the viscera of victims who were still alive; the confinement of people who had been mortally tortured, in agony for days; the opening of the wombs of pregnant women, and other similarly atrocious acts—such acts were not only actions of extreme cruelty against the victims, but also morally degraded the perpetrators and those who inspired, ordered, or tolerated these actions. (Comisíon para el Esclaracimiento Historíco 1999, English edition, Conclusions, #87)

The United Nations report found that from a military point of view, there was no need for such inhumane measures to overcome the insurgency.

> The magnitude of the State's repressive response, totally disproportionate to the military force of the insurgency, can only be understood within the framework of the country's profound social, economic and cultural conflicts. Based on the results of its investigation, the CEH concludes that from 1978 to 1982 citizens from broad sectors of society participated in growing social mobilizations and political opposition to the continuity of the country's established order. These movements in some cases maintained ties of a varying nature with the insurgency. However, at no time during the internal armed confrontation did the guerrilla groups have the military potential necessary to pose an imminent threat to the State. The number of insurgent combatants was too small to be able to compete in the military arena with the Guatemalan Army, which had more troops and superior weaponry, as well as better training and coordination. It has also been confirmed that during the armed confrontation, the State and the Army had knowledge of the level of organization, the number of combatants, the type of weaponry and the strategy of the insurgent groups. They were therefore well aware that the insurgents' military capacity did not represent a real threat to Guatemala's political order.

(Comisíon para el Esclaracimiento Historíco 1999, English Edition, Conclusions, #24)

The report concluded that the majority of victims of the army's two "100%" solutions were civilians.

> The State deliberately magnified the military threat of the insurgency, a practice justified by the concept of the internal enemy. The inclusion of all opponents under one banner, democratic or otherwise, pacifist or guerrilla, legal or illegal, communist or non-communist, served to justify numerous and serious crimes. Faced with widespread political, socioeconomic and cultural opposition, the State resorted to military operations directed towards the physical annihilation or absolute intimidation of this opposition, through a plan of repression carried out mainly by the Army and national security forces. On this basis the CEH explains why the vast majority of the victims of the acts committed by the State were not combatants in guerrilla groups, but civilians. (Comisíon para el Esclaracimiento Historíco 1999, English edition, Conclusions, #25)

Religion was a special target of the army's abuses, especially Catholics. Falla's analysis of this aspect of the atrocities supplements the U.N.'s findings.

> The army marginalized, deported and killed priests . . . perceiving them as promoters of the liberation of the poor or potential supporters of the guerrillas. The army restricted and infiltrated the religious ceremonies, deceived the leaders of the Church, and threatened the members. It killed catechists and massacred religious groups, surrounding them during church services and burning them in their temples, particularly in Evangelical chapels. The army considered the religious activities deceitful, not so much a front for nonbelievers as an authentically religious screen that fomented subversion while projecting the false image of innocuous prayers and singing. The army used religious symbols to punish the population, hanging or crucifying people and leaving corpses in kneeling positions. It destroyed churches and religious symbols, including those of saints, to make people's lives difficult. Only later . . . did the army make a distinction between the churches in an attempt to manipulate them, using religion to curtail the force of liberating faith. (Falla 1994: 187)

Catechists insisted that Catholic Action itself, even in the strict sense, was never subversive (Stoll 1993: 128). However, the army was not interested in distinctions. The "ashes strategy" was effective. It put an end to guerrilla successes. The EGP survived as a remnant force hiding out in the jungles of northwestern Quiché until 1996, but it was no longer a military threat of any kind.

Summary

Guerrilla warfare in the Maya area began in the late 1970s, reached a climax in 1980–82, and continued with low intensity until 1996, when peace accords were signed, although many of their provisions have never been implemented. During that time, there were an estimated 190,000 deaths, of whom 90 percent were Maya.[1] Various government forces were responsible for 93 percent of the deaths; the guerrillas for 3 percent; and unknown forces for 4 percent. Of the government forces, the army alone inflicted 85 percent of the deaths. Many deaths at the hands of the army were preceded by torture and rape.

Charges of genocide were made against the army. The United Nations commission (CEH) did an in-depth investigation of these charges in four areas: northern Huehuetenango; the Ixil area of El Quiché; the Joyabaj-Zacualpa-Chiché sector of El Quiché; and among the Achí of Rabinal, Baja Verapaz. The investigation concluded that the government and military of Guatemala had conducted a campaign of genocide against the Maya in these areas, and probably in some other areas that were not investigated in-depth (Comisión para el Esclaracimiento Historíco, 1999, English edition, Conclusions, II, #108–#123). There were many additional cases of torture and rape not followed by death, as well as numerous kidnappings, unjust detentions, and attacks resulting in bodily harm or destruction of property. At least sixteen Catholic priests and nuns were murdered by government-aligned groups (Conferencia Episcopal de Guatemala 2007: 51–375, 411–13). Government and military officials as well as some members of the national press blamed the Catholic Church for the rebellion and the dreadful consequences. This will be discussed in chapter 26.

The years of the Guatemalan insurgency and counterinsurgency were among the bloodiest in Guatemalan history. There are many personal histories of suffering. Portions of these histories are recounted in the numerous testimonies documented by two investigatory commissions: the Recovery of Historical Memory Project (REMHI), administered by the Guatemalan Catholic Church, which published *Guatemala, Never Again!*; and the report of the United Nation's Comisíon para el Esclaracimiento Historíco (CEH), entitled *Guatemala: Memory of Silence*. The events in Guatemala were to have an important influence on the later insurgency and counterinsurgency in Chiapas.

In 1999, President William Clinton visited Guatemala and publicly apologized for the role of the United States in the murders and genocides of Maya during the counterinsurgency. "It is important that I state clearly that support for military forces or intelligence units which engaged in violent and widespread repression of the kind described in the report [CEH] was wrong, . . . and the United States must not repeat that mistake. We must, and we will, instead continue to support the peace and reconciliation process in Guatemala." (Clinton 1999)

19

Militarization in Guatemala

When the EGP had been defeated, the military turned the Maya areas into militarized zones. The goal was to prevent a reoccurrence of insurgency by rooting out any remaining guerrilla bands and by asserting its control over the civilian population. This chapter examines Catholicism among the Maya during this period.

After reducing many communities to ashes in the early 1980s, the army took over all civil functions in the western departments. The army's objective was "to regain the indigenous population and rescue their mentality from the guerrilla.... For the first time in Guatemalan history both the guerrilla and the army sought to gain the hearts and minds of the indigenous population" (Schirmer 1998: 38). Initially, no one was allowed to return to their home areas. All civilians and refugees were placed in model villages, some of which would become poles of socioeconomic development. This would separate the civilian population from the guerrillas, cutting them off from their support system. These settlements were to form their own armed forces—civil patrols for self-defense against guerrilla incursions as well as a means for the army to keep surveillance on the male population. The leaders of these patrols were given special privileges so that they would bond with the army and the government. This would pit Maya against Maya.

Former members of the EGP were to be reeducated. "When the displaced were brought in during this period, they would be lined up in front of the army compound. . . . Since they had walked for several days or months and had high levels of malnutrition, everyone usually was first given hot food. They were asked by the S-5 specialists if any were former soldiers. . . . Those who were combatants in the revolutionary cadres, if not considered 'lost' to subversion, would be taken aside for 'informal conversations' and given 'special attention' as one Kaibil lieutenant . . . explains: everyone was politically 'reeducated' but those who had positions of responsibility within indigenous organization were kept 'close at hand,' sometimes living with their families within the army compound in a house close to the officer in charge" (Schirmer 1998: 69).

This was an effort to turn the Maya into ladinos. As published in the official army journal, the plan was for the army "to put all its efforts into Civil Affairs Units to complete its assigned mission of intensifying the ladinization of the Ixil population until it disappears as a cultural subgroup foreign to the national way of being. . . . By ladinization one must understand it to mean *castellanizar*, to pressure the population to use Spanish language and culture, to suppress the distinctive *traje*, indigenous dress and other exterior displays of differentiating oneself from the group. Given that the constitutional concepts are in the official language of Spanish, these measures will facilitate communication. Without these differentiating characteristics, the Ixils would stop thinking as they do and accept all the abstractions that constitute nationality, patriotism, etc." (104). Under these conditions, Maya Catholics sought to go on with their lives in a number of different ways, some under army control and others seeking to escape it.

Army-Controlled Villages

In 1982, the new president of Guatemala, General Rios Mont, offered amnesty to all subversives, provided they present themselves to the military and collaborate with them. Many took advantage of the amnesty. By that time, it was clear that the guerrilla movement was in trouble. But for many, their communities had been destroyed. Others did not want to return to their communities, as they feared both the army and the guerrillas. The army established a number of resettlement villages. Over seventy communities, with schools, health clinics, and markets, were constructed. Wooden houses with electricity and potable water were built close to each other for surveillance purposes. These communities met the military needs of the army and the felt needs of the Maya in these circumstances (Stoll 1993: 156–61). Any type of Action Catholicism was not tolerated.

Another tactic of militarization was to rebuild some of the destroyed villages (Manz 2004: 155–82). The strategy was to repopulate each with diverse Maya groups that would militate against a unified community that might cause trouble, somewhat reminiscent of the Spanish *congregaciones* of different *parcialidades* in a town. A core group were the few original villagers who had escaped death (*antiguos*). New settlers from diverse areas (*nuevos*) were brought in and given land to make them loyal to the army. Later, returning refugees would resent both groups, who had taken over their former lands and home sites. Santa María Tzejá was resettled in this manner, yet it managed to overcome the divisiveness envisioned by the army. Much of this cohesiveness was due to the quality of the community's leadership, which had enabled the refugees to maintain their organization while they were in Mexico. Also, their community leaders were able to place themselves in key organizational positions for arranging the return. "Why was this village able to succeed when so many others have stagnated or failed? Underlying the success was the strong sense of cooperation and community that Father Luis and the villagers forged in the highlands and tempered in the early years in the village. This 'consciousness of community' . . . was infused with liberation theology, which provided a moral center that even the onslaught of the military could not break. This spirit survived and was able to provide the energy and framework for the reconstruction" (Manz 2004: 208).

In every community an important mechanism of control was the establishment of civil patrols supervised by the local military commissioner. A group of men, armed with old rifles, was required to be constantly on duty to guard against attacks from the guerrillas. Every male was forced to serve a certain number of days each month without pay. The loyalties of the civil patrol members were divided. Some were sympathetic to the guerrillas and secretly aided them. Others rejected the guerrillas and actively pursued them. Some patrols became an abusive political faction, causing many problems in numerous communities (Kobrak 2003: 99–113).

Sexual abuse of women was ever present as part of the military's strategy of psychological destruction and vengeance. On the local level, sexual abuse was a common practice by military commissioners, leaders, and members of the civil patrols, especially where the patrols gained control of a village. The existence of various kinds of abuse was one of the untold stories of the war, even after the signing of the peace accords. The guilty did not want the issue discussed, and the women were too ashamed to recount the traumas they had suffered. The REMHI and CEH reports noted that of all the human rights abuses they researched, it was most difficult to obtain information about sexual abuse. Matilde González (2002: 375–424), as a result of her extended experience in San Bartolomé Jocotenango, obtained many testimonies about the various coercive methods used against women. As she relates, some communities had been reduced to sexual-slave camps.

Joyabaj: A Community Attempts to Recover

In El Quiché, the community of Joyabaj was investigated as a site of genocide by the CEH study, which confirmed the accusation (Comisión para el Esclaracimiento Historíco 1999, English summary, Conclusions II: #110–#126). After the murder of Father Villanueva in 1980 (see chapter 23), it was not until 1988 that the church returned to Joyabaj. The new pastor was Father Vázquez, who wanted to restore the community's faith in the church and help them to overcome their fears and face the future with some confidence. He visited all the communities of the parish, staying overnight in each one as a sign of his commitment. He "wanted people to know that celebrations and masses were taking place again, and that people could come to marry, and baptize their children. He went from village to village for days on end, celebrating masses and revitalizing local Catholic Action groups" (Remijnse 2002: 187). He organized choirs and ensemble groups in most of the sixty villages and hamlets of Joyabaj, as well as among the town's ladinos. The priest saw these gatherings as "emerging spaces for people to gather self-confidence and rebuild relations of mutual trust between community members . . . where the word of God could help console people for what had happened" (189), since for almost ten years it had been impossible for the people to gather in groups, including for any religious services.

But the work of this very active priest eventually got him into trouble. Catholic Action was not happy about his close relationship with the traditionalists. He came under constant surveillance by the military, the military commissioners, and the civil patrols, especially with respect to his activities to "wake up" the people. In 1992, the parish hall was torched, along with the instruments of the many musical groups Father Vásquez had founded. In the same year, Vázquez made the annual procession during the feast of the Sacred Heart into an act of remembrance for Father Villanueva, who had been murdered on the feast day in 1980. A large photograph of Father Villanueva was carried at the head of the procession. It was one of the few attempts to remember the violence of the past in a formal way.

In 1994, Father Rudi became the pastor of Joyabaj, and he remained until the spring of 2000 (Remijnse 2002: 257–89). The violence of the past had abated. With the signing of the peace accords in 1996, the problem now was how to deal with the psychological trauma caused by the war. "The Catholic Church clearly connects truth with reconciliation, as does Padre Rudi. In their view reconciliation is possible only when one knows what has happened and thus what one has to forgive. There cannot be forgiveness and reconciliation without truth" (259). Father Rudi visited the communities of the parish. His sermons in some communities would discuss the atrocities of the past, the role of the military and the civil patrols, and how to come to terms with them by reconciliation and forgiveness. These were communities that had experienced great suffering. But in other communities, he would hardly mention the past. These were communities where tensions were still high and needed more time to recover. Discussing the past would have caused more trouble. Other communities had witnessed few atrocities and so were not interested in any discussion. In still others, the more-pressing problem was the infighting and division within Catholic Action, which needed immediate mediation.

One of Father Rudi's last projects in Joyabaj was to arrange for the reburial of Father Villanueva, who had been buried in Chichicastenango. His remains were brought back to Joyabaj for burial in a chapel within the church. An immense procession, mostly Maya, accompanied the casket to its final resting place. Father Rudi's stay in Joyabaj also involved two of the Catholic Church's efforts at reconciliation: the REMHI project (Recovery of Historical Memory Project), and participation in the exhumation of mass graves through the church's Oficina de Paz y Reconciliación (Office of Peace and Reconciliation). Joyabaj provides a look at these projects at the local level.

REMHI, the Recovery of Historical Memory Project, collected testimonies about what had happened to families during the war: who were the victims, the atrocities they suffered, and who committed these crimes (Remijnse 2002: 267–70). This effort to learn the truth was expected to be the beginning of forgiveness and reconciliation. Father Rudi was responsible for recruiting and training the interviewers. There was a reluctance among the people to be interviewers. Finally, eight people volunteered, mostly literate and bilingual Maya, who were trained in interview techniques and given tape recorders. At the Sunday masses, Father Rudi introduced the interviewers, explained the project, and encouraged the people to tell their stories. Many were reluctant to do so because they saw the project as perhaps causing more trouble. Catholic Action in some communities refused to cooperate. In the end, forty people out of a population of twenty thousand were interviewed, most of them from villages that had suffered the most during the war. When the REMHI report was published, in 1998, only a few fragments of the Joyabaj interviews were used. The rest were lost in the editing of the flood of available material. Many interviewees were disappointed; they had agreed to tell their stories precisely because they wanted them to be published. A typical testimonial account is illustrated in the introductory photograph, which shows a therapeutic painting done by a child after the war. In the painting, a young girl is sitting between two graves, presumably those of her parents. The toppled cross of one signifies its desecration. She stares at a globe of the world pierced by a rifle from whose mouth pours out blood, symbolizing the atrocities. Some of the blood trickles to one of the graves, symbolizing the cause of her parents' deaths. The young girl sits, grieving and trying to make sense of her horrible experiences.

Because of kidnappings, and for other reasons, many people never learned what had happened to their family members. There were known mass graves in numerous communities, but no attempt was made to exhume the remains unless specifically requested by members of the community. Some requests were made, to bring closure to people's grief and to help reconciliation. But for fear that it would cause more strife within the community, some preferred to let the graves be. During the war, the army had occupied the church, the parish rooms, and the priest's quarters in Joyabaj, from 1981 to 1986 (Remijnse 2002: 270–85). They had used the rooms for interrogation and torture of their victims. When their victims died as a result, the bodies had been dumped into a deep ravine in back of the church that also served as the town dump. In August 1999, heavy rains caused a mudslide that uncovered some of the remains in the ravine. Something had to be done with them. But no one came forward to ask for exhumation. Most of Catholic Action was reluctant to get involved. Since it was church property, Father Rudi made the request. Thirty-two victims were found, but only one could be identified. It was suspected that more remains that had not been uncovered by the mudslide were buried farther down the ravine.

Escaping Army Control

Fearful of the Guatemalan army, other communities—Comunidades de Población en Resistencia (Communities of Population in Resistance, or CPRs)—avoided contact by retreating to one small area in the mountains still held by the EGP or by hiding out in the jungles of northwest Ixcán. The EGP and some of its collaborators retreated to Sta. Clara, Cabá, Xecoyeu, and later to the Mount Sumal area to the north of Nebaj (Stoll 1993: 132, 146–53, 173, 290–94; Falla: personal communication). A few catechists continued to hold meetings and function as best they could under the difficult circumstances. In 1988, a community hiding in the Ixcán sent three catechists from the Ixil area on a dangerous journey through army-controlled territory to visit the Mount Sumal communities. After their arrival, "they brought the people together for catechism courses, for large worship services, and for baptisms so large that some people climbed trees in order to see them. . . . | In three months they performed a thousand baptisms" (Falla 1998: 85). On their return, they were accompanied by two Ixil catechists. The catechists stayed for several months learning how to form a pastoral team, and then returned to the Mount Sumal area. The army initiated a blockade of Mount Sumal, and people began to leave. In the late 1980s, the area was overrun.

It is estimated from 1800 to 6000 people (Falla opts for 1800), most Catholic Action Maya, hid out from the army in the jungles of northwest Ixcán, along with remnant EPG forces, but independent of them (Falla 1998: 56). Many had been members of the original Ixcán colonies organized by Maryknoll at the end of the 1960s. To escape detection, they formed themselves into thirty mobile bands. There were catechists in all the bands holding Word of God meetings to reflect on the Bible as an aid for surviving their situation. They modeled themselves after the first Christian communities described in the Acts of the Apostles. "How do people sustain their morale? They constantly organize celebration of the Word of God and that is their only consolation. That's true, the life they lead is incredible. Everything is shared. Like the life of the first Christians. Nobody keeps corn or whatever apart. Everything is administered by those responsible in the community" (Falla 1988: 246).

Religious life was well organized, with a team of lay pastoral workers and medical assistants who periodically visited the communities. There were three ministries of the church: "its ministry of accompaniment (being with the people), its ministry of evangelization (knowing, loving, following Jesus) and its ministry of socio-political formation (interpreting reality from within and from the outside)" (Falla 1998: 101). The priest-anthropologist Ricardo Falla was a member of the team for several months in 1983, and again for five years from 1987 until 1992. In 1983, the team consisted of three pastoral workers and two health promoters. "We planned to be two or three days in each community, celebrating mass, meeting with people and trying to encourage them, asking about their needs, and in my case, interviewing them about their past experiences. . . . Two young women who were health promoters accompanied us and were able to respond in a small way to the needs of the people" (25). Falla notes that the team was helped considerably by the continuing respect for the memory of Father Bill Woods, who had helped organize a number of the communities and was later killed by the military (chapter 24).

Falla's first stay was cut short. Because the army had stepped up its search for these communities, the refugees were constantly moving their camps to avoid detection and the sporadic bombings by warplanes. Concerned for his safety and despite his objections, in 1983 the guerillas ordered him out of the area they controlled. After many requests, the guerillas allowed him to return in 1987. This became his permanent assignment providing that, for two months of every year, he come out to renew his contact with the Jesuits. During Falla's second period in the jungle, the pastoral team developed a coordinated structure that involved all the catechists in the various bands. The bands were divided into zones, each zone comprising five neighboring communities, with a coordinator who conducted meetings for the catechists of his or her zone. The coordinators also distributed the Eucharist, keeping the consecrated hosts in small structures used as chapels. A pastoral team consisting of two lay coordinators and Falla visited all the camps. There Falla led reflections on the Word of God and taught with paintings, and later slides, about biblical scenes. "With the help of explanation, they understood what was being presented on the screen. That was the first step. The second step was for them to find its application for their lives of resistance, so the projector showing would not just be entertainment but would bring new insights to everyone's lives" (Falla 1988: 82). The Noah story was one of those used for reflection. Catechists and their study groups interpreted it in much the same way as noted in figure 13.1. "The rains of the flood were the massacres by the Army; the Ark was the holy jungle that protected us; Noah's faith was the faith of the people who fled into the jungle; the raven was the bad messenger who never returned and stayed in refuge in Mexico; and the dove was the messenger who returned with the good news of hope. The application of the text not only looked to the past, but also into the future, and Noah was an example of how we would leave the jungle one day just as he had left the Ark alive" (83). The use of the Noah story followed a traditional method of biblical interpretation. First the Word of God is studied, and then a search is made for help in interpreting the group's current situation. Later, the pastoral team inverted the steps, similar to the method described in chapter 14. Falla also used for reflection "the book of Judith, a woman who was the model of resistance" (33). Chapter 13 described how it employs both variants of the biblical pattern as found in the Exodus event. In 1986, Padre Luis Gurriarán joined the Ixcan resistance camps. He notes: "Until now my main work has been to visit the camps and take care of the religious necessities of the people in a very adaptive and liberating manner. Mass is celebrated with tortillas and the juice from sugar cane. During every mass I administer general absolution. We intend to leave 'ordained' in each community, lay ministers for baptisms, reconciliations, and the Lord's Supper. That is to say, we are trying to create a real 'Church of the Poor' with the least amount of dependence on those who have been said to be indispensible. But you may object, this is not very orthodox! Perhaps, but it is very Christian and within the framework traced out by Vatican Council II during its good moments of inspiration by the Holy Spirit" (Santos 2007: 297). After his recovery, he worked in the Guatemalan refugee camps in the State of Quintana Roo, a part of the Yucatan Peninsula. After a year, Luis became sick and left for Mexico and Spain to receive medical treatment.

Falla left the communities at the end of 1992 after the army found out about his presence and accused him of being a guerrilla combatant. He wanted to publicly clarify that he was there as a priest under Jesuit jurisdiction serving mobile communities seeking survival; that the communities were in contact with the EGP but not under its control. He was refused permission to return by the bishop of the Quiché Diocese, who feared the army's retaliation. If Falla was allowed to return to the jungle groups, there was the danger, with precedent in the diocese, that in his place the army would kill another pastoral worker. In spite of continual sweeps by the army looking for these groups, they were able to avoid capture. With the signing of the Peace Accords in 1996, the jungle groups finally formed their own permanent villages and reintegrated themselves into Guatemalan life (Falla 1998).

To survive the horrors of the counterinsurgency, an estimated 100,000 to 200,000 Guatemalan Maya fled to Mexico, most to Chiapas, where they lived in fifty-six refugee camps spread along the Mexican-Guatemalan border (Manz 1988: 147). This was a dangerous trip, as the Guatemalan army considered any group on the move to be guerrillas seeking to establish bases in Mexico. At times, the Guatemalan army crossed the border and raided refugee camps on Mexican territory. For some, the decision to move began with a dream that their patron saints had left the community church and fled to Mexico, a sign that they should do the same (Montejo 1999: 48, 159). Some communities carried the statues of their patron saints with them during their flight and built small chapels for them in the Mexican camps (113, 121, 160). Mexican priests visited the refugee camps to administer the sacraments. Catechists carried on without assistance from pastoral workers (140, 161, 193). The San Cristóbal Diocese provided substantial material assistance. "Don Samuel Ruiz Garcia worked tirelessly and without compromise to aid the refugees. In the initial period, a total debacle was avoided by the refugees own tenacity and spirit of survival, combined with church efforts as well as government aid. Two factors allowed the relief to be quickly organized: many refugees came as communities . . . and the church had an existing network of village organization throughout the Diocese. The effort required a high degree of commitment, volunteerism, and a strong sense of solidarity with the refugees, qualities the Diocese could quickly mobilize" (Manz 1988: 147). When he recovered his health, Padre Luis Gurriarán worked in the camps, which in the late 1980s were moved from Chiapas to the Mexican states of Campeche and Quintana Roo. Later, some refugees returned to Guatemala on their own initiative, while others did so under the repatriation accords between the Mexican and Guatemalan governments, whose negotiations began in 1987 and were implemented in 1992.

Summary: Decline of Action Catholicism

The insurgency was attributed to Action Catholicism and its biblical influence. The defeat of the EPG created a dangerous situation for catechists wishing to conduct reflection sessions for Maya empowerment and nonviolent action. Such activities were considered subversive and reported to military authorities, with some type of retribution bound to follow. Action Catholicism lost many of its catechists. In the Ixil area, about 70 percent had either died, fled, joined Pentecostal churches, or started Catholic charismatic groups that sharply distinguished themselves from Catholic Action (Stoll 1993: 173, 191, 194). A few catechists, such as those in Cotzal, went underground. Although they could not hold meetings, they kept the organization alive by individual contacts. They emerged in 1984 with the reappearance of the Catholic priest in the area (191). Many Catholic Action groups evolved into parish organizations concerned with helping the priest organize sacramental activities and with the resettlement of refugees. The names Catholic Action and Liberation Catholics were no longer used, and they became known simply as Catholics.

20

Chiapas

The Role of the Maya in the Worldview
of the Zapatista Insurgency

In the early morning hours of January 1, 1994, five thousand armed combatants of the Zapatista Army of National Liberation (EZLN) surprised and seized four municipal centers within the boundaries of the San Cristóbal Diocese, including San Cristóbal de Las Casas itself. Two brief encounters with and defeats by the Mexican army disrupted their strategic plan. The EZLN units melted back into their bases and support communities. Twelve days later, the army declared a unilateral cease fire. The dead included seventy Zapatista soldiers, most of whom were Maya, thirteen Mexican army soldiers, thirty-eight Chiapas state police, and several hundred civilians

(Womack 1999: 43–44). The majority of the EZLN combatants, both officers and their battalions, were Maya Liberation Catholics, including some catechists and deacons. The uprising received national and international publicity. In many quarters, the Catholic Church and Bishop Samuel Ruíz were seen as the instigators. The labels "Maya" and "liberation theology" became closely linked with "insurrection" in many people's minds.

Overlooked was the fact that only a small minority of Maya Liberation Catholics in the San Cristóbal Diocese were involved in the insurgency. Twenty percent of the EZLN came from the Tzotzil highlands, 5 percent from the northern Chol sector, while the main body of the insurgents, 75 percent, came from the Cañadas, a relatively small area of the diocese and only one of the four sectors comprising the Lacandón (map 4). When Bishop Ruíz learned about the preparations for the insurgency, he, along with a number of pastoral workers and catechists, had launched a campaign of dissuasion. Meyer (2000: 93) estimates that the EZLN lost 40 percent or more of its armed force due to these efforts. Tello Díaz (2001: 220) places the number at 60 percent. In 1993 alone, the EZLN's numbers declined from 12,000 to 9,000 (Womack 1999: 42). In the Lacandón, "revolt was not inevitable. . . . There inside by then were probably 1,800 colonies and communities. . . . The great majority of the rural poor who did not belong to the Zapatista Army of National Liberation, did not rebel, while the minority who did belong, did rebel" (23). None of the communities from the Jesuit zone of influence (see chapter 14) were involved. When Rudolfo, the deacon of Magdalenas (chapters 6 and 15), was asked to persuade his community to join the rebellion, he replied: "The Apostles carried swords. We know this from the bible. But when one of them tried to cut off the ear of the people who came to arrest Jesus, he told them to put it back. So we don't think about using arms to attack the government. We won't be oppressors. But if the government attacks us, we'll defend ourselves" (Chojnacki 2004: 55). Rudolfo clearly distinguishes between defensive and offensive use of arms. The catechists in nearby San Pedro Chenalhó also refused the invitation to violence (Kovic 2005: 160).

The remainder of this chapter examines some events leading up to the Chiapas insurgency and the church's role in them. It has parallels to the Guatemala situation, which had a major impact on these later events in Chiapas. The treatment here emphasizes the role of worldviews. Other authors give more in-depth accounts of the political and economic factors in this highly complex situation.

The Contending Forces

The four issues raised by the Maya Congress in 1974 continued to be problematic. The Maya of the Cañadas were demanding titled recognition of the lands they had settled and the removal of the constant threat of eviction. They desperately needed health and education facilities for their communities. For help in confronting these problems, the Maya became involved with three occasionally cooperating but often competing sectors of Mexican public life: the government, at both the national and state levels; the Marxists of the Mexican Left; and the Maya's ongoing relationship with the Catholic Church, previously described.

Mexico had been ruled by the dictatorship of the Partido Revolucionario Institucional (PRI) since the 1930s. One of the PRI's branches, the National Peasant Confederation (CNC), was the party's labor union. It controlled the rural work force. Governmental corruption, bribery, and the buying of votes were rampant. The government's primary focus was on industrial development that would be controlled by and provide financial rewards to Mexican elites. This development was urban-centered; the neglect of the rural sector resulted in an economic crisis in the rural economy. The peasantry had lost land in spite of constitutional provisions for land reform, one result of the 1910 revolution.

During a 1968 student protest against the dictatorial government in the Tlatelolco sector of Mexico City, federal troops had been ordered to fire on the protestors. It is estimated that two hundred or more of them were killed, although the government disputes these figures. Outraged at the massacre and the beatings, many students joined leftist organizations. Plans were made to go to the countryside, insert themselves into peasant organizations, and slowly build up political-military support before engaging in some type of public action against the government, either as legitimate political parties or as subversive armed forces.

As a result, the Maya crisis in Chiapas attracted an influx of Mexican leftist groups, all seeking to organize the Maya in their struggle for rights, and at the same time to incorporate them into a national movement seeking to overthrow the PRI-dominated Mexican government. These leftist organizations included the Central Independiente de Obreros Agrícolas y Campesinos (CIOAC), part of the Mexican Communist Party; the Mao-inspired Línea Proletaria (LP), which later absorbed the Política Popular (PP) and the Unión del Pueblo (UP); the Organización Campesina Emiliano Zapata (OCEZ), aligned with the teachers' union centered in Mexico City;

and the clandestine Frente de Liberación Nacional (FLN), which included its public political group, the Alianza Nacional Campesina Independiente Emiliano Zapata (ANCIEZ), and its secret military branch, the Ejército Zapatista de Liberación Nacional (EZLN). To instill their worldviews in the minds of the Maya, the leftist parties fought each other, the government, the mainstream political parties, and the church-based organizations.

National and State Crises

The Mexican nation was going through a period of economic and political turmoil. During the 1970s and 1980s, a national agricultural crisis resulted in significantly less demand for agricultural labor in Chiapas. Historically, work in agriculture had been the survival outlet for land-poor Maya.

> It appears that the first stage in this shift was triggered by a decline in the profitability of agriculture. Stagnant and falling commodity prices, rising costs of fertilizer and fuel, scarce and expensive credit for farming, unfavorable exchange rates for exporters—as the 1970s went on, the cumulative impact of these conditions led large landowners to stop investing in their land and scale back production. Indeed, many began to pull out of agriculture altogether and either sold their land or converted it to cattle raising. . . . Finally, by the 1990s, most large landowners had abandoned the countryside altogether, to be replaced by small owners or *ejidatarios*, or communal holders. While these new farmers did begin slowly to increase production again in the late 1980s, they did not need as much hired labor as the plantations they replaced. The effect of all these changes was that demand for agricultural laborers probably reached its peak in the early to mid-1970s and then declined through most of the 1980s before creeping back to former levels at the end of the decade. As of the mid-1990s the demand appeared to be almost exactly what it had been twenty years earlier. (Rus, Hernández Castillo, and Mattiace 2003b: 6)

By itself, the conversion from agriculture to cattle raising, which required significantly less labor, had a devastating impact.

But the recovery of agriculture to the original level of the early 1970s did not relieve the problem. In the thirty-year period from 1970 to 1990, the number of Maya in Chiapas had doubled, from 100,000 to over 200,000. In addition, 100,000 or more Guatemalan Maya had sought refuge in the state and were now also seeking agricultural work. "Essentially Chiapas's indigenous peoples, who for almost a century had been maneuvered into relying on seasonal, often migratory agricultural labor to maintain themselves, suddenly found that the agricultural economy did not need them" (Rus, Hernández Castillo, and Mattiace 2003b: 7). In 1989, the International Coffee Agreement was not renewed. This caused the price of coffee to drop by 50 percent, driving many producers from the market.

The state of Chiapas had always functioned as a semi-autonomous region, poorly integrated with the rest of Mexico. It retained characteristics of the colonial era, and its economy had been scarcely touched by the Revolution of 1910. It was an economy based on large agricultural plantations that later were used for cattle raising and tree harvesting. The plantations were owned by a local oligarchy who ran the state for their own benefit. In many respects, they were the ideological descendants of the Spanish *encomenderos*. They controlled the interaction between the people of Chiapas and the national government. Their plantations were lands that had belonged to the Maya but had been taken away from them, as described in chapter 4. The Maya represented cheap labor, either living and working on their former lands as plantation serfs, or supplying seasonal labor when needed while trying to eke out a bare subsistence in their traditional communities. The results were summarized in chapter 6.

Against this background, the state government was taken over in 1982 by a new governor, and virtual dictator, General Absalón Castellanos, one of the richest cattlemen in the state. Castellanos came from a clan in Comitán that owned extensive plantations in the Lacandón that were politically protected. In 1980–81, he had been in charge of the Chiapas military zone. During that time, whenever confronted with land claims or labor disputes by the Maya, he had provided armed support to the plantation owners. In 1982, his orders were to prevent a civil war similar to what was happening in Nicaragua, El Salvador, and Guatemala.

> Among all the many oligarchs who had governed Chiapas, General Absalón Castellanos had proved by 1986 that he was one of the greediest, most corrupt, most rigid, and most violent. . . . At 59, he had taken office determined to stamp out "inciters of anarchism, pokers of the fire, fishers in troubled waters, who are not far from the trash the current carries," that is, the CIOAC [Communist Party], the OCEZ [a labor union], and the Union of Unions (UU) [a federation of Word of God ejido and other communities]. On federal funding of U.S. $100 million for his Program of Agrarian Rehabilitation, he paid landlords in full for land claimed by landless communities,

[and] eventually 200,000 acres [were] granted to 159 communities [in return for being] enrolled (in all but a few cases) in the . . . PRI. By 1986 he had also given landlords hundreds upon hundreds of exemptions covering nearly three million acres. His brother, head of the State Forestry Commission, was stripping large tracts of the rain forest for private profit. The OCEZ had lost more than 20 members killed. In 1985 the CIOAC's state secretary-general and its main lawyer, a Socialist deputy in the federal chamber, had both been murdered, the latter, according to Amnesty International, by hit men for the governor's brother. Between 1982 and 1988, when the general's term expired, the number of political assassinations would run to 153, making the governorship one of the most gruesome in Mexico's modern history. (Womack 1999: 199)

During the late 1980s and early 1990s, the Mexican government was negotiating the NAFTA trade treaty with the United States and Canada. One of its agreed-upon provisions was the abolition of small, uneconomical production units dependent on government subsidies. This included ejidos. This required repeal of Article 27 of the Mexican constitution, which was the basis for government grants of unused lands as ejidos. All applications for ejido grants were canceled. In addition, existing ejidos would be privatized, which meant their land could be sold. This opened the way for more land acquisition by non-Maya. For many, this was the final straw of frustration. To them, it proved that reform was impossible.

The Catholic Church and the Marxists in the Cañadas

As described in earlier chapters, community reflection on the Bible had brought Maya communities to a wider consciousness of the injustices of the Mexican political-economic system in which they were embedded. In the Cañadas, the Ocosingo pastoral team and catechists drew up a twofold plan to encourage Maya initiative: composition of the Tzeltal "Word of God" catechism, described in chapter 16; and a social action program, Ach Lecubtesel (a new way of living better).

Prior to the Indian Congress of 1974, catechists in the Cañadas organized groups to discuss local community problems. To assist the catechists, the Ocosingo pastoral team asked the help of a group of ladino community workers who had been working in Oaxaca. This group were members of the clandestine Unión del Pueblo (UP), a leftist organization advocating the overthrow of the Mexican government (Legorreta Díaz 1998: 62–78). It employed a Maoist strategy of gradually bringing the peasantry to a revolutionary consciousness, which would eventually result in a socialist government. To accomplish this, the UP did not rule out the use of armed violence, but it did not see violence as the only possible means to attain their goal. For strategic purposes, the UP would periodically cooperate with the government. A former director, Hector Zamudio, said, "We decided to create an armed organization to defend itself and to answer any aggression by repressive forces. It was not a military organization, but a political party. However differences arose and a militarist faction arose within the party" (63). The UP welcomed the opportunity to work among the Catholic groups in the Cañadas. They saw them as potential recruits for the UP's revolutionary consciousness. As the UP was a clandestine group, the pastoral workers of the Cañadas did not initially realize they were being infiltrated, and when they did, some had no objections, stemming from their Christian Marxist perspective. The UP assisted the communities by supplying information and helping them discuss their problems in a broader context than had been their custom. To prepare for the final presentations during the Congress itself, they trained a group of translators, with whom they formed a close bond.

During the Indian Congress of 1974, communities worked together and realized they all had the same problems. With the encouragement of pastoral workers and the assistance of UP, ten to fifteen communities of the Cañadas joined together to form agricultural and consumer cooperatives. In 1975, joined by delegates from eighteen additional ejidos, they formally organized as Quiptic ta Lecubtesel (United for Our Progress) (Vos 2002: 251–56). Soon, twenty-five additional communities joined. Quiptic ta Lecubtesel was modeled on the catechetical structure of the diocese and was called "the little brother of the Word of God." It represented an implementation of the worldview expressed in the Tzeltal catechism (Iribarren 1985: 65).

Javier Vargas led the Ocosingo pastoral workers and supervised the activities of the Cañadas catechists. He collaborated with the UP workers, introducing them to the catechists, who in turn introduced them to their Word of God communities. At community meetings, catechists would speak first, then a member of the UP would highlight local problems and politicize the communities by proposing the UP's solution as the answer: a socialist state. UP workers were supplying a specific implementation of "act" for the "observe-judge-act" triad of Action Catholicism.

Catholic Communities Turn to the Left

The UP workers began to take over community leadership from the catechists. They started to organize community meetings without the catechists to study the national law of agrarian reform, the history of Mexican revolutions as an example of campesino class struggle, elements of historical materialism, and the theory of political-economic organization (Tello Díaz 2001: 74–79; Legorreta Díaz 1998: 70). The UP also politicized a number of pastoral workers. At a 1979 meeting, pastoral workers concluded that "Christianity was the prophecy of justice and equality among men; and socialism is the concrete and material reality to bring it about" (Vos 2002: 235). The collaboration of pastoral workers with the UP, and later the PP, took place in the Dominican-administered Cañadas and in several communities of the Chol and Tzotzil areas.

In 1978, Bishop Ruíz, impressed by their community-development efforts in northern Mexico, had invited members of Política Popular (PP) to assist any diocesan groups wishing for their help. They were welcomed in the Ocosingo parish and Quiptic. Soon afterward, the UP and PP national organizations merged to form the Organización Ideológico Dirigente (OID), also known as Línea Popular (LP) because of their Maoist worldview. A reorganization took place, and the original UP personnel were transferred to other regions, leaving only PP personnel to work in the Cañadas communities. Based on their Maoist philosophy of decision-making by the entire community, the PP began to criticize the church, the catechists, and the predeacons as being part of a vertical elitist structure, an authoritarian organization. They accused the catechists and the predeacons of using their religious positions to amass political power. This was true in some cases. In 1979, the pastoral workers reacted by withdrawing their cooperation with PP personnel. They asked Word of God communities to cease feeding them. Most were forced to leave the area, as there were no other available food sources (Tello Díaz 2001: 86). However, some pastoral workers continued to work clandestinely with a few PP members who had returned.

In 1980, Quiptic was absorbed into the Unión de Uniones Ejidales y Grupos Campesinos Solidarios de Chiapas (UU), an umbrella organization of smaller unions sponsored by the LP Maoist movement. The UU represented 12,000 families in 180 communities from 15 municipios. The Unión successfully negotiated with the government for long-promised financial assistance and other benefits for new agriculture settlements. Negotiating was a temporary tactic, as they remained committed to ultimately overthrowing the government. Quiptic, whose members comprised about 60 percent of the UU, objected to the way the UU was negotiating with the government and withdrew from it. The leadership of both the LP and UU, as well as those representing the government, was ladino. Thus, ladinos were dealing with ladinos about Maya concerns. Quiptic rejected this and disliked collaboration with government entities, saying that it involved frustration and bribes, and that it ultimately led to being controlled by the government.

Biblical reflection had helped to bring the Cañadas communities to social consciousness, and their successive inclusions in UP, PP, LP, and UU had narrowed their outlook to a Marxian political philosophy. With Quiptic's withdrawal from the UU, there was no remaining political party in the Maya communities with an agenda to specify the "act" portion of the "observe-judge-act" formula. It was no longer being provided by Action Catholicism, whose work in the communities stopped short of forming a political party. Although there were national reform parties, none were present among the Maya in any significant way, as was the case for the Christian Democrats in Guatemala.

The Zapatistas: Forces of National Liberation (FLN)

With the reappearance in 1983 of a radical leftist organization, the FLN, the political vacuum began to be filled (Legorreta Díaz 1998: 183). The FLN was founded in 1969 as a clandestine Marxist organization based in Monterrey. The FLN had made two previous attempts to establish a guerrilla presence in Chiapas, in 1968 and 1972, but had failed when it was detected by the government. In 1983, the FLN returned again to Chiapas. Their strategy was similar to that of the Guatemalan Left, described in chapter 17. They would recruit combat and support forces in a remote area for an extended period of time, initiate guerrilla warfare, build a peasant army, and take over a region as a liberated zone. From there, they would join leftist forces from other regions and capture Mexico City. In contrast to the Guatemalan Left, their goal was to initiate a democratic process that, hopefully, would lead to a government committed to social justice.

The Maya populations of southern Mexico were an ideal place to renew their efforts. In the first two years, about twenty-five FLN members arrived in Chiapas, including a Captain Marcos. Starting in northern Chiapas near the border with Tabasco, the FLN were successful in gaining some Chol and Tzotzil recruits and building a base of support in the area around Sabanilla. They were assisted by the region's sympathetic parish priest. From there, they moved to the encampments of the poor around San Cristóbal de las

Casas and recruited more members (Tello Díaz 2001: 93). Next, they went to the remote Cañadas of the Lacandón, where the socially conscious and already politicized Liberation Catholic communities provided ideal conditions for recruitment. As in Guatemala, indigenous issues would be used to recruit, although they were of secondary importance to the long-range objectives of the FLN.

Zapatista Communities and Military Training

Given the social conditions in Chiapas and the political vacuum in the Cañadas, it was not difficult for Catholic communities to become Zapatista communities. The moral codes were similar. Both sought reforms that would bring social justice; both took strong stands against the use of alcohol; both insisted on respect and equality for women. All decisions were to be made by an assembly of the community itself, where all had the right to express their opinions, including women. All who disagreed with Zapatismo either voluntarily left the community for fear of possible army retaliation or were evicted (Le Bot 1997: 278; Tello Díaz 2001: 238). (The introductory photograph shows Marcos, on horseback, and his staff entering the Zapatista center of Realidad for a meeting in 2007.)

Once a Zapatista community was established, young men and women left their communities and went to live in the mountains to be trained as *guerrilleros*, armed combatants. They ranged in age from seventeen to twenty-six years, with a few as young as eight or nine (Leyva Solano and Ascencio Franco 1996: 227). The EZLN built a sizeable military force in a few years. For many families in the Cañadas, the family ejido land was too small to be divided for agricultural purposes. Without any inheritance and without any opportunity for schooling in these jungle communities, what was the next generation to do? The Maya crisis had produced a generation of frustrated youth seeking a future. Many decided it was with the EZLN.

In the mountain camps there was intensive military training and further indoctrination in the Marxist worldview. Military discipline set the tone of the camps. Upon rising in the morning, the trainees took long hikes to find food, carry it back to camp, and prepare it. This was followed by an hour-and-a-half of study—Spanish for the many Maya who needed it. Then there was a period of instruction about the worldview of the FLN. This was followed by the evening meal. Afterward, the group divided into small cells to study books dealing with political subjects and the history of Mexico. These included biographies of Mexican rural revolutionaries: Pancho Villa, Emiliano Zapata, Father Miguel Hidalgo y Costilla, Father José

María Morelos, and others. Each member of a cell took his or her turn reading a paragraph from the book. In this way, many learned to read. These recruits were absorbing the worldview of the rural Mexican revolutionary tradition, first, led by the priest Hidalgo to liberate Mexico from Spain, and then, in 1910–17, the Mexican Revolution to liberate the country's rural areas from the servitude of the hacienda system. They also studied U.S. army manuals. Because of their leadership training and experience, many catechists and predeacons became officers in the ranks of the EZLN. While in the mountains, the recruits were supplied with food and goods from their home communities, which became support bases. The support bases also formed armed militias, a reserve group that would be mobilized only to defend the communities if attacked (Le Bot 1995: 176).

The EZLN initially recruited women in secret meetings away from the men because of woman's traditional position in Maya communities. The Zapatista program was attractive, as it openly discussed women's problems within Maya culture, as described in chapter 5. "A revolution within the revolution" took place as women were given the same status and opportunity as men. Women's new place was summarized in ten laws passed by the central Zapatista coordinating committee, the Comité Clandestino Revolucionario Indígena (CCRI), and affirmed by the local communities (Rovira 1994: 231–36). Women volunteered to go to the mountains and be trained as armed combatants. It is estimated that they made up 35 percent of the EZLN army, some of them becoming high-ranking officers (61). If an EZLN couple wanted to marry, they had to receive permission from their military commander, to verify that neither was already married or separated. The EZLN had its own marriage ceremony. Women were not allowed to have children while with the EZLN, due to their workload and the frequency of military travel. Contraceptives were widely distributed. If a woman wanted to have a child, she could take a leave of absence, return to her community, have the child, leave it with her mother or mother-in-law, and then return to the EZLN. Women also became members of the militias. They were the backbone of the support bases preparing and sending food to those training as combatants in the mountains. Marcos and others cite the importance of women in recruiting their husbands, sons, daughters, and in-laws into the EZLN (Le Bot 1995: 165; Rovira 1994: 216).

Expanding Zapatista Influence

The next phase was initiated when the now-trained guerrillas and their instructors descended from the mountains and began the political-military indoctrination of the Cañadas communities. They began by discussing the miserable conditions in the Cañadas and the reasons for them. The message was clear: "Be prepared to fight. At any moment we must rise up with arms, either to defend ourselves or to obtain what we are lacking" (Rovira 1994: 47). The need for armed defense was confirmed in 1986 when a group of masked plantation paramilitaries destroyed an ejido community. This confirmed the new element of resistance introduced by the FLN—the necessity of fighting and taking steps to obtain arms and prepare for war. The development of a social consciousness had already been achieved by the Catholic Church, and infusing that consciousness with a Marxian political philosophy had already been done by the Marxian predecessors of the EZLN.

During this period of expanding Zapatista influence, Subcomandante Marcos, as the head of military planning for the EZLN, organized military shows to build morale and give the impression of a military force of great strength. In 1988, he brought together members from seventy or more communities to celebrate the anniversary of the founding of the FLN. He did the same in 1992 to celebrate five hundred years of Maya resistance to the Spaniards and Mexicans. On both occasions there was a parade of EZLN combatants, one thousand the first time, and five thousand in 1992. These spectacles gave the impression of an armed force of great strength, and boosted the morale of all (Vos 2002: 339–42). The EZLN stocked some weapons and made plans to resist with force any incursion of the army into EZLN communities.

Originally, the FLN had been working clandestinely within the Unión de Uniones (later called ARIC Unión de Uniones). But in 1991, when it saw the UU collaborating with the government, the FLN withdrew and formed its own political organization, the Alianza Nacional Campesina Independiente Emiliano Zapata (ANCIEZ). In 1992, it staged a number of public demonstrations against various government actions that obstructed land reform. During a demonstration in San Cristóbal, ANCIEZ pulled down the statue of the Spanish conquistador of Chiapas, without government retaliation. These demonstrations gave the impression to many Maya that the EZLN was a viable organization capable of fighting government forces.

The EZLN and the Catholic Church

The EZLN was significantly helped by a number of pastoral workers in the Ocosingo parish because of their Christian Marxist orientation. Initially, they had used the Marxian economic model as a tool of analysis, but as they came to understand the situation in Chiapas, they went further, seeing a strategy of armed violence as the only viable means to bring about social justice. The Ocosingo pastoral workers had chosen to work with the UP and PP, introduced them to the catechists and Word of God communities, and encouraged cooperation. The politicized foundation laid by these groups remained after their departure. When the EZLN reentered the region, they found immediate support from a number of the same pastoral workers, catechists, and predeacons. An observer told Meyer (2000: 110) that a majority of the pastoral workers in the Cañadas supported the EZLN. After the uprising and the ensuing deaths, some Maya recalled with resentment that "they [the leftists] entered with the priests. Because of this the people readily believed what they said. They were saying that we will not gain anything unless we fight" (Tello Diaz 2001: 117).

As explained in chapter 11, pastoral workers and catechists could influence communities by the biblical passages chosen for reflection and the questions asked to focus the ensuing discussion. Those in favor of the EZLN's agenda of armed violence chose passages from the Book of Revelation (the Apocalypse) that describe the great battle between the forces of good and the forces of evil preceding the Messianic Kingdom. As described by a former EZLN combatant: "People joined because the Bible was interpreted in light of the Apocalypse. It taught that the people of God must fight with arms. This was the result of the reflection each Sunday and in each community under the influence and leadership of the catechists who had become members of the EZLN" (Legorreta Díaz 1998: 203). "With the Questions, the church members began to evaluate their position within the broader Mexican society. Comparing this with conditions that Jesus had condemned during his time on earth, they have found that their own living conditions were against the will of God. Converts told me repeatedly that it was with the advent of the Questions that the political reflection began" (Moksnes 2003: 143–44). The EZLN catechists morally justified the use of violence by their own interpretation of the Exodus event. "They were arriving at the conclusion that the people of God should struggle with arms, not because the Bible said so, but they were drawn in this direction by the questions they asked about arms, about what those followers of Christ did when they fought in Egypt, when they sacked Egyptian communities where they had been ground down as servants, as slaves. How were they able to free themselves? . . . Because they believed in God, they believed in armed struggle. We can accomplish the same thing if we do what they did. If we don't do it, then we are not followers of Christ. This is what the discussion of the questions was saying" (Legorreta Díaz 1998:

203–4). In this EZLN interpretation of Exodus, Christians were involved in the Exodus and the Maya were responsible for the plagues inflicted on the Egyptians.

The EZLN buildup was also helped by the initial waverings of Bishop Ruíz about armed rebellion—and the interpretations of his waverings. As will be seen in chapter 26, one of the conditions for the moral legitimacy of a just war is the possibility that the use of arms will succeed. The bishop had been deeply moved by conditions in Maya villages and frustrated by the machinations and attacks of the Chiapas elite backed by governmental corruption and lies. He was the one who originally invited the PP into the diocese. It is not clear when the bishop realized that the PP were radical Marxists, and how long it took for him to reject them once he did. There was a period of dalliance as he pondered the moral question. He visited Nicaragua and was impressed by the Sandinista victory. There, in 1987, he wrote, "We must leave Egypt to arrive at the Promised Land. . . . But this involves political interventions, it does not come about by prayer alone. . . . Christians can take part without misgivings of conscience in the Sandinista movement, even in the armed struggle so that a new society may arise. . . . Christians can be there, for there is the Kingdom" (Tello Dias 2001: 106). For the bishop, the Sandinistas had shown that armed violence was a viable possibility, and therefore he confirmed the morality of its use in Nicaragua. Javier Vargas, who was a close advisor of the bishop and prominent in the activities of the Ocosingo pastoral team, shared the bishop's view. Vargas also visited Nicaragua. He returned highly enthused about what he had experienced (119).

This dalliance was interpreted by some as approval of an armed uprising. A former Zapatista recounted his understanding of the reflection periods. "On Sundays it was the custom for all to go to church . . . but we did not go to learn the Bible of God, that is to say, the Bible itself, but some questions that the people of the diocese had composed. The intention of the bishop was to order that these questions be sent . . . to the catechists of each community for discussion during the church meetings. But in reality it was not the Bible of God nor His word but the word of the Bishop and his politics. This is what was happening in terms of the struggle. . . . There were questions of how to liberate yourself, how to defend yourself, what was the intention of the government, of how all of us campesinos are ground down. This is what was discussed in church" (Legorreta Díaz 1998: 203). These accounts indicate how some communities and pastoral workers interpreted the bishop's condemnation of the existing conditions in Chiapas and his silence about the use of armed violence as backing for the position of the EZLN, thereby giving it legitimation and moral justification. In 1993, the Jesuit Mardonio Morales, who had worked among the Maya in the diocese for thirty years, said that the bishop had been naive in introducing the Marxists into the Word of God communities, that the church had been undermined by a politics of violence (Rovira 1997: 118).

In summary, the social and political consciousness of these communities, their confusion over the political vacuum that had developed, their frustrations with and distrust of the government, their anger over the proposed repeal of Article 27 of the constitution, the similarity of the Zapatista moral code with that of Catholic Action, the silence of the bishop about the use of armed violence because of his dalliance—all these factors combined to induce a number of Word of God communities with their catechists and predeacons to adopt Zapatismo as the embodiment of the "act" part of "observe-judge-act" formula.

Some Catholics Turn against Zapatismo

By 1989, the bishop had rejected the strategy of war advocated by the EZLN. He realized that it could not succeed against the strength of the Mexican army. At that time the brutality and magnitude of the Guatemalan counterinsurgency was beginning to be understood. The bishop came to the conclusion that armed violence could not bring about social justice and would only lead to a repetition of what had happened in Guatemala.

To diminish the prestige of the EZLN in the Cañadas, the bishop and Slohp, a group of catechists and deacons who worked closely with the bishop, emphasized that the EZLN leadership consisted of Marcos and other ladinos. They asked, Why should Maya communities reject the Maya leadership of those catechists who opposed the EZLN, in favor of ladinos? Playing the ethnic card was an effective strategy and led to desertions from the EZLN, which began in the late 1980s (Legorreta Díaz 1998: 224).

In 1993, as the EZLN hastened its preparations for war, the bishop presided over a meeting of Slohp. It opened with reflection and reading from the Bible, Matthew 4:1–11 on the triple temptations of Jesus. The first temptation regarded food, and the third was political power over all the world. Another reflection was on John 15:1–10, the parable of the vine and its branches. The biblical passages were used to express Slohp's message: remain faithful to the vine of the Word of God and your indigenous culture. In other words, remain faithful to the flame of the Tzeltal catechism; do not let yourselves be tempted with illusions of food and power by outside ladinos whose ideas are contrary to indigenous culture (Vos 2002: 343–48; Tello Díaz 2001: 119, 139, 219).

Aware of the situation in the Cañadas and taking advantage of Pope John Paul II's visit to Yucatán in March 1993, Bishop Ruíz issued a lengthy pastoral letter, "In This Hour of Grace." The first sections reviewed the past and contemporary history of economic, political, and cultural oppression in Chiapas. Then he reviewed the attempts of the diocese to cope with these problems. The letter warned against violence and closed with a plea for dialogue. "Creation itself protests against the way of plunder and concentration of wealth that our social system generates, signaling the urgency of a change of course for humanity, lest our planetary survival itself be jeopardized. Why not start a different path without waiting until the social structures change because of the desperation of those who have been crushed since antiquity? . . . Dialogue, which is a condition for fraternal relationships, supposes a prior willingness to listen and has as its platform the acceptance of the other, without presuming him to be of excessively bad faith" (Ruíz García 1994: 601–2).

But as the year continued and reports from the Cañadas and other areas revealed the stepped-up preparations of the EZLN, the bishop continued his efforts to dissuade the Word of God communities from taking up arms. Although he knew the tyrannical conditions in Chiapas seemed to justify a war for social justice, he also knew it would be a costly and unsuccessful effort against the state and national governments with the various armed forces at their command. He knew only too well the sufferings of the 100,000 or more Guatemalan refugees in Chiapas, as well as the stories of those who had died within Guatemala. He and numerous pastoral workers visited Maya communities in the Cañadas and other areas, strongly urging them not to support the armed struggle. Their efforts were influential in Marcos's losing at least 40 percent, possibly 60 percent, of his armed forces (Meyer 2000: 93; Tello Diaz 2001: 220).

The EZLN claimed to be neutral with regard to religion—that it did not favor or discriminate against any type of religious belief. Most, if not all FLN leaders were Marxian atheists. In spite of their public stance, privately, they denounced the biblical basis of liberation—and on occasions, publicly. To counter the bishop's efforts, one time Marcos addressed his battalion: "You go around saying that God is going to support you in the struggle. This isn't correct, compañeros. God, yes, he came into the world, but they killed him, so there is no God now. Now God is worth shit. Now this Bible is worth shit; it is the book of a coward. Our only God now is a gun. This is our God [as he held up his gun]" (Legorreta Díaz 1998: 223).

This kind of talk alienated people. "I left the EZLN in 1989 because Marcos began saying that 'the word of God was not worth shit.' Some of those who left felt deceived by the Zapatistas. They began to realize that the Zapatistas were only using the word of God for their own purposes and were not telling the truth. The majority had joined the Zapatistas convinced that it was in accord with the Bible, but then Marcos himself began speaking against the word of God, that God did not exist and things like that. The people became very confused" (Meyer 2000: 92). Marcos always played down the role of the church in Maya empowerment.

Some Zapatistas thought the EZLN was not strong enough to fight the Mexican army, and left for that reason. More left when the army discovered a training area with arms of the EZLN. A substantial minority of EZLN communities disapproved of the decision to wage war for other reasons. Some dissenters feared that army retaliation against their communities would inevitably follow any such attempt. Others thought that since the vote was favorable by only a small margin, the EZLN communities were not united behind the effort (Le Bot 1995: 192; Vos 2002: 341).

The Pressure for War

Marcos feared there would be more desertions unless the EZLN was able to prove itself soon by offensive action. Marcos himself agreed that the EZLN was not ready for war. "Originally we were not an army prepared to take the offensive. Since we were a political-military organization, we were ready for a war that would be waged against us at any moment, but not for one where we would go in search of the enemy. After our contact with the communities, we had conceived our military role in terms of defense. We assumed that if the army came into the region or discovered us, it would be necessary to fight and resist, first against the state police and then the national police and finally the Mexican army itself. Therefore we had a defensive military strategy that covered all the Cañadas and the most important places in the highlands. The majority of our troops were located in these two areas" (Le Bot 1997: 193). But to prevent more desertions, Marcos had to speed up his timetable and train an offensive army.

The Decision to Fight

In addition to mounting desertions from his army, Marcos faced another problem: the internal politics of the FLN. The national directorate had come to the conclusion that, given the deteriorating situation of the FLN in other regions of the country, its best strategy was to cease efforts at armed insurrection and to become a legal party of the Left. This threatened Marcos's Chiapas strategy and the sacrifices the EZLN communities had made to buy a few arms and build up the organization. To thwart the proposed policies of the national directorate, Marcos needed his armed force to go into action (Vos 2000: 352–54; Tello Díaz 2001: 184–87; Henck 2007: 151).

In January 1993, the first reorganization meeting of the FLN was held in the ejido community of El Prado. In attendance were forty officers and representatives from the weak FLN units in other states, the national directorate of the FLN, the officers of the EZLN, and about three hundred representatives from Zapatista Lacandón communities. On the second day, Marcos launched a vehement attack against the changes proposed by the national directorate. He said it was necessary to undertake offensive action in Chiapas as soon as possible. When the other delegates saw the support of the three hundred Maya representatives for Marcos's position, they voted in favor of it. The question of going to war against the government was then taken to the Zapatista communities for a vote, since they were the ultimate authority for all Zapatista decisions. A slight majority of those voting, perhaps fifteen thousand, voted in favor of initiating the rebellion (Tello Díaz 2001: 187–89). There is some controversy about this vote. Some say that Marcos manipulated the vote by the selection of communities that voted, or used other means so that it was not a legitimate voting process. Others say that the government treatment of Maya communities created a state of impatient rage, so that many welcomed the possibility of an armed uprising. Both factors were involved. It is not clear which was the more important (Henck 2007: 151–57).

To gather, interpret, and execute Zapatista community decisions, a coordinating committee, the Comité Clandestino Revolucionario Indigena (CCRI), had been created (Rovira 1994: 37–45; Vos 2002: 352). Because of the large number of communities involved, intermediate layers were established. The community leaders brought their community decisions to regional leaders where they were further discussed; then representatives of the regions brought the results of these discussions to the leaders of the four language zones, who were also members of the CCRI, along with other important officials. They included Subcomandante Marcos as head of military planning, which gave him a more prominent role in the war effort; Tacho, a former catechist; and David, a former seminarian from San Andrés Larraínzar (Vos 2002: 348–49; Rovira 1994: 257–59).

In accordance with the bare majority vote from the Zapatista communities, the CCRI commanded the military branch, the EZLN, to make preparations for war (Le Bot 1997: 195–96). As outlined by the committee, the war was to be:

1. an indigenous war, not of one language group, but of all four of the principal groups in Chiapas.
2. a war not only for indigenous demands, but for national ones as well.
3. a war not only in the local area, but in the whole territory of Mexico, or at least in all the states where the EZLN exists.
4. an offensive war, no longer one of self defense. This means using our forces as offensive units, and preparing the communities to attack rather than simply resisting an attack.
5. A war for a change to a democratic system, not a fight to take over the government.
(Vos 2002: 352)

In spite of the slight majority vote in the EZLN communities, Marcos ordered his army to go on the offensive. On January 1, 1994, the armed uprising took place as five thousand EZLN combatants surprised and seized four municipal centers. It was the day that the provisions of the NAFTA treaty were to take effect. Surprise was the EZLN's key tactic. They killed state policemen guarding public buildings. According to Womack, from the four municipal centers, their strategy was to take over the highlands of Chiapas, "attack and capture the state's military headquarters in Tuxtla, after which they would turn north toward Mexico City. Meanwhile, having prepared the northern and central fronts, the secretary general and commander in chief would be coordinating operations in Chihuahua, Michoacán, Puebla and Vera Cruz" (1999: 43; see also Vos 2000: 335).

The Uprising

During the first day of the attack, the EZLN did not encounter any units of the Mexican army. This was fortunate because the entire EZLN force had only 200 automatic weapons and 2,000 rifles; the remaining thousands of combatants had only .22 caliber shotguns, pistols, and wooden lances (Womack 1999: 42). An Ocosingo unit had no automatic weapons. To obtain modern arms, the EZLN attacked but failed to capture the army post outside San Cristóbal. In Ocosingo, the army arrived on the second day and trapped a small force, whose withdrawal on the first day had been delayed by a shortage of trucks. The army blocked their retreat to the Cañadas using helicopters and airborne infantry. The EZLN suffered most of its casualties in these two encounters. With these setbacks, the EZLN forces melted back into their bases and home communities.

This initial effort of the EZLN revealed its weakness, that it was not a military threat to the government. To protect its international reputation and its ability to attract foreign investment, the Mexican government did not want a repeat of the murders and genocides that resulted from the Guatemalan counterinsurgency, although it could have easily done so. It proclaimed a unilateral cease-fire after twelve days. Without admitting it, the Zapatistas welcomed the cease-fire. A series of negotiations between the government and EZLN ensued. These eventually led to the San Andrés Accords on Indigenous Culture and Rights, in February 1996.

The bishop and the liberation movement were blamed for the uprising by many local ladinos, government officials, and newspaper columnists in Mexico City. The bishop opposed the EZLN's decision but sympathized with those who had taken up arms. "We understand the subjective situation of many of our brothers that have opted for a path that we consider wrong. The cry of anguish of those who give up their lives in search of better conditions deserves all our understanding, if not for them, at least for their children. As a diocese we declare that violence obstructs the path of true solutions, and from this rejection of violence, we want to accompany the people and their social organizations to defend their rights and better the true conditions of their lives" (Floyd 1997: 100).

Summary

This chapter has examined why some Maya Catholics in Chiapas took up arms against the Mexican government. Based on their frustration with government corruption in their efforts to overcome their crisis, and taking advantage of the influence of the Bible in empowering them and bringing them to social consciousness, radical Marxian leftists convinced some Word of God communities that "act" in the formula "observe-judge-act," meant they should become Zapatistas and prepare for an armed uprising to achieve social justice. They formed a poorly armed military force that initiated an armed insurgency. It was short-lived. The Zapatistas lived on as a political-social movement that will be described in the following chapter.

21

Militarization in Chiapas

When the EZLN retreated to their communities, the army did not immediately pursue them. It brought in additional military units, militarized the state, and developed a counterinsurgency plan that was initiated in 1995, more than a year after the uprising. During this period, additional Word of God communities became Zapatista supporters.

Zapatista Autonomous Zones

The failure of the Zapatista uprising showed that they were not a serious military threat, although they did not renounce the use of armed force. Instead, Marcos declared that the Zapatista communities would be "in resistance" against the state and national governments. Thirty-eight EZLN municipios identified themselves as autonomous, with their own governments completely independent of the state and national governments. This independence included the refusal to accept or participate in any government public-service programs, including education, health, or agricultural assistance. In their view, any dependency on the government meant surrendering to inefficiency, giving bribes, submitting to other forms of corruption and discrimination, and forfeiting the ability to confront injustice—ultimately leading to a self-defeating situation. They would govern and provide for themselves in communities modeled on the original EZLN communities in the Cañadas (Burguete Cal y Mayor 2003: 191–218). In communities where the Zapatistas were only a part of the community, there were two parallel governments and systems of public services within the same community. The introductory photograph shows an independent Zapatista market.

San Emiliano

In the 1950s, San Emiliano was a newly founded community in an isolated part of the Cañadas (Barmeyer 2009: 66–74, 78, 80, 92–93, 103–4, 109–23, 133–35). It built a church, installed a statue of its patron saint, became a Word of God community in the 1960s, joined the Dominican-encouraged Quiptic ta Lecubtesel in 1975, was absorbed into the Unión de Uniones in 1980, became a Zapatista support community in the late 1980s, contributed combatants to the uprising in 1994, and received some land from the *fincas* taken over during the uprising. The majority of the community withdrew from the Zapatistas in 1998.

The community began by carving out house sites and fields for milpas from the raw jungle. San Emiliano received a land title for its original settlement. As its population grew, community members needed more land, but the government refused to give them titles after 1970. The community constantly struggled to provide for its needs. With the arrival of pastoral workers and catechists in the 1960s, the community began using the "observe-judge-act" formula and the Bible for reflections on their situation. This influenced their involvement in Quiptic and the Unión de Uniones. In the mid-1980s, the Zapatistas appeared in the community, talked at the assemblies, and gradually convinced the community to become a Zapatista support base while continuing to be a Word of God community. This was a relatively easy matter, as the Zapatista code of conduct was similar to the one already existing in the Word of God communities. The Zapatistas provided the framework for the "act" part of the community reflections. The main difference between Word of God communities who became Zapatistas and those who did not was the acceptance or rejection of possible armed violence as a means to obtain social justice.

The Zapatista Community Structure after 1994

In theory, the heart of the Zapatista system was the local support community. All decisions, except some military matters, were to be discussed and approved by the assembly of the entire community. The Zapatistas retained from the Maya tradition and the Word of God meetings the consensus method of arriving at decisions during their weekly meetings—the tijwane method described in chapter 14. The actual governance of the communities was by committees. The main ones were the Ejidal, Land Control, and Police committees, each with its own internal hierarchical structure of a president, etc. There were a number of additional standing or temporary committees for specialized functions, such as education, health, and installing a potable water system. These committees formed the Council of Authorities, the group entrusted with the overall governance of the community. In communities with a militia, there was a *responsable*, a person who reported directly to and took orders from the majors of the standing EZLN army. The responsable was selected by the military branch and presented to community members for their endorsement.

The local community and its committees were coordinated by a similar organization at the municipio level. It was composed of the same committees that governed the local communities. The members of the municipio committees were appointed by the local communities. There was also a *regional*, the counterpart of the responsable at the community level. This person reported to and took orders directly from the military. In 2003, a more-inclusive organization was added to the structure that encompassed all the municipios within a designated region (see the map in Burguete Cal y Mayor 2003: 192).

The military branch of the Zapatistas consisted of the Comité Clandestino Revolucionario Indígena (CCRI), the standing army, and the militias. The members of the CCRI were elected by each Zapatista support community in which there was a militia—a trained, part-time military force that lived in the support communities. The militias were called upon to supplement the standing army in times of emergencies. The standing army was led by three subcommanders, each in charge of a regiment of 1,500 soldiers, both full-time military men/women and militias. Below them were eight majors, each in charge of a battalion consisting of 40 combatants and 500 militias. Under each major were captains in charge of 15 soldiers, and below each captain, 3 lieutenants in charge of 5 soldiers each (Henck 2007: 135). The three subcommanders were also members of the CCRI. Since the CCRI was elected by the local community assemblies and the subcommanders were subject to the CCRI, there was civilian control of the military. But because each community's responsable and each municipio's regional were chosen by the military and took orders from the major in charge of the area's militias, rather than the local community, the military had a quasi-independence of its own. Since the structure was a gradual development, a Zapatista support community inserted itself into a version of this framework depending on when it joined the Zapatistas. It was not static, as adjustments were constantly made. Both communal and individual membership in the EZLN was often temporary and motivated by pragmatic considerations.

Biblical Reflection

All assemblies of the community, no matter what the topic for discussion might be, were preceded by biblical reflection.

> "Being a community with strong ties to the diocese, the village of San Emiliano had its own *comité iglesial* made up of the *tuhunel* (deacon) and a team of catechists who cooperated with the *consejo de autoridades* (governance committee) in choosing Bible passages befitting the topics to be discussed in the next communal assembly" (Barmeyer 2009: 92).

Many discussions during the Sunday assemblies were motivated by selected Bible citations. These often referred to very secular themes, such as the installation of a vegetable plantation by the community's women, the distribution of paid work with a road construction company among the men, a land conflict with a nearby army camp, or a contract on marriage exchanges with neighboring base communities. Less frequently, these were consultations (*consultas*) by which the command of the EZLN gauged the mood among its supporters. Issues included deciding if the organization should transform into a political party, whether it should begin or discontinue negotiations with the government, or whether to embark on a nationwide campaign to press for the recognition of indigenous rights to autonomy (Barmeyer 2009: 93–94).

This illustrates how, after the uprising, Zapatista communities continued to be or became Word of God communities. They continued to find in the Zapatista organization and its agenda for social justice the implementation of the "act" part of Liberation Catholicism's triad of "observe-judge-act."

Counterinsurgency against the Zapatistas

In 1995, the government launched a counterinsurgency to bring Zapatista communities under government control. It consisted of a twofold strategy. The first part was an attempt to destroy the few possessions the communities had, reducing them to dire poverty. When the army entered the Cañadas, many inhabitants of Zapatista communities fled to the mountains. In a village near San Emiliano an army unit set up camp. "When the soldiers entered the deserted villages, they broke into the houses, stole tools and destroyed the furniture. . . . For eight days, . . . [they] slaughtered and ate the cows, pigs and chickens the villagers had left behind. Houses where the soldiers found military clothing or 'books by Che Guevara' were burned. . . . When they moved on, they destroyed the maize granaries and roaming pigs and horses, and finished off the communities entire food reserves. . . . In the mountain hideout there was not enough food for everyone and several children and old people died of disease and exposure. Babies were born under difficult circumstances but with everyone's support" (Barmeyer 2009: 67–68). Eight days after the army's departure, the inhabitants returned to their decimated community. These already poor communities were reduced to desperate poverty. The failure of the EZLN army to defend their invaded villages alienated many Zapatistas.

This alienation was furthered by the second prong of the government's counterinsurgency. Building on their desperate poverty, the government allowed Zapatista communities to resume obtaining government subsidies and services for health, education, and agriculture that had been cut off at the time of the uprising. This was tempting bait to win over Zapatista communities in dire poverty and who were still required to support the Zapatista standing army. In San Emiliano and other communities, the government's strategy was successful. To sustain themselves, many turned to government assistance and submitted to government authority. Zapatista groups that remained in resistance were maintained by assistance from NGOs and other forms of outside aid (Barmeyer 2009: 130–236).

Zapatismo and Liberation Catholics in a Tojolabal Lacandón Community

Antonio Sánchez Cruz, a Tojolabal-speaking Maya, was born in 1926 near Margaritas (Earle and Simonelli 2005). A year later, he migrated with his grandmother's family to a distant area of the Lacandón jungle, where the family worked on cattle ranches. Antonio matured, married, and cultivated unused land. In 1951, he joined forty-five others to form an ejido group that petitioned the government for title to 906 hectares (2,238 acres) of land, each member being responsible for about 20 hectares (49 acres). "We had been working the land for many years, but the people from the Church told us we needed a paper. We began the petition in 1951 and made out first trip to Tuxtla [the state capital]. In 1952, the engineer came for the first time, but without money you don't get the title. So we worked with the church; they gave us direction even then" (Earle and Simonelli 2005: 77). The petition was granted, and the ejido community of Orilla del Bosque was formed.

Twenty-eight years later, in 1980, the government built a road in the region. The ejido group decided to relocate adjacent to the road since it provided better access to markets for their products. They formed a new ejido community, Cerro Verde. The new community included Maya, mestizos, and immigrants from other states, which made it difficult to unify the community. Antonio, now a grandfather, was active in the move, which included his daughter Luz, her husband, Rodrigo, and their young daughter, Ana. A pastoral worker, Javier Vargas, described the role of the Bible in the lives of these people.

> Luz and Rodrigo were a young couple when I first met them. They had one daughter only, Ana, who had a difficult childhood, and they sought medical care with the nuns in Comitán. Like the other young adults, they were uneducated, without shoes even, trying to survive on what their milpas produced, and anxious to learn production techniques that might increase their yields, and give them something to sell. We began working with some of the members, and they learned to read with the Bible. . . . So we taught a few, and they taught each other. In this way they learned to speak, to feel *capaz*, capable to interpret the Word of God and the world around them.
>
> The communities were in charge of the process. They formed committees and some became catechists. It was dynamic, they had such wisdom, and the reflection and learning gave life to this. Then faith began to live in their daily life. The young men and women, with

their families, began collective work, especially in health and education, providing for themselves what the government failed to provide. Where there were no medicines, they began herbal gardens, set up a community pharmacy, learned to be health promoters as they learned to make the environment they lived in healthier. All this was born from the Word of God. It was a great horizon of hope. (Earle and Simonelli 2005: 85–86)

Late in 1994, probably as a result of the uprising at the beginning of the year, they heard about some people who could help their community. "This is how our work began, then, knowing what the Church taught us, to respect our heritage and identity as indigenous, to trust in our dear god. But we heard that there were others further into the jungle who could teach us more. We talked among ourselves and asked . . . Where were these brothers working, and where could we find them?" (Earle and Simonelli 2005: 87).

A Word of God Community Embraces Zapatismo

A group from Cerro Verde made three trips before the people in the jungle would talk to them—Zapatistas on their guard against strangers. They decided to form their own Zapatista group within the Cerro Verde community. "For the families of Cerro Verde who ultimately made the decision to enter into [the] resistance, it was a sign that there was another way to confront the endless poverty and hunger, that they were not powerless. The Zapatistas had taken up the armed struggle, and without doing so themselves, the families could become part of a united front and provide a base of support. Their assistance to the movement began in the form of peaceful civil resistance to government policies and programs, those age-old forms of pork-barrel handouts that served to foster division and envy, one PRI program after another that served to create dependency and make a sham of democratic practice by buying votes in exchange for aid" (Earle and Simonelli 2005: 88).

This was a Zapatista group within a mixed community, not the whole community as in San Emiliano. They made a formal declaration of their intention to the ejido committee of Cerro Verde, saying that they would continue to cooperate in all local agricultural matters and maintenance of community trails, but not in health, education, or any other project that involved PRI government programs. One of Antonio's daughters remarked about the role of Zapatista women: "In the small pueblos we have different areas of work—health care, education, representatives for the Church. And we share Catechism with the children, and share our work with the children. This is what we have at one level of the group. Some of our women do health care. Since 1995 we became integrated with the others in health care [municipio level]. We became more capacitated to help our own families. This came about by necessity. We looked for people to help us become capable" (Earle and Simonelli 2005: 91). Later, the municipio group that included Cerro Verde Zapatista community joined with other Zapatista municipios to form the autonomous region, Tierra y Libertad (Land and Liberty), with its administrative center in the community of La Realidad.

Counterinsurgency and Defections

As part of the counterinsurgency, in 1995 the army occupied Cerro Verde because of the Zapatista presence. They set up programs of social services to counter those of the Zapatistas. Anyone who did not use the army's services was considered a Zapatista. Antonio's daughter continues, "There were police, Federal troops, and people were being asked to watch us. It was partly because we were in resistance, and partly because we are Catholic and there are a lot of evangelical church groups here. The village is very divided, and it has gotten worse since the struggle began in 1994. Everyone in the village was Catholic before the struggle. The sects divide the people; the evangelicals receive food and money from the government.... And we refused to go over to the side of the government, to eat government food, because it is tainted. We needed our own strength, and we kept strong by working together.... And that's when we knew we had to look for our own authorities, to coordinate us, to organize us, and to help us to educate ourselves. Our enemies made us strong" (Earle and Simonelli 2005: 91–92).

In April and May of 1998 the army and the state governor renewed the campaign to destroy the autonomous zones. Using any pretext it could find, the army again occupied Zapatista communities, destroying anything of economic value they could find (Earle and Simonelli 2005: 104–6). In 2000, men affiliated with the government's political party (PRI) attempted to force the Zapatista groups to join the PRI or be expelled from the local ejidos. In Cerro Verde, these were the original founders of the ejido or their descendants. The given reason was their refusal to work on government health and other projects. Therefore, it was argued, they were no longer meeting their ejido obligations. After a long legal battle and economic problems, a federal ejidal judge ruled against the expulsion of the Zapatista group (147–53).

In the midst of all this turmoil, the Zapatista community continued with their Word of God meetings and services. "Many of them headed up the hill to the tiny Catholic church for a meeting to hear another version of the word of God. Cerro's religious background remained a critical underpinning to its resistance, and they remained practicing Catholics. . . . The seeds of liberation theology were sown in the communities. They had been left to fend for themselves, seeing a visiting nun perhaps once a month. The sacraments reached them twice a year. They carried on with Bible readings interpreted in the face of struggle and conflict by trained catechists. The church was truly autochthonous. But it seemed easy to understand the growth of other religious sects in the area. These were not dependent for an important part of their ritual on the availability of ordained clergy" (Earle and Simonelli 2005: 200). In 2003, the community joined with 22,000 Zapatista supporters during a peaceful march in San Cristóbal commemorating the ninth anniversary of the 1994 uprising (226).

But the hard work to maintain their independence and the continual opposition of the rest of the Cerro Verde community began to wear down the Zapatistas. "Half the people who entered the Movement a decade ago in Cerro Verde left, pressured by others, facilitated by religious conversion, or in some cases bolting over the issue to drink and its aftermath. Don Antonio lost one of his sons to the government side over soccer because it placed his son in the company of non-Zapatistas, whose habit was to drink after a game, to celebrate if they won, or to forget if they lost" (Earle and Simonelli 2005: 283). And as Alma [Antonio's daughter] pointed out, "concerning all of this suffering and persecution of our community, envy and gossip is the worst thing we faced within the larger community. But at the same time, it fills us with bravery, motivation to persist" (283).

Zapatismo and Liberation Catholics in San Pedro Chenalhó

After the uprising, a Zapatista autonomous zone was declared in the San Pedro Chenalhó municipio, with its center in Polhó (see the map in Burguete Cal y Mayor 2003: 192, no. 25 on the map). As word of the Zapatista uprising and the reasons for it became known, a Word of God community in San Pedro became a Zapatista support group that included some Evangelicals. Among its members was a Catholic couple, Domingo and Antonia. Domingo worked long hours as a Zapatista representative. Antonia recalled the beginnings of a Zapatista support community. "After a few months in 1994, we began to learn that the Zapatistas were indigenous people struggling for justice and in support of the poor. Given that we, too, were poor we began to make connections with them. Support groups began to form in Chenalhó. Many Catholics joined the Zapatista support base, but some didn't want to. The priests and nuns who came to our community to give courses said that what the Zapatistas were doing wasn't a just struggle. Some Catechists listened to them, but later many Catechists began to join the bases. That's when a split began among the Catholics, between those who wanted the Zapatistas and others who didn't want them" (Eber 2011: 70).

Antonia compares Zapatismo to the Exodus. "The Zapatistas are liberating us now the same way that Moses liberated the Israelites in the past. Although the Zapatistas rose up in arms, it was like calling out with a trumpet so that the government would hear us. Using only our voices, the government didn't pay attention to us. That's why the Zapatistas rose up in arms. But not all the Zapatistas use arms. We in the bases support the army, but we don't use arms" (Eber 2011: 70).

Antonia explained how Marcos had advanced the quest for social justice over the previous efforts of diocesan-inspired organizations. "The government began to listen a little when Marcos arrived and the war began. Before Marcos and the Zapatistas we were always standing at the door knocking, but we couldn't open the door. It was as if we didn't have the key. But it seems that Marcos had the key. When he opened the door many people went walking with him" (Eber 2011: 72).

"Knocking at the Door"

This phrase of Antonia's is referring to efforts prior to the Zapatista uprising by a number of human rights organizations with close ties to the liberation movement of the diocese and its biblical roots.

Xi' Nich' (The Ants)

This human rights group emerged from the organization of a protest march in 1992 (Floyd 1997: 197–201; Kovic 2004: 64–65). In December of 1991, the inhabitants of Palenque seized its town hall in protest over their treatment there by state officials (Harvey 1998: 194–95). Their demands were (1) the right to select their own officials; (2) the employment of Mayas to translate between Spanish and the three Maya languages used in the town halls of Palenque and Ocosingo; (3) an end to the violent treatment and detention of protestors; and (4) a program of reforestation to replace the trees lost to clear-cutting by ladino timber interests.

The group was violently evicted by state police, who wounded many, including women and children. One hundred three were detained, including a priest pastoral worker. Ninety-four were freed after a few days, but the nine who remained in detention were beaten, tortured, and confined in the state prison in Tuxtla.

The protests at the town hall continued for two months and were met by more violence. When local efforts were unsuccessful, the protestors decided to take their grievances to the president and the federal government. They organized a forty-eight day, seven-hundred-mile "March for Peace and Indigenous Rights," from Palenque to Mexico City. Over four hundred people from 118 communities made the trip, including Bishop Ruíz and a number of pastoral workers. It began on March 7 and ended on April 25. Federal authorities, embarrassed at the group's presence, met them at the outskirts of Mexico City and agreed to their demands. The protestors then went to the shrine of Guadalupe, where the bishop offered a mass of thanksgiving. Returning home, the group continued as a human rights organization in the Palenque area. Although the march contributed to a growing awareness of the Maya demands, many of the government promises were never fulfilled. The door remained shut.

Pueblo Creyente (The Believing Community)

This organization began as an informal group among some pastoral workers who wanted to coordinate efforts dealing with the problems of the poor at the diocesan level, rather than limiting themselves to local communities. They held bimonthly meetings of catechists and community representatives, which included biblical readings and study (Kovic 2005: 164–76, 1–3, 204). A conflict in Simojovel in the northern highlands was instrumental in expanding the role of the group. It arose over the government's expulsion of Maya from their homes and lands sought by wealthy ladinos (Harvey 1998: 153, 172; Tavanti 2003: 131). The priest in the region, Padre Joel Padrón, protested the government's action. As a result, he was charged with robbery, conspiracy, damage to property, having an illegal weapon, and inciting the peasants, was arrested, and then jailed in the state capital, Tuxtla Gutiérrez. Without any urging from diocesan authorities, five hundred Maya began an eighty-mile march from Simojovel to protest the government's arbitrary action in violation of the priest's human rights. Along the way, other Maya communities joined the march, so that eighteen thousand Maya arrived in Tuxtla. They remained for three days of peaceful protest, prayer, and penance. Although the government later freed the priest, the government's treatment of the marchers showed that the door was still shut.

Marches such as these were frequently used by the human rights groups to protest government actions. Antonia participated in these marches. "In the time of the believers we demonstrated and went on pilgrimage to Tuxtla. I walked with [my daughter] Rosalva on my back in the heat. But the government didn't pay attention to us. It treated us like animals, like monkeys, like flies that fill up the streets. It never listened to us because it knew that the believers didn't have guns and it wasn't afraid of us" (Eber 2011: 74). This was keeping the door shut.

Realizing their ability to organize as a result of the march, understanding that they could pressure the government when united, drawing on their biblical reflections in the Word of God communities, seeing that a structure was needed for the discussion and election of representatives, in 1991 the group became a formal organization dedicated to social justice and human rights. Its goals, "working for life, equality, unity, social change—are explicitly political but come from their faith" (Kovic 2005: 170). The group was nonpartisan and experienced explosive growth. Protest marches with up to twenty thousand people participating, often against government policies, were among its most important activities (173).

Pueblo Creyente also corrected a glaring deficiency in the organization of the diocese. Theoretically, "accompaniment" by pastoral workers meant that they would be alongside the Maya in their struggle against oppression. But being better educated, there was always the temptation to be paternalistic, to perceive their pastoral role as teacher to student, to be the voice of the voiceless. The most glaring manifestation of this tendency was the structure of the diocesan assemblies. They were composed of pastoral workers, ladino advisors, and the bishop, but no Maya representatives. The Maya had indirect input, through their advice to the pastoral workers, but no presence or direct input. Some pastoral workers had complained for years about this contradiction in a diocesan policy that said it opted for Maya self-determination. The organizational ability and strength shown by Pueblo Creyente in their spontaneous support of Padre Padrón swept away all misgivings. The diocese asked the communities involved in Pueblo Creyente to elect representatives who would become members of the diocesan assemblies and an integral part of the decision-making of the diocese. In Kovic's apt phrase, diocesan accompaniment, instead of being the voice of the voiceless, now became "the ear of an earless church" (2005: 164–65, 205).

Las Abejas (The Bees)

Originally, this was the Pueblo Creyente group in the municipio of San Pedro Chenalhó. It had participated in the protest marches over the imprisonment of Father Padrón. In 1992, it formed its own organization to work only within the municipio. It was named Las Abejas, the Bees, to signify their work in unison (Tavanti 2003). Their goals were social justice and defense of human rights. But they rejected armed rebellion as the means to obtain these goals. This put them in a no-man's-land between the government and the Zapatistas.

Abejas originated against the background of a family land dispute (Chanteau 1999: 55–58; Moksnes 2003: 111–13). At the death of his father, a son claimed the family land according to the Maya custom that excluded his two sisters because they were women. But they wanted a share of the land for their three sons. A bitter dispute ensued. A community meeting was called to resolve the issue. It was decided to divide the land equally between the son of the deceased and the sons of his two daughters. In retaliation, three women related to the sisters were robbed and raped in their homes. The same day, the three grandsons of the deceased were ambushed and shot, and one died. A catechist and a Presbyterian found the victims, and with the assistance of three others, carried them to a road, where they were transported to San Cristóbal for medical treatment. For their efforts, the five were arrested for the murder and imprisoned in San Cristóbal. The arrest was an attempt by the mayor from the government party (PRI) to discredit the Catholic group, which was anti-PRI. Incensed Pedranos from twenty-two communities rallied to the five's defense. They mobilized 1,500 people to assemble in San Cristóbal. For five consecutive days they marched to the prison where the men were held. After a few weeks, the men were released. This was the political factor that occasioned the foundation of Abejas.

But there was also a biblical factor (worldview) that contributed to its origin. According to one of its founders,

> Thanks to the pastoral work of the church we have learned the story of the people of Israel when they were living in slavery in Egypt as described in the Old Testament. We have also learned that these stories of liberation and salvation are alive in today's stories of peoples searching for freedom and justice. We have also learned that God cares for our cultural traditions that are essential to really understand the vitality of today's stories of salvation and liberation. Indigenous theology has been particularly helpful to see ourselves with our own

Old Testament stories. It was through the different approach of indigenous theology that we learned how to go back to our cultural traditions and refer to it for our spiritual life and moral decisions. Our ancestors did not have the Catholic church but they knew how to respect the elders and therefore there was no robbery, no bad words and they knew how to pray. Indigenous theology does not depend on the directives of the government of the *mestizos*. (Tavanti 2003: 119)

The founding of Abejas illustrates the dialectic between political factors and worldview in accounting for social action. The indigenous theology movement will be discussed in chapter 29. As of December 1999, the organization had 5,083 members from twenty-seven communities of the municipio. Although the majority of its members were from Word of God communities, it was an ecumenical organization, as 10 percent of the membership were traditionalists, 5 percent were Presbyterians, and there was a small group of Pentecostals (Tavanti 2003: 19, 121). Each of the twenty-seven communities elected a representative. Governance was by a six-member board of directors who also served as members of four standing committees. The first three committees were a judicial committee to mediate conflicts within communities or the municipio; a committee to investigate cases of those forcibly displaced from their homes and lands for possible recuperation; and an agrarian committee to handle land conflicts. These committees used traditional Maya methods of conflict resolution rather than relying on Mexican laws drawn up with no understanding of indigenous culture (Collier 1973: 91–108). The fourth committee assisted the Abejas families who were the victims of the Acteal massacre, to be described below. The Acteal victims were the martyrs of the organization, and showed that nonviolent protests do not guarantee immunity from the effects of violence. The founding fathers of the organization remained as advisors to the board of directors, an echo of the role of the principales in the traditional system.

But a number of Abejas communities fissioned when some of their members became Zapatistas. Antonia describes the split that occurred in her Word of God community when some became a support group for the Zapatistas.

With this [the uprising], the religious ones in Chenalhó [Word of God Catholics] divided amongst themselves. Before this we were united just by being in the Word of God. But after the uprising, we began to divide, because people didn't know which way to go, what was good, what was bad. We were confused. Although we had learned about

The Word of God, the Zapatistas and the Bees began to criticize each other. . . . The two groups didn't understand each other . . . I don't know how many years it took, but later the Bees understood that what the Zapatistas were doing came from the Word of God, too." (Eber 2011: 70)

Abejas Caught in the Middle: The Acteal Case

After the agreement in 1994 to respect Maya cultural rights, the government initiated its two-prong counterinsurgency. Defection from the Zapatistas and reintegration into Mexican civil society usually meant backing the PRI, the government's political party, in return for participating in government-assistance programs. Abejas agreed with the Zapatista's refusal to accept public services from the government when services would only be provided if the communities joined the PRI. They also agreed with the Zapatistas' demands for social justice. But they rejected the Zapatista use of violence to attain it, and the setting up of parallel Zapatista governments and systems of public services. But not bothering with distinctions, the government considered Abejas communities to be Zapatistas. As part of the counterinsurgency, it launched a campaign of retribution, evicting Abejas members from their communities and placing them in refugee camps.

This campaign came to a climax on December 27, 1997, when government-backed paramilitaries attacked Abejas refugees in the Acteal refugee camp. For five hours, a group of about sixty men, many armed with AK-47s, poured bullets into the section of the camp to which they had been assigned, shooting anyone they saw. When they finished, there were forty-five dead—fifteen children, nine men, and twenty-one women, five of whom were pregnant. Among the survivors, twenty-five were wounded, nine of them seriously. A number had survived by hiding behind or beneath the bodies of the dead. (For the bloody details, see Tavanti 2003: 9–14, and the sources cited there.)

Father Miguel Chanteau, the pastor of San Pedro Chenalhó, preached at a commemorative mass for the victims of the massacre. "Brothers and Sisters, consider ourselves as the community of God on our way to the Promised Land. For we are on our way to a place where we will have enough land to feed our families. But suddenly we pass from Exodus to Exile. But remember that during the Exile, God raised up some prophets to keep alive the hope of the community that one day they would return to their home land. Therefore I commission you, brothers and sisters catechists, to be today the prophets of your communities sustaining their hopes that one day you also will return to live in peace in your villages in justice and dignity" (Chanteau 1999: 105).

The massacre galvanized the Abejas organization and resulted in a huge increase in its membership to more than five thousand adherents. Biblical theology was called upon to sustain communities psychologically devastated by the Acteal murders. The Abejas interpretation of the tragedy saw the dead as martyrs for the cause of social justice, as contemporary biblical faithful who by their living the word of God, suffering, and dying, would ultimately bring about the liberation of their people (Tavanti 2003: 91–95).

"Opening the Door"

Three different human rights groups had been unable to make substantial progress against governmental injustice. Antonia saw these groups as implementing Bishop Ruíz's insufficient approach to the problem of social justice in Chiapas. The government was simply ignoring his efforts and those of Liberation Catholics. Antonia saw Marcos and his strategy of armed rebellion as having been successful in breaking through the government's indifference, disrespect, and continual enforcement of injustice. Here, she compares it to cutting through a log. "When I joined the Word of God it was my way to learn about injustices. It was the light on the path, so I could see where to walk. We learned how things are and what we can do in the future. But on our path we came to a big log, the government and powerful people, that blocked our way. We couldn't go any further until the Zapatistas came and cut the log" (Eber 2011: 72). Antonia sees the Zapatistas in biblical terms. "God isn't angry at those who defend their rights. It's like David did in the time of Jehovah. Do you know that there was a prophet David who defended his rights? There was this soldier named Goliath. I don't know in what text of the Latin American Bible it tells about this, but David confronted the government soldier. David was poor and little, a soldier of the poor people. The same thing is happening right now. We are acting like David. What else can we do?" (80).

But as time went on, Antonia became discouraged when many Zapatista communities were barely able to survive or defected as a result of the counterinsurgency.

> In the beginning I always participated in the meetings because I wanted to hear what the Zapatistas had to say. Although I was very excited at first, later as the years passed and they were saying, "We're going to win! We're going to win a better life!" I didn't see any

triumph. I began to think, "Ah, the triumph won't come soon. All we can do is to struggle and struggle more and not give up."
We struggle for our grandchildren, the children of the future, not for ourselves. Although I have seen that many have died, I see no gains. I could die in a week, or in a few months, so it's better that I not focus on triumph. It's better just to struggle so that something might change in the future. I no longer think that in a few months or years we're going to win, that we will have, as stated in the Zapatista demands, decent housing, schools, and other things. The government made an agreement in San Andres with the Zapatistas but they didn't fulfill the agreement. Who knows when we will see what the Zapatistas are demanding? That's why I don't think about triumph right now. I just want to continue to struggle. Change takes a long time. The government is only beginning to open its eyes and its ears. That's why it's necessary to remain calm. (Eber 2011: 72)

An Evolving Marcos

Subcomandante Marcos emerges as a pivotal figure in the Zapatista insurgency and, later, "the resistance." Knowing something about him throws some light on these key events involving the Maya of Chiapas. He has remained a controversial figure, about whom there had been much speculation. However, this much is clear. He came out of a Catholic background, became a doctrinaire Marxist, and as a result of his experience with the Maya, developed a blend of Marxism and Maya philosophy. Marcos says of his evolving worldview, "It's a process . . . you begin to take steps—first becoming interested in a situation, then understanding that there is injustice, then trying to understand the roots of this injustice. . . . You begin by helping out in small ways, taking logical steps" (Henck 2007: 23). This has similarities to the evolving worldviews of some Catholic pastoral workers, which will be discussed in Parts VII and VIII.

Marcos divides his logical steps into three stages. "There are three Marcoses: Marcos of the past who has a past, the Marcos of the mountains before the first of January [1994], and post January 1 Marcos" (Henck 2007: 9). The past of Marcos, the nom de guerre of Rafael Sebastián Guillén Vicente, began with his birth, on June 19, 1957, in Tampico. In 1963, he began school at the Jesuit-run Colégio Felix de Jesús Rougier, which his three siblings also attended; in 1969, he began his secondary studies at the Jesuit-run Instituto Cultural de Tampico; in 1973, he entered college at the same institution; in 1977, he began advanced studies in philosophy at the Universidad Nacional Autónoma de México (UNAM) in Mexico City; in 1979, he began teaching Marxism at the Universidad Autónoma Metropolitana (UAM) in the same city, probably in the same year he became a member of the FLN; in 1980, he finished his thesis at UNAM; and in 1984, the FLN sent him to Chiapas with the rank of captain in the military branch of the organization, the EZLN (Henck 2007: passim).

His background in Catholicism consisted of his family's influence and his education in Jesuit-run institutions from grade school to his graduation from college at the age of twenty (Henck 2007: 1–28). During these formative years, little is known about the role of religion in his consciousness. But his fourteen years of education in Jesuit schools must have had some kind of influence. An important question is: What type of Catholicism was he exposed to in the Jesuit institutions? Given the pronounced conservatism of the Mexican church and of many Mexican Jesuits of that period, and given the dates of his formative years, he was probably exposed to Tridentine Catholicism. Although some have speculated that he was exposed to liberation theology, Medellín and the liberation movement did not begin to take root until the 1970s, and never did in many dioceses. In 1957, the year Rafael was born, Castro and Che Guevara were initiating the contemporary Latin America culture of revolution. Rafael was a good student, inquisitive about the world about him. By his own testimony, he was strongly attracted during these years to questions of social justice. He mentions reading newspapers. The logical progression led to his admiration for the Latin American revolutionary movement. Early on, Che Guevara became one of his heroes. The 1968 student massacre at Tlatelolco took place when he was eleven. When he was twelve, Don Quixote became one of his heroes. He took a copy of the book with him to Chiapas (21). In contrast, the passivity of Tridentine Catholicism, its emphasis on ritual and belief, and its silence about social order, would indicate that Rafael did not find any meaning in it. He told Le Bot (1997: 53), "The last religious service I attended was when I took my First Communion. I was eight years old."

His family was Catholic, but not the Latin "muy catolico" (very Catholic) type. His parents valued learning and encouraged their children to read literature and Mexican history, with its strong revolutionary strain. They had a strong sense of morality. "My parents taught us that, whatever path we chose, we should always choose *el camino de la verdad*—the path of truth—no matter how hard it might be, whatever it might cost. That we shouldn't value life over the truth. That it was better to lose your life than to lose truth.... We were taught that all human beings had rights, and it was our duty to fight against injustice" (Henck 2007: 19). This morality was not based on sectarian religion, Catholic or otherwise. "We [Rafael's family] were very independent of religion. It was a very humanist tradition, and not attached to any particular line" (21).

In summary, it appears that during his formative years, given the type of Catholicism he was exposed to, given his family's moral stance, given his interest in social justice, and given the developing Latin American revolutionary movement, the young Rafael just drifted away from Catholicism without any strong rejection or denunciation. It had no meaning for him. It was an easy transition, scarcely an evolution. During his Mexico City years, while attending and teaching at the very political and Marxian-influenced universities, he became a doctrinaire Marxist and a member of the FLN (Henck 2007: 29–52). The FLN was the structural implementation of his Marxian worldview—a natural progression given his concern for social justice and a missionary zeal to implement it. This period lasted ten years.

In 1984, the FLN sent him to Chiapas to organize a military force among the Maya that would lead to the establishment of a political order based on democratic elections. He was plunged into Maya culture, so different from the Mexican culture he knew in Tampico and Mexico City. As he encountered the Maya worldview, it confounded and upset his doctrinaire Marxism and he began to evolve. This was his mountain period. With the appearance of Viejo Antonio and his teachings, Marcos began to absorb aspects of the Maya worldview and came to respect it. He realized its importance and the developing Catholic openness to it in the communities he was attempting to convert to Marxism. He realized that doctrinaire Marxism was not going to win converts, that he had to develop an ideology that took Mayan culture into consideration. It was a conversion experience. Marcos inserted a humanistic vocabulary into his Zapatista discourses, speaking of love and other such "soft" topics, rather than dwelling on the constant refrains of class warfare. Attempting to summarize this stage of Marcos's evolution, Henck (2007: 5) is "prepared to accept (with at least 5,000 indigenous who follow him) Marcos's claim that the cross-pollination of Marxism with indigenous ideology and social structures produced a new thought system."

Marcos has said that the reason the Zapatistas refrained from any mention of religion was the fear that the Maya would see the Zapatistas as just another religious group trying to missionize them. If they did, the war for liberation would become a religious war over religious belief systems (Le Bot 1997: 326). Marcos began to make distinctions between the types of Catholicism he encountered in Chiapas. "We cannot deny that the moral authority of the progressive church [liberation movement] in the highlands, in the northern sector and in the Lacandón is very great and has much to do with the moral authority of Samuel Ruíz. The fact that his words are listened to—I don't say agreed with, but listened to—in the communities, in many of them, is very significant. There has been formed a group of [pastoral] workers that have very close contact with the communities, almost as close as the Zapatistas. This enables them to know about problems, disagreements, frictions, differences, concerns in the communities that other churches are incapable of because they are more distant from the people. This group of the diocese, whether they are in the communities or not since they come and go, have very close contact with the communities. And what is great about it, is that they are there to be of service to the communities!" (Henck 2007: 328). Marcos had a tendency to take credit for much of the preparatory work done by the diocese on which he built the EZLN. Here he shows some respect. Indeed, he proposed Bishop Ruíz to be the mediator of the dialogues with the government after the uprising, and the government agreed.

Summary of Part VI

The last five chapters have examined the role of the Maya in the worldview of Marxist insurgents and of government counterinsurgencies in Guatemala and Chiapas. Initially, Marxists viewed the Maya as a rural underclass whom they would use to initiate an armed struggle to overthrow the national government in Guatemala or to institute a democratic political process in Mexico. Using local problems created by the Maya crisis as their springboard, using the biblically inspired Maya social consciousness and resolve to confront systemic injustice, using their frustration with trying to obtain social justice by nonviolent means, urban ladino Marxists went to Maya areas and, with the assistance of some pastoral workers and catechists, were able to insert armed rebellion as the only realistic "act" to attain social justice. They enlisted Maya Catholics as supporters of or as combatants in military units. These units then initiated failed insurgencies against government forces, in which they were decisively defeated. In the aftermath, the Guatemalan and Mexican armies militarized insurgent Maya zones, maintaining tight control over the population. As a result, in both countries, Action Catholicism went into decline.

VII

The Impact of the Maya Crisis on the Worldviews of Pastoral Workers

The contact of Catholic non-Maya pastoral workers with the Maya was not always unidirectional in its impact. Some pastoral workers were raised as and remained Tridentine Catholics as they carried out their ritual duties in Maya parishes. But for others, the Maya experience triggered their evolution to Catholic Action, and some continued evolving to becoming Liberation Catholics. This evolution may have been accompanied by a second type of evolution, a breaking out of the ethnocentrism associated with Western Catholic theology to an appreciation of a non-Western Maya theology as a basis for a Maya Christianity.

The pastoral workers who remained Tridentine tended to limit themselves to sacramental duties and religious instruction and to abstain from involvement in non-religious community structures and, consequently, from any participation in the insurgencies. They saw the sacraments as contributing to personal reform that would help overcome alcoholism and some of the degenerate aspects of the traditional Maya ritual system. They perceived their efforts as strengthening the moral fiber of the community and in that way contributing to its efforts for social justice. Individual cases will be examined in chapter 23.

In addition to administering the sacraments, others insisted that the meaning of the sacraments needed to be shown by personal involvement in community activities that directly confronted the Maya crisis and the problems of social injustice. They worked with Maya catechists to empower the Maya with the aid of biblical reflection. They promoted health programs, or taught literacy, or collaborated in founding various types of economic cooperatives. As empowering, these activities involved political issues that posed some delicate questions as to the limits of the church's involvement. Individual histories will be described in chapter 24.

A third type of pastoral worker evolved from the preceding type. This type initiated or worked with the empowerment programs but found their efforts and those with whom they worked frustrated by the region's continual, systemic governmental corruption and repression. In spite of their efforts for social justice, such workers felt their efforts were useless in the face of an entrenched system of economic and political injustice enforced by violence. They saw that the only realistic remedy was to resort to armed violence themselves. When offered the opportunity to do so by radical Marxists, they either became guerrilla combatants or support personnel under the jurisdiction of the EGP commanders. This type will be the subject of chapter 25. The final chapter in Part VII examines the moral question posed by the resort to armed warfare.

22

Two Pastoral Workers Evolve

This chapter describes the Guatemalan experiences of two Americans of the Maryknoll congregation, shown in the introductory photograph. Their personal histories trace their evolutions through all three types of commitment.

The Dynamo Nun

Sister Marian Peter's autobiography is dedicated to Bishop Helder Cámara of Brazil, a model of Latin American Liberation bishops; to the guerrilla priest Camilo Torres; and to three leaders of the Guatemalan insurgency mentioned in chapter 17: Luis Turcios, César Montes, and Yon Sosa (Melville and Melville 1971). The dedication indicates the end-point of her evolution.

Figure 22.1. The dynamo nun in uniform.

Born in Mexico of American parents, Marian Peter's religious formation in Catholic schools and Maryknoll had been Tridentine. In 1953, Sister Marian arrived in Guatemala, and in spite of her desire to work with the poor, she was assigned to teach in a Maryknoll high school for upper-class girls in Guatemala City (Behrens 2004). The director of the school explained its purpose. "We were asked to establish in Guatemala a school where English would be taught so the daughters of the so-called 'better families' who would not be sent away to the States to study. Our purpose in founding and continuing the school is to educate apostolic Christians with a sense of social responsibility. It is hoped that a large percentage of our graduates will become active leaders in the community" (204). In 1956, Sister Marian founded a Girl Scout troop at the high school. One girl refused to join when informed she would have to make her own bed while on scouting trips.

During two vacation periods, the Sister worked among the Maya in Jacaltenango, Huehuetenango, learning the realities of the Maya crisis, so different from her experiences among the elite in Guatemala City. Before returning to Guatemala City, she told the pastor, "You've given me a new reason to return to the City and work there. You can encourage the Indians and help educate them. Yet, if the politicians and plantation owners who live down in the capital don't see the need and begin to help, the Indians will always be exploited and held down. I'm going to work for them from the other end . . . with the young girls who are the future wives of politicians and wealthy landowners" (Behrens 2004: 201). This was her movement to Catholic Action, and it set her agenda for the next thirteen years of working for social justice on behalf of the Guatemalan Maya.

In 1960, she helped found a teacher-training program that prepared students for school administration. Student teachers did internships in a school for the poor that was opened next to the high school, and in Maryknoll schools in Huehuetenango. One student in the program was Yolanda Colom, later an EGP guerrillera, who discovered the Maya crisis while doing her student teaching in Cuilco, Huehuetenango. Marian Peter also started a group to prepare catechists for teaching in the public schools of Guatemala City, and she herself began teaching in them and learning more about Guatemala's poverty. She also helped initiate an adult education program in a slum area of the city.

Expanding the Social Consciousness of Students

In 1962, a Venezuelan Jesuit visited Guatemala to give a series of lectures—Cursillos de Capacitación Social (Short Courses about Social Empowerment). Their purpose was to develop the social consciousness of students whose social position and opportunities for education would enable them to take political, economic, and social leadership roles in their countries. Hopefully, they would work to bring about a process of reforms and social changes based on a Christian ethic that at the same time would have erected a barrier against Communism (Porras Castejón 2009: 30). Sister Marian attended, then was invited to Venezuela to learn more about the Cursillos. She returned to Guatemala, and in 1962–63 helped arrange five Cursillos by the Jesuits for high school and university students. Most of the participants came from Catholic high schools for the elite. Later, in the town of Huehuetenango and in Maya communities of that department, the Cursillos were given by the students themselves.

Student Cursillos were conducted over an eight-day period. Cursillos conducted for Mayas were one-, two-, or five-day adaptations because of time constraints. A typical day would be scheduled in this way (Bonpane 1985: 31–32):

6:00—Rise, calisthenics, wash, dress.
7:00—Meditation, usually on a passage from the New Testament.
7:30—Breakfast.

8:30—Inquiry: Given my background, where do I belong in the overall Guatemalan social structure?

9:00 to 12:00—Three classes on topics about the Guatemalan reality. See below.

12:00—Group discussions.

12:30—Sports.

1:00—Lunch.

2:30—More Inquiry: For example, are demonstrations effective? Is faith compatible with revolution?

3:00 to 5:00—Two more classes on topics about the Guatemalan reality.

5:00—Required extemporaneous talks by participants on given situational topics. For example, as a leader of a student demonstration, what instructions do you give to demonstrators if the army opens fire?

6:30—Folk Mass oriented toward liberation. The sermon is a dialogue with all the participants.

7:00—Dinner.

8:00—Entertainment: songs, games, dances by participants.

9:00—Debate: for example, is armed struggle a necessity?

10:00—Review of the day and self-criticism of programs, of conduct of individuals, acknowledgements of lapses by those who have not fully participated.

10:30—Retire.

The topics covered in the classes were: The social problem; The family; Liberalism; Communism; Capitalism; The social doctrine of the church; The formation of a Christian social sense; Love; Marxism; Socialism; Agrarian reform; Revolution; Why must we change the structures of society?; The dignity of the human person; A Christian philosophy of life; Guatemalan social and political statistics; Social conditions in El Salvador, Nicaragua, Honduras, Costa Rica, and Panama; The common good; and Responsibility and the need for action. The classes were oriented toward a nonviolent Christian socialism, a middle-of-the-road philosophy between laissez-faire capitalism and communism. They drew heavily from the social encyclicals of Pope John XXIII: "Mater et Magistra," "Pacem in Terris," and "Populorum Progressio."

A former student describes a result of the Cursillos.

> With the inspiration, enthusiasm and support of a Maryknoll nun, Marian Peter, we founded El Centro de Capacitación Social (CEDECAS)

(The Center for Social Empowerment). We rented a house in the center of the city that we baptized with the name Cráter. This became a meeting place where young Christian socialists could develop. But it went beyond that. . . . By selling meals for twenty five centavos (cents), we began to develop relationships with public school students and those from non-elite high schools run by religious groups. For many, this opened up unknown worlds as Guatemala was divided into sectors, each one closed in on itself without knowing the others, but prejudging them. Fraternizing with these boys and girls, coming to know their worlds and points of view, we began to change our own views. Shortly afterward we went to work in Huehuetenango, to the indigenous communities where the Maryknoll priests had their missions. Coming to know the dignity and depth of the worldview of these people who lived in extreme poverty, who suffered abuse and profound disrespect, and in addition were the object of violence when any attempt was made to change the situation; all this and other questions they raised, resulted in a radical change in the lives of many of us. (Porras Castejón 2009: 36)

The author of this statement, Gustavo Porras, after repeated frustrations in bringing about peaceful social change, joined the EGP as a staff member of the National Directorate. He resigned over policy differences (chapter 26), and later became a close advisor to President Álvaro Arzú, playing a prominent role in arranging the Guatemala Peace Accords of 1996. In his memoir, he repeatedly comes back to the seminal role of the Cráter in his life (2009: 209, 231, 309). Ninety students from the Cráter began working in thirty villages in Huehuetenango and Quiché conducting Cursillos for Maya peasants (Bonpane 1985: 25).

Moving Beyond Catholic Action

One day, a group that included the nun, some of her high school students, a former university student who was a frequent visitor at the Cráter, and an unknown friend of his, took a hiking trip to the volcano Agua (Melville and Melville 1971). When they reached the top, the unknown friend revealed his identity—he was Luis Turcios, the leader of the FAR-EGP. The university student had gone to the eastern mountains, where he had become Turcios's bodyguard. Turcios and the nun had an extended discussion about the plight of Guatemala's poor. At that time, the Guatemalan army, the secret police, and U.S. military intelligence were blanketing Guatemala searching for him.

Turcios was from a Catholic background, had gone to Catholic schools, and had entered the novitiate of a teaching congregation, the Christian Brothers, intending to become a member. He stayed a very short time. "The false piety he had witnessed in the novitiate had disgusted him. His teachers couldn't explain to him what life was all about" (Melville and Melville 1971: 205). He appears to have been a victim of Tridentine Catholicism. At that point, he entered the military academy to prepare for a career as an army officer.

The encounter with Turcios expanded the nun's already growing social consciousness and the need to do something about the problem of Guatemala's desperate poverty. About this time, she met a Maryknoll priest working in the Western Highlands, Thomas Melville. She explained to him the work at the Cráter and the students' volunteer efforts in the Maya areas. His reaction: "I had no idea there were so many persons here in Guatemala City really interested in doing something about the plight of the peasants. I think myself that the guerrillas are the only ones who have the answer." The nun's reaction: "He had just voiced a growing conviction of my own that I hadn't dared as yet to admit to myself" (Melville and Melville 1971: 208).

Turcios was killed in an automobile accident in 1966. His place was taken by César Montes, who had been baptized a Catholic, become a Mormon, and by 1966 had given up on religion altogether. He had been a teacher, then began law studies only to abandon them to join the guerrillas in the Eastern Highlands. While helping the same former university-student-turned-bodyguard who had brought Turcios on the volcano hike move some books, he happened to meet the nun. They discussed the problems of Guatemala. These discussions continued when the nun periodically acted as chauffeur for Montes and his bodyguard during their clandestine visits to Guatemala City. They included a covert visit to Turcios's tomb, and another to the Mexican border on the first leg of Montes's trip to Cuba. At that time, Montes, as Turcios before him, was the most sought after man in the country, by Guatemalan security forces and U.S. intelligence units. Here, the rest of the nun's story becomes entwined with that of Thomas Melville, which will be discussed below.

The Nun's Motivation

From what depths did this young woman draw the courage to immerse herself in the problems of Guatemala's poverty that would initiate the evolution of her worldview? It was the Bible, especially its Christology.

During one of the Cursillos, the nun was leading a discussion of Matthew 25:31–45, which depicts the scenario of the Last Judgment when the just are told: "I was hungry and you fed me, thirsty and you gave me to drink, I was a stranger and you received me into your home, naked and you clothed me, I was sick and you took care of me, in prison and you visited me." The King is asked: "When did these events occur?" The King answers: "I tell you, indeed, whenever you did this for one of the least important of these brothers of mine, you did it for me."

Her meditation on this passage brought about a conversion experience. "As we considered the words of Christ, a stronger meaning began to show through to me. My heart began to beat very fast. . . . Suddenly I understood. And my literal, geographical concept of the Kingdom of Heaven and the Lake of Eternal Fire fell away before the onset of a new revelation: The real heaven is a state of communion, of compassion, and loving concern, with all our brothers and sisters, all of them incarnations of the Living Christ; the real Hell is a state of alienation, a being out of communion with yourself and with other people and with God, distrustful, competitive, prejudiced, seeing others as things, sources of profit and gratification only. This was a completely new insight for me, and I was thrilled with its meaning. I saw Christ struggling to be born in other faces" (Melville and Melville 1971: 149). The importance of this biblical passage for Action Catholicism was noted in earlier chapters.

An Australian couple who had worked in China wrote an article that gave her further insights to reconcile Guatemalan reality with her search for meaning. Part of its message was: "There is no need to add that they [Christian evangelists in China] came from corrupt societies where Christianity was very powerful—that is to say, very corrupt. . . . The result was a violently anti-Christian China, which is not as disastrous as it sounds. Anti-Christianity is not always anti-Christ. A society can have Christian features without knowing Christ. Certainly a pagan society can be more Christian than one which has betrayed Christ. In other words, a rejection of pseudo-Christianity is not necessarily a rejection of Christ himself" (Neale and Deidre Hunter 1966, in Melville and Melville 1971: 232).

Thomas Melville, and others, were trying to discourage the nun's meetings with the guerrillas that were becoming known within Maryknoll. He told her that she was showing her lack of patience to work inside the church to bring the Spirit of Christ to life. Her reply: "Christ is almost dead in the Church, and you know it better than I do. Besides you told me you thought we all might have to work with the guerrillas someday." Melville insisted, saying, "When you tie yourself up with these people, you'll be crucifying Christ all over again." Her reply: "You know who's crucifying Christ nowadays" (Melville and Melville 1971: 216). Christology as a part of biblical theology was providing the frame of reference for these Maryknollers attempting to find meaning in the fog of abject poverty, racism, structural injustice, and an archaic church. At this point, the nun's evolution joins with that of Thomas Melville.

A Tridentine Warrior Begins His Journey

Maryknoll priests were invited to staff the parishes in the predominantly Maya Department of Huehuetenango, and a few in Quetzaltenango. Thomas Melville has also written about his days in Guatemala, in the same book with Sister Marian, and dedicated to the same people (Melville and Melville 1971). Brought up in an Irish American neighborhood in the Boston area, he had many of the characteristics typical of this background—he was hard-working, street-smart, had a blunt oral style, a sympathy for the ordinary worker, and little patience with trivia.

He had arrived in Guatemala in the late 1950s after finishing his courses in Tridentine theology. His first assignment was the parish of San Miguel Acatán in Huehuetenango. He expressed his frustration trying to teach the Tridentine catechism:

> We had to try to teach the catechism in Spanish and to see that the people memorized the questions and answers. They might not know the meaning of the Spanish words, but they could say them and were thus assured of salvation. . . . One day a woman came in to take part in a baptism as a *madrina* (godmother). I insisted she learn the catechism questions: "You promised last time you'd learn the catechism. You didn't learn it. I'm not going to give baptism this time unless you do."
>
> She became angry. . . . She went to the mayor to complain. The mayor was Indian, and for the Indians, religion is a way of life inseparable from politics.
>
> The mayor called me over to his office and insisted I baptize the child. . . . What I failed to understand was that the mayor's responsibilities extended to all the concerns of his fellow Indians—as leader, he was as much a spiritual father as a political elder. . . . The mayor flatly demanded that I perform the baptism. Meanwhile the father of the child had gotten drunk. When he returned to the mayor's office

and found that I was still refusing to baptize the child, he shoved me in the chest. I knocked him down right there and stormed out. . . .

I was determined that these people were going to learn about Catholicism and that I was going to save their souls in spite of themselves, even if I had to knock people down or get killed myself in the process. I had the special mission of taking God's undiluted word to the heathen. They would hear me, and I would settle for nothing else. For I was convinced that most of the Indians, even those who took part in Catholic services, but who did not learn the teachings in the catechism, were in peril of everlasting hellfire. I thought I had no right to compromise on questions of faith and that I had to insist, before giving the Eucharist or before baptizing a baby, that all the participants—and most certainly the parents and godparents in baptism—study the catechism. (Melville and Melville 1971: 38–39)

Melville was completely unaware of the Maya perception of baptism as a ritual that makes infants less susceptible to sickness (Early 2006: 211–25). This was the reason behind the Maya insistence on the ritual.

Melville was transferred to San Juan Ixcoy, a very traditional town in Huehuetenango (Melville and Melville 1971: 52–62). During the long absence of a resident priest in the town, the church had come under the jurisdiction of community authorities for the use of local shamans and other members of the community. Melville mistakenly thought that the church was under his jurisdiction. His first problem was to gain control of the church. The shamans kept the keys, opening and closing the church in accord with their ceremonies, especially during the late night hours. They refused to surrender the keys to him and allowed Melville entrance only at the times acceptable to themselves. This enraged Melville, who broke the old lock, replaced it with a new one, and proceeded to set the schedule for the use of the church. This, in turn, enraged the shamans, who complained to the authorities. They intervened on behalf of the shamans, but Melville refused to surrender the keys.

Having gained control of the church, he found an oven in the middle of the nave where the shamans were accustomed to sacrifice chickens, spreading the blood on the floor as a feeding ritual to the Maya gods. Melville proceeded to destroy the oven and replace it with pews for a congregation that never used pews. He then proceeded to remove some of the many saint statues that crowded the interior of the church. This further enraged the community, who again complained to the authorities and were on the verge of physically assaulting the priest. All of this was climaxed by Melville's storming into a private home where one of the most sacred objects in the community was kept to receive offerings of candles and incense. This was the *caja real*, a box containing documents handed down from the community's ancestors. He committed the ultimate sacrilege by opening the box and removing the documents, to the horror of all present. Actually, they were old sacramental registers from the colonial period. The priest took the documents and retreated to his quarters for protection against an enraged mob. His imminent death was predicted as punishment for the sacrilegious act. However, the shaman leading the opposition to him died a short time later, and the community began to have second thoughts about the power of the padre. This was a Tridentine priest who believed that saving souls was not simply a matter of orthodox belief and baptism, but also required the destruction of any signs of contrary beliefs.

The community demanded Melville's removal, and the Maryknoll bishop readily complied. "I had been kicked out of San Juan Ixcoy because in my zeal to do what I thought we were supposed to do—to destroy the Indian's belief in their pagan gods—I had over reached myself. . . . I resented the criticism implicit in [the removal] . . . that I was too imprudent to be allowed to act as a genuine missioner, and so was now to become a sort of clerical caretaker" (Melville and Melville 1971: 92).

The Warrior Begins to Evolve

As his Tridentine worldview was being undermined, Melville was sent to Cabricán, a parish in the Department of Quetzaltenango. "My state of mind now permitted me to be open, vulnerable, to what I found in Cabricán. I no longer felt any compulsion to prove anything, either to the Indians or to myself. I listened in a different way to my parishioners now, for my reasons for listening had changed. And that listening changed me. I had set out, like Columbus, with one goal and, like him, had landed in another place, with a different set of conditions and with pressing questions about my original destination and purpose" (Melville and Melville 1971: 92).

Melville describes how he learned about the lives and beliefs of the Cabricán Maya, and how this in turn led to his admiration and love for them as a people and as personal friends, not as potential converts to Tridentine Catholicism. They exchanged views about God, with the result that "we began to realize that our God was the same—He creates us, wants us to live together and share our lives; he prepares our end. We discovered on both sides that we weren't too sure about our beginnings or the inevitable end. And although we could never be certain we actually understood each other's concepts, we seemed to agree that people should find themselves through relationships based on love and mutual respect. We said it in different ways, but the essential elements were the same: love and trust" (Melville and Melville 1971: 94). His ethnocentrism was being eroded.

Melville continued the displacement of his Tridentine worldview as he kept pondering on the meaning of his current experiences. "Why were these people, who had so little, so unselfish? Why were people like me, who owned so much, so careful of possessions—although Christ has promised our sharing will be rewarded a hundredfold? The contradictions struck me more and more deeply. I began to wonder and question. For perhaps the first time in my life, I began to think for myself" (Melville and Melville 1971: 94). Melville goes on to describe how his Tridentine worldview collapsed under the stimulus of this intense cross-cultural experience.

Evolving to Catholic Action

When he asked the people what else he could do for them beside reciting the prayers of the mass, baptism, and chants for the dead, they sought his assistance in purchasing a truck to carry limestone and wood. Cabricán had large limestone deposits from which the people made lime. At that time, they knew nothing about how to purchase a truck, prices, maintenance, etc. Melville agreed to help with the lengthy undertaking of introducing them to Guatemalan businessmen and enabling mutual understanding, given their very different worldviews (Melville and Melville 1971: 98–102). The cooperative in Santiago Atitlán had a similar experience buying a tractor. The truck was finally obtained. Melville then helped to formalize and develop the incipient lime cooperative into a vibrant organization. It later bought two wheat threshers and spun off three credit cooperatives. One managed to accumulate $3,000 in savings, and by 1966 had loaned $8,000 in small amounts to its members (163). Melville assisted in the formation of credit cooperatives in surrounding communities, and was instrumental in the formation of the National Federation of Credit Unions, serving as one of its elected officers. But he was not satisfied. "Though expansion was the watchword, we were still reaching no more than a small percentage of the people. Only with more land could we begin to effect any substantial change" (163). Land was available in remote areas. After previous involvement in helping to plan the Maryknoll colonization projects in Ixcán (Morrissey 1978: 151–57), in 1966 Melville, with fifteen Cabricán families as a trial group, initiated the Pope John XXIII colony at San Juan Acul on the

Figure 22.2. The truck of the Cabricán cooperative. Thomas Melville is leaning out of the window on the passenger side of the cab.

Pasión River in southern Petén. As a social activist, Melville had come a long way from his days as a Tridentine warrior in San Miguel Acatán and San Juan Ixcoy.

But Melville was becoming discouraged. Earlier, he and other priests interested in social action had met to discuss the plight of the people in their parishes. They issued a report that concentrated on internal migration and land problems. Their parishes involved about 75 percent of all migrants in the country. The report was submitted to the bishops of the four dioceses in which the priests worked. Soon, a Tridentine reply came back. The bishops accused the priests of mixing in politics when their real job was to preach the Gospel. "If we [the editorial 'We'] were sincere, we would preach to the landowners, and they, on hearing the message of charity and justice and peace, would put it into practice in their own good time" (Morrissey 1978: 226).

During a visit to Guatemala City, Melville met Sister Marian Peter accompanied by César Montes and Nestor Valle of the FAR-EGP. They questioned Melville about the colonization project in the Petén and congratulated him on his work, but pointed out that it would be limited to only a few thousand people at the most, a very small percentage of those in need. They said the government would use the project as a showpiece, as an excuse to cover up its repression of the many. The Maya would still be forced to supply cheap labor for the commercial plantations of the economic and military elite. They went on to explain that they saw the essence of their own work as education, that they were armed teachers. They claimed that if the government left them alone to educate the peasants about their rights, there would be no need for arms. They emphasized that the guerrillas were attempting to do the same as he was, but his was an unarmed version of limited reach in the midst of a large Maya population. The arguments of the guerrillas were upsetting Melville. "My own troubles with the hierarchy didn't help me resist the revolutionary approach" (Melville and Melville 1971: 220–23).

In the exchange with the guerrillas, one of them quoted a biblical passage about Christ. Melville resented it, pointing out that the guerrilla didn't believe in Christ. The guerrilla replied that he accepted Christ as a historical human being, that Christ was not Melville's personal property, that he belonged to every human regardless of religion. Again Melville and his identification with Christ was being challenged. "I had trouble with my own conception of Christ. I call myself Christian, but what does it mean? . . . For me, Christ had meant celibacy and reading the Psalms of the breviary every day. The bishop was His voice, 'Do it. Don't do it. Do this, don't do that. . . . My admiration for the revolutionaries was slowly growing. They got to me by their direct questions, questions that had churned in my gut but that I had tried to keep out of my consciousness" (Melville and Melville 1971: 220–23).

As Melville and Marian Peter plunged ahead with their social action, they sought an answer to the many social and personal confusions that had overtaken them. A pressing question was, What should they do in the face of the continuing oppression of the Maya? They turned to biblical theology for assistance. A Maya leader, probably a catechist, gave his own explanation of the biblical parable of the Good Samaritan, a *locus classicus* of the liberation movement. He took a hypothetical view of the parable and wondered about the outcome if the Samaritan had had a faster mule. If he had arrived earlier, he might have found the thieves beating up the man from Jericho. The question: Should he have waited until they finished the beatings so he could exercise charity toward the victim? Or should he have scolded the thieves, and told them and their victim to love each other? Or should he have attempted to rescue the victim before he was badly beaten? In this case, "rescue" could only mean "fight." This hypothetical use of the famous parable carried only one message—fight with the guerrillas (Melville and Melville 1971: 263).

Further Evolution: Liberation with Arms

At Melville's suggestion, representatives of a small group of Catholic social activists held a secret meeting (Melville and Melville 1971: 261–64). Its purpose was to decide what specific action they should undertake in order to bring about a meaningful revolution. Those present included Sister Marian Peter, Art Melville (Thomas's brother, also a Maryknoll priest in Huehuetenango), a nun from Art's parish, three students from the Cráter, a FAR-EGP representative, the leader of the Cabricán colony in the Petén, and Padre Luis Gurriarán from Quiché (see chapters 12 and 24). The group decided on a "Christian revolution based on an organization of self-defense, to reoccupy the lands that had been stolen from them [the Maya] by generations of white men" (263). Such a move was bound to bring military retaliation and initiate a combat situation.

They wished to retain a Christian identity; therefore, they would not join the FAR-EGP, but would collaborate with them. For the immediate future, they would organize a base of support among peasant leaders and prepare themselves for the ordeal. The time frame was quickly altered when they realized that if they did not go to the mountains soon, they would be unable to do so later, as news of the decision was bound to leak out. Peasant leaders in Huehuetenango and Quiché were asked about their interest in joining the revolution, and they readily agreed to do so. A later meeting was held in Escuintla in which the plans were discussed with a group of students. Twelve of them went to the Petén to find staging areas for military actions and hiding places for food and weapons. Without revealing their plans, Marian Peter discussed the Guatemalan situation with a Maryknoll priest who opposed any use of violence. Marian Peter insisted that revolution "was the only path left to achieve social justice" (Melville and Melville 1971: 267).

In the weeks following the Escuintla meeting, Luis Gurriarán developed doubts about the plan's feasibility and feared the consequences for foreign missionaries in the country. The government had threatened their expulsion if his own congregation did not order Luis to leave the country in 1965. He revealed the plan to a Maryknoll priest, who in turn notified the Maryknoll authorities and the U.S. ambassador. Melville and his brother, Sister Marian Peter, and several others from Maryknoll were expelled from the country. Some went to Mexico for four months, in an effort to keep their movement alive from there, but realizing its futility, they returned to the United States.

There is another aspect to the evolution of these pastoral workers. Father Melville and Sister Marian Peter fell in love during this period, became engaged, and married while in Mexico. They saw this as another aspect of their liberation, from a celibacy that had set them apart from the people with whom they were working—a hindrance to their identification with the poor (Melville and Melville 1971: 280–81).

In the United States, they became part of the Catonsville Nine. This group forced its way into a draft board in Catonsville, Maryland, seized its records, and burned them on the front lawn as they stood in a circle singing hymns and waiting for the arrival of the police. The Melvilles joined this group to publicize the tyranny in Guatemala. Others were demonstrating against the Vietnam War. Later, they obtained doctorates in anthropology and pursued academic and business careers. They continued their advocacy of human rights, especially in Thomas Melville's book, *Through a Glass Darkly* (2005), which seeks to draw lessons for the United States and the Catholic Church from the Guatemalan genocides.

A Later Account

The Melvilles felt betrayed by Father Luis Gurriarán's revealing their plans. Their joint book was published in 1971, only three years after the Escuintla meetings and still a dangerous time in Guatemala. Gurriarán's more-detailed account of his meetings with the Melvilles was not published until 2007 (Santos 2007: 82–88). Because of the Guatemalan students involved in the Melvilles' plans, their 1971 account may have omitted some details to protect the students.

Gurriarán says that in 1967, shortly after returning from his expulsion, he was visited by Sister Marian Peter and another Maryknoll nun who invited him to a meeting with the Melville brothers. He accepted, and at that meeting they told him about the Cráter and its relationship with the FAR-EGP. Melville invited Luis to visit the colonization project in the Petén. There Luis discovered it was located at an abandoned military training camp that had been used by the United States in preparation for the invasion of Cuba in 1961. Luis was told that a guerrilla cell was already active there, and that in conjunction with the Cráter, they were building a revolutionary movement.

Later, there was another meeting in Luis's parish with members of the FAR-EGP in attendance, then a third one at the Maryknoll high school, and a fourth one on a plantation near Escuintla. The plantation belonged to the father of a young woman from the Cráter who was also a FAR-EGP combatant. Those present were: Luis, with some Quiché cooperative leaders; some students from the Cráter; and members of FAR-EGP, without their usual face coverings. The FAR-EGP representatives presented their agenda. They claimed that conditions were ripe for an immediate uprising and takeover of the Guatemalan government. They said that many were involved. The center of operations would be the colonization project in the Petén. Melville would fly in arms from Cuba. The Cuban experience would be repeated and Guatemala would be free at last.

At first, both Luis and the cooperative members were excited by what they heard from the guerrillas. But as Luis began to think over the problems of involving the network of Quiché cooperatives, doubts began to arise. In his view, the FAR-EGP was a weak organization. It had been defeated earlier in Eastern Guatemala. He questioned its ability to obtain arms from Cuba. His question was, "Is it morally right, decent, and a normal way of acting to bring about the death of Christian leaders, cooperative officials who are being very effective in the work they are doing?" (Santos 2007: 85). Luis did not say anything to the Melvilles about his change of heart. After several weeks, he went to a Maryknoll priest and revealed what was taking place. Looking back on the episode, he summarized his position: "All of it was an illusion, a determined but immature plan disconnected from reality" (87).

Summary

This chapter has traced the personal evolutions in the worldviews of two pastoral workers, from Tridentine Catholicism to Catholic Action to a version of Liberation Catholicism that opted for armed rebellion as the only solution to the Maya crisis. A similar journey lies behind the lives of some other Catholic pastoral workers. Blase Bonpane, an ex–Marine Corps officer, was a Maryknoll priest in Guatemala who supervised the students from the Cráter working in Quiché and Huehuetenango. He was also deported from the country with the Melvilles. He kept a rambling diary (Bonpane 1985) of the kaleidoscope of questions and reactions the experience triggered. It resulted in the undoing of his faith in the worldview of both the United States and the Catholic Church. The following chapters will describe some other pastoral workers and the impact of their cross-cultural experiences among the Maya.

23

Social Justice by Sacramental Observance

The theological formation of many priests in the Maya area had been Tridentine. It prepared them for teaching the Tridentine catechesis and administering the sacramental rituals. For some, it could involve many hours of travel by horse or foot over winding mountainous trails to reach scattered Maya communities. This type of pastoral worker would help individuals who asked for assistance. But going out into the community, becoming close to the community, helping to build needed community structures to empower them to confront their critical problems—these were not seen as the work of a priest. Instead, they were to concentrate on belief and ritual, following the Tridentine paradigm.

Counterinsurgency in El Quiché

During this period, three Sacred Heart priests in the Ixil area were murdered (Conferencia Episcopal de Guatemala 2007: 177–200). Two of them traveled throughout the Ixil area preaching and administering the sacraments. They were not known as social activists in the mold of Father Luis Gurriarán.

Padre Faustino Villanueva from Navarre, Spain, was living in Joyabaj after returning from Nicaragua, where he had been convalescing from health problems (Santos 2007: 220–21). He was apolitical and enjoyed playing card games. On the evening of July 10, 1980, two young men appeared at the parish house, asked for the priest, and when he appeared, put two bullets into him. They were later identified as having come from the police office in Santa Cruz. According to a former colleague, "He was a good priest, very popular, very much a friend of the people, but not a leader in the community, he was not a restless pastoral agent talking about social revolutions. On the contrary, he was a typical traditional priest humble, simple, loved good company, but not a man of initiatives or an organizer of people" (Remijnse 2002: 97). If he noticed people giving political messages in the church, he would ask them to leave. He was frequently heard to say, "Politics is OK, but outside the church." Chapter 19 described his reburial in Joyabaj.

Juan Alonso from Asturias, Spain, was the pastor in Lancetillo (Santos 2007: 231–32). He was an ardent anticommunist, a strident defender of the Tridentine Church, as well as of the army's omnipresence in the Maya area. He was an expert horseman and marksman who never took food with him on his visits to outlying communities, but hunted it along the way. Because of his extreme views, he clashed with the other priests of the diocese, spent a few years working in Indonesia, then returned to Guatemala, to a parish in the Petén. After Bishop Gerardi closed the Quiché diocese, the Vatican disapproved of it and named an administrator who recruited priests to return and administer the sacraments. Juan Alonso volunteered. Shortly after his return, he started out on a motorcycle for Cunén to celebrate mass, was allowed to pass through an army roadblock, and then was shot with three bullets in the back of the head by the soldiers.

The Oklahoma Effort Continues in Santiago Atitlán: A Murdered Priest

Father Stanley Rother (see introductory photograph) was the first of the second generation of pastoral workers that gradually replaced the original group beginning at the end of the 1960s. Born in 1935, ordained a priest in 1963, he arrived in Santiago in mid-1968 with no preparation for the cultural situation in which he was to become involved. To determine where

Father Rother should be placed within the various ongoing programs of Maya empowerment, he was evaluated by the first-generation Oklahoma priests. "A nice enough guy but not so bright, conservative and basically a 'Mass priest.' There were doubts about his ability to master Spanish well enough to be effective at the mission" (Monahan 1984: 3). He became the Santiago pastor in 1974, and soon was the only priest serving the parish. He remained until his murder by military operatives in 1981. Reviewing his thirteen years in Santiago also tells much of the later history of Catholicism in this Mayan community that was described earlier in chapters 8, 12, and 14.

Internal Dispute and Decline of Action Catholicism

The broad vision of the original Oklahoma effort began to dim in the late 1960s. Among the first generation of pastoral workers disagreement arose about the social distance between the Oklahoma group and the Atiteco community. The team had expanded to a sizable group of eight to ten people housed in a large building where they lived a modest American lifestyle that was vastly different from that of the struggling Atitecos. Atitecos were amazed at the differences. One of the Montessori teachers had a pet beagle that triggered the Atiteco comment, "They even feed their dog meat." Meat was a rare luxury for most Atitecos.

As a result, a division developed within the group between a culturally sensitive faction and those who were still working with an American mentality. The culturally sensitive faction thought that the group should split up into smaller units, living among the Atitecos and following a lifestyle closer to theirs. They rented quarters in Atiteco family compounds and became more involved in everyday Atiteco life. In accordance with the policy of developing Maya leadership, this faction formed a Guatemalan corporation that took over ownership of the radio station from the parish, with the Atiteco literacy teachers as the board of directors. The Oklahoma diocese continued financial support (Nick, n.d.).

All this was threatening to the American mentality faction, whose views were summed up by the editor of the Oklahoma diocesan newspaper: "The Catholic Church in Oklahoma would come to serve the people of Santiago Atitlán with priests, nuns, lay people, money and Yankee know-how. ... The Oklahomans found a gentle people living patiently, even passively, in great want" (Monahan 1984: 2). The paternalism and ethnocentrism behind this statement also expressed the mentality of Oklahoma diocesan authorities. They disapproved of the culturally sensitive faction, with

Father Rother in agreement. As a result, Father Stafford was ordered back to Oklahoma and the remainder of the faction eventually resigned.

The literacy program matured and thrived under Atiteco direction (Asociación La Voz de Atitlan 1998). For several years, the program retained the original curriculum, whose discussions were oriented toward traditional Atiteco conditions of the 1950s and 1960s. In the 1970s, they developed their own version of it, *Pensemos Juntos*. It was more adapted to their changing situation by focusing on Atiteco's problems arising from increased contact with the national society. They made new posters (cartelas) to focus the discussions (Konefal 2010: 45–48). This further development by many of the original teachers fulfilled the goals of the pastoral workers who had founded the radio station and its literacy program (214n55, 215n62). During the insurgency, some personnel of the radio station became supporters of ORPA and were killed by the army. But the station recovered, and its literacy work continued.

But the literacy program was the exception. To summarize a complicated history, the 1970s saw the decline or disappearance of the other social-action programs—the weaving, credit and agricultural cooperatives, the Montessori school, the orchestra. Difficulties arose with the second-generation American supervisors of the hospital and its medical program. It was taken over by Project Concern. The Oklahoma effort eventually dwindled to a single priest, Father Rother, and several Guatemalan nuns presiding over a Tridentine parish. As pastor, Father Rother concentrated on church services and the renovation of the church. Beyond this, he enjoyed driving what had been the cooperative tractor on former cooperative land now owned by the diocese. He loaned out parcels that helped to feed three hundred families. This was welcome charity, but not community development.

Father Rother had a reputation of being *bravo* (easy to anger). At times he quarreled with both Catholic Action and traditionalists when he felt his authority was threatened. He and other second-generation priests were involved in a number of disputes with various Atitecos and town organizations (Tarn 1997: 325–53). In 1975, a group of traditionalists wrote a letter to the Bishop of Sololá listing their complaints (324).

> 1) that Atitecos did not know under what conditions the American padres entered into Atitlán; [There was a twenty-five-year contract with the diocese of Sololá; the Atitecos were never consulted.]
> 2) that the padres did not carry out fiesta ritual as expected traditionally by the town and said mass in their own way "contrary to church custom"; [See Early 2006: 74, 219, for the importance of

fixed rituals and prayer formulas in traditional thought; some of the changes were introduced by Vatican II, and others were attempts at liturgical adaptation to Atiteco culture, initiated by the Fathers Carlin and Stafford.]

3) that much money had been sent from the United States which had not benefited the town in any way and that the Americans were merely drawing their salaries and doing nothing for the people; [This was their impression of the second-generation pastoral workers.]

4) that the Americans had set bad example in that four priests had left the priesthood and married; [Two were first-generation priests who returned to the States, served in parishes there, and then resigned. One had been recalled because of the dispute about social distance, which was a factor in his later resignation. The other two were second-generation priests who resigned while in Santiago.]

5) that the Americans were bringing about the entry of *los japes* (hippies). They should be replaced by Guatemalans or Spaniards. [There was a large American hippie colony living across the lake in Panajachel. They visited Atitlán and were known to be friendly with one of the second-generation priests, who was somewhat of a hippy himself and later married.]

While the list contains some misunderstandings, it does capture the change from the type of Catholicism promoted by the original pastoral workers in 1964 and the difference with the second generation. It was a gradual return to Tridentine Catholicism. "By 1979, Father Stanley Rother seemed to be following a policy of minimal interference, to be attending cofradía rituals when invited . . . , and to be continuing the program of enriching church ritual and liturgy. . . . We heard several times . . . that Father Rother spent too much time cloistered in his rectory or repairing and maintaining his equipment and not enough going around the village to visit his parishioners" (Tarn 1997: 322, 324). This description stands in contrast to remarks about the first-generation pastoral workers. "From the indigenous viewpoint, a *Catequista* official told us in 1979 that, originally, Atitecos had been very satisfied with 'these American priests who cared about poverty, health, education and so forth and who said hello to everyone whatever their religion might be.' They [the Atitecos] sided with the Americans against the local ladinos who were very jealous of them since many of their little trading deals had been upset by the American generosity" (315). This was especially the case when the cooperative store changed the traditional high price, low volume policy of rural merchants to low price, high volume marketing that greatly benefited the community.

As the insurgency-counterinsurgency unfolded, Tarn's evaluation of Father Rother was: "As far as Father Rother is concerned, our experience in 1979 was that he was an honest and dedicated man doing a difficult job to the very best of his abilities. On the other hand, it cannot be denied that he was also feisty and pugnacious, with a reputation for authoritarianism at times, for picking fights and for attacking tradition head-on when diplomacy and patience might have been better advised" (Tarn 1997: 323). This does not mean that he attacked Atiteco tradition, but he would get upset about minor matters with traditionalists, as well as with Catholic Action and other town factions.

Following the initiatives of Fathers Carlin and Stafford, he took an interest in incorporating Maya symbols into the Catholic liturgy. While renovating the church, he encouraged two local woodcarvers to incorporate traditional Maya symbols in the altarpiece, pulpit, and priest's chair (Christenson 2001). "As reconstructed by the Chavez brothers, the central altarpiece in the church . . . represents a translation of contemporary Tz'utujil theology into material form. . . . The artists included European Christian elements in the narrative panels at the suggestion of Father Rother and used them as the starting point to expound visually analogous Tz'utujil concepts. The Catholic components are therefore integral to the over all message of the monument" (212). The Catholic components were defined by the Maya worldview, contrary to Catholic orthodoxy (Early 2006, and chapter 2). This went far beyond Father Rother's understanding of the significance of the carvings.

Occupation by the Guatemalan Army

The volcano Atitlán, especially its southern and eastern flanks, became a region of operations for ORPA (Organización del Pueblo en Armas). Their command post was located on the upper reaches of the volcano. They carried out clandestine recruitment in Santiago, and occupied it briefly for open recruitment. Much of its support came from Atitecos. They were so dependable in difficult circumstances that the ORPA *comandante* took "Santiago" not only as his wartime alias, but later legally assumed it as his first name, out of respect and gratitude to them (Santa Cruz Mendoza 2004: 38, 58, 93, 149 and passim).

Figure 23.1. The shrine to Father Rother in the church of Santiago Atitlán.

As a result, the Guatemalan army built an encampment on the outskirts of the town and implemented the first phase of its counterinsurgency—kidnappings, tortures, and murders. Father Rother complained to the military. For him, the climax came when his chief catechist was kidnapped from the porch of the church as he looked on. This conservative priest complained bitterly to the military and roundly denounced them in public. The Atiteco community loved him for it, and this event became the defining moment of his life. For his own safety, he left the country, but returned after a few months. By the rules of counterinsurgency, his fate was sealed. On July 28, 1981, in his forty-sixth year, Father Rother was murdered in the parish rectory. This was one more example of the military carrying out their counterinsurgency by indiscriminately eliminating priests whom they considered to be Catholic Action types. It also served to intimidate the community. The Atitecos were outraged, as this was one more in a series of 500 to 2,000 murders of Atitecos during the counterinsurgency (Tarn 1997: 353–63).

Figure 23.2. The shrine's sculpted image of Father Rother.

Figure 23.3. The inscription at the foot of the shrine reads: "Stanley Francis 'Aplás' Rother, martyred priest." It continues with the dates of his birth, ordination, arrival in Santiago, his assassination, and mentions his thirteen-year residency in the parish. It closes with the verse from John 15:13: "There is no greater love than this: to give up one's life for one's friends."

Some traditionalists interpreted Rother's death in shamanic terms. They view the Catholic priest as a shaman, a person endowed with powers to communicate with the saint-gods of the Maya covenant through their knowledge of prayer formulas (Early 2006: 211–25). Shamans protect Maya communities by their rituals and prayers. Famous shamans can morph into various human beings and reappear at trying moments of Maya history. "Even today, Maya confraternity members who are otherwise wary of the intentions of Catholic authorities recall Father Rother with some fondness and suggest that he may have been one of the many manifestations of their great culture hero, Francisco (Pala's) Soquel" (Christenson 2001: 60).

After Father Rother's death, Atitecos insisted on removing his heart and preserving it in a jar, now buried beneath a shrine inside the church. The Oklahoma City diocese is seeking the canonization of Father Rother as a martyred saint of the Catholic Church. An appendix to this chapter discusses this effort.

Renewal of Catholic Action

In the late 1980s, the Oklahoma dioceses (Oklahoma City and Tulsa were now centers of separate dioceses) returned to Atitlán. Father John Vesey, an authoritarian Tridentine warrior, became the pastor for four turbulent months before being transferred to the neighboring parish of San Pedro La Laguna (Perera 1993: 191–95). There he became the center of a seven-year controversy (Early 2006: 49–56).

He was succeeded in Atitlán by Father Thomas McSherry, who revitalized Action Catholicism. He organized numerous projects of health, education, and housing; helped build a public library and a park, marking the spot where a number of Atitecos were murdered by the army during the counterinsurgency. He also helped build a small market where war widows and others could market their weavings (Carlson 1997: 37). He had a tense relationship with traditionalists and in 1997 moved his residence to Cerro de Oro, an aldea of the municipio. Suddenly, in July 2001, the bishop of Sololá terminated the agreement with the Oklahoma diocese and installed a Guatemalan priest as the pastor. Father McSherry returned to Oklahoma, ending the diocese's somewhat turbulent history of sending priests to Guatemala. The Oklahoma dioceses have continued financial support of the parish.

A Priest Defies the Liberation Movement

The San Cristóbal Diocese's embrace of the liberation movement was not accepted by all its priests. An American, Jim (Diego) Lockett, following service in the United States Marine Corps, decided to become a priest, a missionary, in order "to save souls" (Chojnacki 2004: 39–40, 246–63; Meyer 2000: 171–75). He had a brief stay with the Jesuits, then attended Mexican seminaries in the late 1950s to prepare for diocesan work in Mexico. The bishop prior to Don Samuel recruited him and assigned him on a trial basis to the Tzotzil parish of San Andrés Larraínzar. Don Samuel made the appointment permanent in 1962, just before Vatican II.

The priest threw himself into the work with Marine Corps energy, riding horseback to visit every tiny settlement in his far-flung parish for the purpose of administering the sacraments. Some blocks distant from the traditional priest's room next to the church on the town square, he built a large, completely walled-in compound that became his residence and office. No one could enter the compound except a few Maya who approved of his activities. He raised money in Mexico and the United States to buy land, on which he built a dispensary, an elementary school, and an experimental potato farm. All remained under his tight control, with little or no effort at preparing the Maya themselves to sustain these projects.

He seldom left his compound except to administer the sacraments and visit his projects. When visiting outlying areas, he refused to eat the local meals prepared for him or accept any other gestures of hospitality. He made it clear that he wanted to leave as soon as he could after performing his ritual duties. Wearing his pith helmet, he swirled across his parish on horseback, stirring up dust as he passed foot-weary Maya carrying their heavy loads.

The Padre ignored the catechists and their place in the community, and attacked the diocese's theology of liberation. He did not attend the catechists' meetings, discontinued their translation of his homilies into Tzotzil, and administered baptism and presided over marriages without prior instruction of the participants by either the catechists or himself. He acted as if the catechists did not exist. This, in turn, lowered the respect of the community for them. Rudolfo (chapters 5 and 15) was the deacon-catechist in Magdalenas, a visita of San Andrés. This usually quiet, respectful man observed that the padre "depreciated us. He said we could eat tortillas, but he was accustomed to meat. I heard him say this! He opposed everything we thought was the word of God—the community, collective work. We didn't learn anything in his courses: he just said we should pray, sing songs of praise and not sin. He never talked to us the way I'm talking to you" (Chojnacki 2004: 39).

The padre, with his authoritarian manner, ran into additional criticisms. He clashed with the traditional authorities, who did not like this foreigner buying land in their community for projects that remained under his complete control. This was ancestral land that frequently had been obtained by ladinos through fraudulent means. His close friendship with the town's ladinos was also disliked. He soon came into conflict with the pastoral workers, Mexican nuns. They accused him of paternalism, of trying to dominate all parish activities while ignoring those he could not control. In contrast, "Madre Lourdes . . . [made] frequent visits to isolated hamlets where she encouraged the catechists in their efforts to evangelize families and friends through established patterns of communal life and work. . . . The woman religious spent long hours in conversation in the communities, animating catechists in their work as much by her willingness to share their daily life as by catechetical lessons" (Chojnacki 2004: 251).

In 1980, and again in the 1990s, the catechists requested Lockett's removal from the parish. Don Samuel listened to their petitions and in 1992 terminated Lockett's permission to work in the San Cristóbal Diocese. The dissident priest was then accepted in the neighboring diocese of Tuxtla by a bishop opposed to Don Samuel's liberation ways. This bishop sent Lockett to the parish of Bochil, one of whose boundaries was the dividing line between the two dioceses. Lockett wrote a letter of protest to the Pope and appealed to the Vatican through the Apostolic Delegate, an avowed enemy of Don Samuel and the liberation movement. Over Don Samuel's protest, the delegate gave Lockett special permission to visit San Andrés once a week to hear confession and celebrate mass in his compound.

Lockett "appealed to theological tradition, citing 'the mind of the church' and the will of the Pope as warrant for his conviction that the first duty of the missionary is to 'save souls,' not . . . the bishop's 'badly mistaken strategy' to foment social struggle" (Chojnacki 2004: 251). In a section of his letter to the Pope, Lockett wrote:

> For some years the majority of the catechists (not all) lost their previous devotion and candor because of the teachings they have received in catechetical courses and meetings. They are taught class conflict to bring about a forced leveling of classes, teachings that have been clearly condemned by the Holy Popes, especially by Pope Leo XIII. The catechists who live distant from the town center and closer to Bochil were not contaminated by such teaching because they did not

attend the classes. But the majority of the catechists who did attend were contaminated and because my instructions were not in agreement with those teachings, they have begun to petition my removal. The Bishop withdrew my diocesan faculties [permission to function as a priest in the diocese]. As a result there was a division among the Catholics, one group that in conscience did not accept these teachings and another that did. I told the faithful that we could call our group (which was the majority) the group following the doctrine of the universal church, or the universal doctrine, and the other group, the liberationists because they like to talk a lot about "liberation," a material liberation that they say will triumph in the battle with the rich. (Meyer 2000: 173)

This was a priest fighting to retain a Tridentine Catholicism.

Appendix: The Proposed Canonization of Father Rother as a Saint of the Catholic Church

There is a legal procedure that precedes the canonization of a person as a saint of the Catholic Church. Its purpose is to collect information verifying that the person proposed for sainthood led a life that showed a "heroic practice of virtue." The current version of the process is outlined in the Apostolic Constitution, *Divines Perfectionism Magister,* issued by Pope John Paul II in 1983. Given the high esteem for martyrs in Catholic communities, the Oklahoma City Diocese has initiated the canonization of Father Rother as a martyred saint.

From an anthropological point of view, Father Rother's story is an example of how the lives of some deceased individuals become mythologized. Events of a person's life become idealized to highlight and symbolize a value held in great esteem by a community that wishes it to be remembered and imitated. For this purpose, the mythologizing presents an ideal picture of the person's life, one that departs from its actual reality. This has frequently happened in the past history of political leaders, military heroes, and especially in the cult of the saints. It has included canonizing persons of virtue that have been described only in myths but that never existed in real life. In Father Rother's case, his murder became a symbol of community resistance to the oppression and atrocities inflicted by the Guatemalan military. Because of the symbolic importance of his death, Father Rother's life in Santiago has been mythologized to an elevated status that never existed. Much of the myth making is by individuals who never lived in Santiago and have little or no knowledge of Maya culture. The oral and written versions of such information are part of a literary genre: missionary writing. It is a genre marked by ethnocentrisms and exaggerations that at times stretch to falsehoods. A number of misconceptions about Father Rother are found in the Oklahoma diocesan newspaper, *The Sooner Catholic* (e.g., the issues for July 15, 2001, 3; and November 9, 2008, 10).

The Bishop of Sololá, Bishop Angelico Melotto (1984: 101–4), delivered a homily at a memorial mass for Father Rother that helped formulate the myth. "His first preoccupation was to understand the mentality of his parishioners, and to do this he spared no effort to learn the difficult language of his people, Tz'utujil, because it is mainly through the spoken language with all its subtleties that one comes to know the mentality of a people and its culture." Monaghan (1984: 7), Nouwen (1985: 35), and Christenson (2001: 60) echo the bishop's statement. Rother did attempt to learn the language, but never acquired any facility in it. Prechtel, a resident of Santiago and a fluent speaker of Tz'utujil, knew Father Rother. He notes that Rother's use of it was, for all practical purposes, unintelligible (Prechtel 1998: 167; 1999: 166). Father Rother had a past history of being poor in languages. Previously, he was forced to leave a seminary because of his inability to learn Latin. Rother's difficulty with a Maya language was not unusual. Most foreign, and many ladino priests, never really master them (Early 2006: 170). The translation of the Bible into Tz'utujil was initiated under the direction of Father Carlin, and is mainly the work of Juan Mendoza Lacán and Antonio Ratzán Tziná. A group of Atitecos completed the project after Ratzán withdrew, due to a dispute with Father Rother, and after Mendoza was kidnapped and killed during the counterinsurgency (Rother and Mendoza Lacán 1991).

In his homily, the bishop continues, "For the sick, he built the clinic and the hospital, for the illiterates he started classes, for the workers and farmers, he founded the cooperative" (Melotto 1984: 103). Perera (1993: 194) repeats these claims. Father Rother, as a second-generation member of the Oklahoma effort, had nothing to do with founding these projects. The clinic was started by the nurse and papal volunteer (Catholic Peace Corps) Marcella Faudree in 1965. The well-baby clinic was started by another papal volunteer, Joan Lichtenberg, in 1966–67. The hospital was started by Father Carlin and an Oklahoma doctor. The actual construction was done by Father Westerman, who worked alongside paid Atiteco workers. Father Stafford did most of the electrical and plumbing installation. Father Rother arrived during the final phase of construction and helped with the completion.

Classes for the illiterate were given by the radio school. This project was entirely the work of Thomas Stafford and Elizabeth Nick, with this writer helping out as a consultant. After their departure due to the dispute about cultural sensitivity, a visiting nun from Nahualá served briefly as an advisor to the program. After the Atitecos took charge of the radio station, the Oklahoma Diocese continued to help subsidize it, with Father Rother channeling the funds.

There were three cooperatives with separate histories. The weaving and agricultural cooperatives were started by Stafford, who also organized the Montessori school. The credit cooperative was founded by the papal volunteer Patrick Pyeatt. It languished after his early departure. This writer, during his periodic residencies, tried to revive it. Lacking direction after the departure of the culturally sensitive group, it and the weaving cooperative collapsed as the Atitecos needed more time to sustain them. Father Rother was never involved.

Further confusion about Father Rother is illustrated in David Monahan's (1984: 1–10) introduction to the collection of Rother's letters. He states: "Father Carlin saw potential in the new young man. . . . Ramon Carlin was to be Stanley Rother's mentor and model as a missionary" (3). Father Carlin left Santiago for Antigua shortly after Rother's arrival. In Antigua he founded the Francisco Marroquín Linguistic Institute with Father Lansing of Maryknoll to promote Maya languages. He never returned on an extended basis to Atitlán.

There is a question of how well Father Rother understood the army's counterinsurgency and the danger it involved for himself. In November 1980, during one of the worst periods of the counterinsurgency, Rother was inviting his relatives and friends to visit Santiago with assurances that there was no danger. He thought American citizens, evidently including himself, were not in any danger (Monahan 1984: 37, 46). In December 1980 he wrote the annual Christmas letter for publication in the *Oklahoma Sooner*. In it he says, "This is one of the reasons I have for staying in the face of physical harm. The shepherd cannot run at the first sign of danger. Pray for us that we may be a sign of the love of Christ for our people, that our presence among them will fortify them to endure these sufferings in preparation for the coming of the Kingdom" (Rother 1984: 55; repeated in Nouwen 1985: 48). In spite of the phrase "in the face of physical harm," he does not seem to have taken it as a serious threat. Extrapolating from the second sentence, Monahan used the phrase "The shepherd cannot run" as the title of the book containing Rother's collected letters.

It was only in the first few days of January 1981, when he witnessed the kidnapping of his head catechist, that Rother appears to have had some second thoughts. It was during this period that he strongly protested to the army. On January 12, he received news that he was in danger. He and his temporary assistant, Father Bocel, went into hiding in Guatemala City and then left for the United States on January 28. The shepherd did run. It was a prudent move. There was no disgrace connected with it; no accusation could be made of desertion. Given the rules of the army's counterinsurgency, Rother's protest had made him a marked man, and his presence in Santiago could have had no effect on the army's actions toward the Atiteco community. While in Oklahoma, Father Rother gave several talks about the injustices of the Guatemalan situation that may have been reported to the Guatemalan embassy.

Father Rother would have stayed in Oklahoma, and his subsequent "martyrdom" would not have taken place if a priest in a neighboring parish had not mistakenly told him it was safe to return. He did so in April 1981. In June of 1981, Rother resumed inviting visitors from the States, although he now conceded that it was no longer as safe for those working in Santiago as it once was (Rother 1984: 76). But in July of 1981 he saw the parish returning to normal, and again repeatedly invited visitors from the United States (77, 81, 83, 85, 87). On July 28, 1981, he was murdered in the parish rectory as part of the army's counterinsurgency effort. Reading his letters indicates his lack of awareness of the extent of the counterinsurgency strategy and the seriousness of his situation. He joined at least fifteen other priests and nuns murdered during the counterinsurgency, and an estimated five hundred to two thousand Atitecos (Conferencia Episcopal de Guatemala 2007: 51–375, 411–13; Tarn 1997: 363).

In the opinion of this writer, who had some acquaintance with Father Rother in his early days in Santiago, he was an ordinary human being with some very obvious faults that made it difficult to work with him. Tarn has mentioned some of these. He was a good priest by Tridentine standards, faithfully doing his ritual duties. But contrary to the mythologized picture of him, he was not the great missionary, the learner of languages, the prober of Maya culture, the outgoing friend of all the town's factions, nor the founder of badly needed educational projects, cooperatives, and medical programs to meet the Maya crisis. This is the mythologizing process at work, a process present in many communities throughout history. The great moment in Rother's life was his brief, but strong and open opposition to the atrocities of the Guatemala army. Whether this comes under the category of a life that showed "heroic practice of virtue" is for the upper levels of the Catholic bureaucracy to decide. If it does, there should be many other saints, both priests and Maya catechists, who were murdered for their efforts during the turbulent years of the Guatemalan counterinsurgency.

If the effort to canonize Father Rother should rest on the basis of the picture that the dioceses have presented of him in the past, there should also be some sort of clarification. This could be done by giving credit to those who were responsible for the work of Action Catholicism. The diocese could also acknowledge the injustice of its recall of Thomas Stafford because of his cultural sensitivity, the priest who, more than anyone else, implemented development projects in Santiago Atitlán to empower the Maya to confront their crisis.

24

Social Justice by Maya Empowerment

In addition to administering the sacraments, the type of pastoral worker discussed in this chapter was involved in various community efforts that empowered the Maya to confront their crisis, especially through literacy, health, or cooperative programs. They agreed with the ultimate goal of the insurgents: social justice. They understood the population's frustrations with governmental corruption, the greed of the economic elite, and the degenerate political process that protected it with violence. But they rejected armed rebellion as the means to attain social justice.

Father Luis Gurriarán

We have previously met Padre Luis (introductory photograph) in chapters 12 and 17, which described his work with the K'iche' cooperatives, consciousness-raising programs, and the Ixcán colonization project in Santa María Tzejá. He was born in 1933 to a middle-class family in Galicia, Spain, the home of the Santiago conquistador cult (Santos 2007). Inspired by a family member who was a Catholic missionary, he entered the seminary at age fifteen. Upon finishing his Tridentine theological studies, he was ordained a priest in 1958 as a member of the Sacred Heart congregation. He was typical of many of the Spanish clergy at that time—conservative, anticommunist, pro-Franco. Padre Luis, along with other Spanish Sacred Heart priests and Dominican nuns, was assigned to evangelize the K'iche' Maya.

All of them suffered severe cultural shock from their initial encounters with Maya culture in the late 1950s. They were overwhelmed by the poverty and cultural differences they encountered. It made them reevaluate their Catholic commitment. Some withdrew from the religious life. Others returned to Spain and the Catholic conservatism of that era (Diócesis del Quiché 1994: 49–50). Others remained and worked at establishing Tridentine Catholicism among the Maya, including the two described in the previous chapter.

Padre Luis had a different reaction—a conversion experience that initiated his evolution away from his Tridentine formation. "I came to follow in the footsteps of the traditional Spanish missionaries. I had that conquest mentality—to introduce the Christian faith. . . . Social matters had not been discovered yet. . . . I, at least had not discovered them. . . . I did know that the people were poor. What I didn't know were the causes of the poverty and exploitation, or that the poverty was in great part due to the years of colonialism. From the vantage point of today, I can see there was a certain culpability, not only from the Spanish kingdom, but a certain moral responsibility on the part of the church. . . . The church allied itself with those in power who subjugated or enslaved the peoples of the Americas. That realization was a surprise or an awakening to me" (Manz 2004: 52; see also Santos 2007: 70).

The experience would completely reorient Padre Luis's life. "I had no other alternative than to figure out how I was going to rearrange my ideas. That meant to bring about a radical change in my mind-set and therefore to find the way to aid people in changing their conditions. . . . I came to evangelize the Indians of Guatemala, but in the process of getting to know them, they evangelized me" (Manz 2004: 52).

He recalled his Tridentine formation and its shortcomings.

> The portrait of Jesus that I was taught in the seminary had nothing to do with the historical Jesus that I kept finding in the people of

[Maya] communities. The Bible that I had studied in my theology courses also had no relevance for their understanding of religion. The revolution that was taking place in my mind would lead sooner or later to a partial break with the past. I realized that at times, my companions, people of the church, would say that I was becoming very radical . . . or that I had become an enemy of the traditional church. I answered them that not only was I not an enemy, but continue to be an active member of the Church. But I have changed my concept of the Church, my concept of God, of the Bible. . . . I could no longer speak of religion as an instrument of submission, of complacency, and to accept things as they are because that is the will of God. (Santos 2007: 78)

He came to understand that his most important work was not the administration of the sacraments, but with the help of biblical theology, to empower the Maya to exercise their rights and seek the social justice that had been denied them (38).

Padre Luis, his cousin Padre Javier, and a few other Sacred Heart priests who shared his views, began to organize small education projects and cooperatives. The Maya response was enthusiastic. Luis wanted to learn more about the organization of cooperatives. In 1962, he attended St. Francis University in Canada, where he enrolled in an intensive nine-month program on cooperatives and community leadership (Santos 2007: 53).

Because of the inroads the cooperatives were making on the profits of ladino merchants, and their bribes to military officers and politicians in El Quiché, he was expelled from the country in 1965. The order of expulsion came from the Sacred Heart congregation itself, but under the government's threat to expel from the country all foreign missionaries unless the congregation ordered Luis to leave. In this way, the government could claim that they were not responsible for the expulsion. Later, the diocese declared that the main objective of the priest's expulsion had been the destruction of the cooperative movement that he had initiated and promoted. During the time of his expulsion, Luis attended courses on cooperative administration at the University of Wisconsin and at the Credit Union National Association. Luis was very aware that good will alone was insufficient for his work of empowerment, that technical training was needed for many types of community organizations.

The expulsion strengthened the movement, and it continued to grow. With the election of a new president the following year, Luis was allowed to return. He founded a center in Santa Cruz to raise the social consciousness of the K'iche's about the problems of their communities and the need for social action. Three years after his expulsion, there were eighteen cooperatives—ten savings and credit, six consumer, one industrial, and one agricultural (Santos 2007: 55). In Santa Cruz, the savings and credit cooperative grew from twelve hundred to two thousand members, quadrupling its capital, and by 1971, it was the largest in the country.

In 1970, attempting to alleviate the shortage of land, Padre Luis assisted in founding a new community in the Ixcán jungle, Santa María Tzejá, described in chapters 17 and 18. One day, an EGP guerrilla band visited Santa María to buy supplies (Santos 2007: 130–40; Manz 2004: 75–76, 87–88, 182). One of its members was César Montes (chapter 22). During the visit, Padre Luis received his daily radio call from the army base in Santa Cruz seeking information about any insurgents in the area as well as checking on Padre Luis himself. The guerrillero Montes (chapter 22) recalled, "I sat right in front of him with the rifle pointed to his chest and I had no doubt that if it had been necessary, I would have fired it" (Manz 2004: 75). Montes recalls that Padre Luis was neither friendly nor unfriendly, and that Luis later told him that he would not help the guerrillas, but he wouldn't denounce them either. Luis's main worry was that the guerrilla presence would bring the army into the village. Later, the army did visit the village several times, strongly questioned Padre Luis, and in his temporary absence occupied and searched his house. Based on its stereotypical assumptions of Catholic Action, the army considered Padre Luis "a communist dressed as a priest." As the army continued its counterinsurgency, the schoolteacher in Santa María was murdered and the threats against Luis mounted, including a near assassination. He left the village in 1975.

In 1976, Luis moved to Nicaragua. From there, he lobbied to bring international attention to the continual violation of human rights in Guatemala (Santos 2007: 191–285 and passim). This involved trips to Europe, the United States, and Mexico. But he wanted to return to Guatemala and the Maya. In 1986, he worked for a year with Falla and the Community of the Population in Resistance (CPR, see chapter 19). He left because of sickness and went to Chiapas, where he worked with Guatemalan refugees.

At least twice, Padre Luis was invited to join the insurgency—by the Melville group (chapter 22) and later, by the EGP (Santos 2007: 76). He refused, but as the Guatemalan army committed more slaughters, Luis came to approve of armed resistance although he himself never took up arms (Falla: personal communication). Padre Luis is an example of the evolution from Tridentine Catholicism to Catholic Action, similar to the first phase of the Melville's evolution. Others priests were going through similar evolutions and becoming Catholic Action priests.

Padre Jose María Gran

Originally from Barcelona, Spain, Padre Jose María Gran was the pastor in Chajul (Santos 2007: 218–20). He sympathized with the efforts of Maya liberation and worked with youth groups in the many outlying Quiche' villages. "Padre Jose Gran organized youth groups with monthly meetings in each aldea. In this way young people came into contact with other youth groups in the department. In these meetings they recognized that they had same problems. He instructed them in religious themes related to their every day lives" (Samandú et al 1990: 78). Many became catechists. He had witnessed the army's execution of some Maya campesinos in the town plaza. He sent denunciations of these executions to the monthly diocesan meetings for distribution throughout the area. On June 4, 1980, he and his sacristan were returning on horseback from a distant community when they were ambushed. One bullet pierced his heart and six more were pumped into his body. Two bullets broke open the sacristan's skull. On television the army announced that they had killed two unidentified subversives. Under the rules of counterinsurgency, anyone who informed the outside world about counterinsurgency atrocities was considered extremely dangerous and to be eliminated. The military feared international pressures would interfere with their plans.

The Maryknoll Colonization Communities

Father Bill Woods was a Maryknoll priest trained in Tridentine theology. He arrived in Guatemala in the late 1950s. He worked for twelve years in Barillas, a Maya town high in the Kuchumatán Mountains. There, he came face to face with the Maya crisis, which began his evolution to Action Catholicism. He became the pastor of the five colonization communities organized by Maryknoll in eastern Ixcán. Woods was a pilot and used an airplane to facilitate his travel for religious duties between the five colonies and for his many trips to Guatemala City. In the city, he wrestled in his aggressive Texan manner with government agencies over their reluctance to carry out their promised aid to the colonization communities. These activities brought him under military surveillance. In 1976, his plane crashed in the narrow mountain pass between the highlands and the Ixcán jungle. There were no survivors. The crash took place where Woods had predicted it would if the army decided to eliminate him. There is almost universal agreement that the plane was destroyed by the military. César Montes of the EGP said that Woods had told him that he would never collaborate with the insurgents, nor would he denounce them. In 2002, Woods's body was brought back for burial in Mayalán, one of the Maryknoll colonies in the Ixcán (Morrissey 1978: 150 and passim; Manz 1988: 129, 132, 256; Manz 2004: 87).

Following the death of Father Woods, his place was taken by Father Stetter (Falla 1985: 8–12). Stetter was a member of a German group working in the Quetzaltenango Diocese and had been the pastor of Cantel. Like Woods, he was an airplane pilot, a skill needed to move with any frequency between the Ixcán communities. Woods had been head of the cooperatives, and he had spent much of his time working with them and in Guatemala City fighting for land titles. He would make hurried visits to three communities on a single Sunday for masses and baptisms. Otherwise, religious functions were left in the hands of the catechists. Stetter did not become personally involved with the cooperatives, but worked more closely with the catechists and on development projects. He obtained metal roofing for the communities to replace the traditional thatch that was inefficient in the tropical lowlands, built a power plant to provide lighting, obtained a power rice sheller to increase community production, built parochial schools, obtained a tractor to carve out streets, and obtained a loan to begin cattle raising. All these activities to empower Maya communities made him suspect to the military garrison of the area, who had him expelled from the country in 1978.

Bishop Juan Gerardi

Gerardi was born in 1922 and raised in Guatemala City, where he entered the seminary and received his college education (Conferencia Episcopal de Guatemala 2007: 51–76). He completed his theological studies in New Orleans in the early 1940s, a Tridentine formation, and was ordained in 1946. He served in rural parishes, and was made bishop of La Verapaz in 1967, where he clashed with conservative Spanish priests. He had a reputation for being authoritarian and a man of crazy ideas—probably a reflection of an outlook derived from his reading contemporary theology. He became the bishop of the Quiché Diocese in 1974. In 1980, he was elected president of the Guatemalan Bishop's Conference (Santos 2007: 226–31). He was the first Guatemalan bishop in recent times who was also an intellectual. Half of his house was filled with books he had read, their passages underlined and notations made in the margins.

Arriving in Quiché, he initially restrained himself, in spite of his authoritarian reputation, learned about the local situation from the Sacred Heart priests, and studied the K'iche' language and culture. He supported his priests and their socioeconomic activities, especially the work of the Gurriaráns when they were attacked by the military. Called before the commander of the local military zone, who wanted his support of the army, he instead condemned the military for their indiscriminate killing of civilians. The army made an unsuccessful attempt to kill him in an ambush (Santos 2007: 223). After the army's assassination of two Sacred Heart priests, the bishop closed the diocese and withdrew all personnel in an effort to bring world attention to the murderous counterinsurgency of the Guatemalan army.

The bishop went to Rome for a conference on family life. He also discussed with the Pope the situation of his diocese. The Pope disliked Gerardi's closing the diocese, but denounced the government in a letter to the bishops of Guatemala. Cardinal Casariego took the letter to the Guatemalan president-general, Romeo Lucas García, who was infuriated by its contents. Because of his denunciations of the military, Bishop Gerardi was refused reentry to both Guatemala and El Salvador, forcing him into exile in Costa Rica. With a change in the presidential administration in 1982, he returned, and became the auxiliary bishop of Guatemala City in 1984.

After the war, he oversaw the project to recover historical memory (REMHI) that revealed the army as the overwhelming source of the massacres, including genocide. Its findings were confirmed by the United Nations' Comisíon para el Esclaracimiento Historíco. The bishop was murdered two days after presenting the findings of the REMHI project to the public. Father Luis Gurriarán remembered the bishop with great affection and admiration for his talents and for his help during the trying times of the counterinsurgency.

The bishop, by the force of his own reading and intellectual probing, moved beyond his Tridentine formation, but he never believed in armed violence as a means to liberation. He mistrusted the guerrillas. He saw the CUC (chapter 17) as Communism even before that organization joined up with the EGP, although he never publically denounced them (Carmack 1988: 51). In the midst of the turmoil, the bishop continued to preach the necessity of dialogue and peace (Le Bot 1995: 143).

Padre Ricardo Falla

Falla was from a well-to-do Guatemalan family. He entered the Jesuits, where he was shaped by the *Spiritual Exercises* of St. Ignatius Loyola. It sent

Figure 24.1. Father Ricardo Falla recording the Maya story as told by one who remembers.

him on a life's search for the object of his love and service. In the 1960s, he studied theology at Innsbruck, Austria, one of the centers for the "New Theology." Falla read the Popol Vuh and became intrigued with the Maya. He went to the University of Texas to obtain the doctorate in anthropology. For his dissertation, Falla worked among the Maya in San Antonio Ilotenango, which resulted in the book *Quiché Rebelde*.

Falla was a member of the Zone 5 Jesuit community in Guatemala City along with Fernando Hoyos. Hoyos was discussed in chapter 17 as a founder of the CUC, and will be seen again in the following chapter as the leader of an EGP combat zone. Falla assisted at Hoyos's first mass (Hoyos de Asig 1997: 60). In his memoir about his years among the refugee mobile camps hiding in the Ixcán during the militarization period (chapter 19), Falla recalls being inspired at seeing the Maya faces during one of their flights from an army search party. "Seeing these faces, the words of Fernando Hoyos came to me: that in the people, one feels the presence of God. Fernando gave great importance to these very real faces" (Falla 1998: 38). Both Falla and Hoyos had found the same object for their loving service. Falla (9–10) writes: "His [Hoyos's] example presented a powerful challenge to me—not to abandon the priesthood and leave the Jesuits [as Hoyos did]—but to attempt something similar, if not so heroic, based on my own religious and priestly identity. If Fernando did it, why couldn't I?"

The memory of Hoyos inspired Falla. It meant living among, ministering to, and writing about the Maya so that others would understand the culture and crisis of at least 50 percent of the Guatemalan population. His pen became his weapon for social justice and empowerment of the Maya. He revealed to the world the slaughter of the army's "ashes" program in *Masacre de la finca San Francisco, Huehuetenango* and in *Massacres in the Jungle*. He has also written about his experiences among the refugee communities (CPRs) in his memoir, *The Story of a Great Love*. During a quick escape from an army search team, Falla hid the mass chalice and liturgy book in a cave. The army found these and other pastoral materials. Realizing Falla's presence among the bands, they accused him of being a combat guerrilla, a priest in arms. Falla left the jungle to publicly defend himself against the accusation, as described in chapter 19. He has been described as a member of the EGP (Konefal 2010: 126). While Falla did require EGP permission to enter the zone under the control of an EGP remnant force, where the CPRs were also hiding, he at all times remained under the jurisdiction of the Jesuits and was never subject to the EGP organization.

Padre Mardonio Morales

In Chiapas, Padre Morales was a Mexican Jesuit who had been shaped by the *Spiritual Exercises* and by the study of Tridentine theology. From their center in Bachajón, he and his brother, Ignacio, also a Jesuit priest, served Maya parishes in Chilón and Sitalá for more than thirty years. His arrival among the Maya during their subsistence crisis led him to help reorganize the Jesuit parishes there, as described in chapter 12 (Zatyrka 2003: 171–97). For years, he protested to the Mexican authorities about the injustices against the Maya and helped organize protest marches. He made a Tzeltal translation of the book of Exodus, to be used in reflection groups. Later, he translated the New Testament. He clashed with the Ocosingo Dominicans over the use of Marxist advisors in the diocese. He and his catechists refused to introduce the advisors to their parishes. As a result, there were no EZLN military units in these areas during the uprising. In later years, he said that Bishop Ruíz had been naive in his dealings with the Marxists. Chapter 29 will describe his efforts towards establishing a Maya-inculturated church.

While the priests who worked closely with Maya for their empowerment were highly respected, if not beloved by some, the Maya were always aware of the cultural differences between themselves and the priests. In Bachajón, the Maya saw Father Morales and the other Jesuits as neither ladinos, because they did not exploit Maya, nor real Tzeltales, because they still thought like ladinos. For this reason, the Maya maintained that the priests could never fully understand them (Maurer 1984: 162).

Padre Miguel Chanteau

Chanteau served as the parish priest in San Pedro Chenalhó, Chiapas, for thirty-two years. He was born in Alençon, in the Normandy region of France, on July 6, 1930 (Chanteau 1999). With the death of both his parents, at age six he entered an orphanage, where he spent much of his youth. Wishing to become a priest, he entered a seminary. But due to his undisciplined ways, he was forced to leave. This process was repeated several times until he finally finished his theological studies and was ordained a priest in October 1958. He continued to have trouble with local ecclesiastical authorities. Volunteering for work in overseas missions, he accepted the invitation of Bishop Ruíz García to join the San Cristóbal Diocese. He arrived in San Pedro Chenalhó in June 1965.

In Don Samuel he finally found a like-minded cleric with whom he could work. During his years of service in San Pedro, he carried out the bishop's vision of what a Maya parish should be. He became deeply involved in the life of the community. Continually at odds with the local ladino population, he fought the injustices they inflicted on the Maya. He organized an extensive network of catechists throughout the parish, all based on biblical reflection inspired by liberation theology. The previous references to the community of Yibeljoj reflect the results of his efforts in that community. With the supervision of Mexican nuns, health promoters were trained for many local communities. Courses were given on the organization of cooperatives. Workshops were held about human rights in conjunction with the center in San Cristóbal. During the militarization of Chiapas, Padre Miguel supported the efforts of Abejas (see chapter 21). He was accused of being an EZLN collaborator, and in 1998 was expelled from the country by the Mexican government. He has since returned to Chiapas, to a parish in San Cristóbal that ministers to Maya refugees.

Yet despite the padre's efforts for Maya empowerment, the Maya were always conscious of the cultural differences between themselves and the priests. "What seems to distance many Pedranos from Padre Miguel is not so much his important position and gender, but his unmarried status and his acceptance of [moderate] drinking. Although Padre Miguel is a godfather to some Pedrano children, he is not a true father and husband. Pedranos told me, 'Priests don't work the land and don't have families to support. They eat good meals and can even get drunk, without anyone telling them that they shouldn't. They have cars and can take off to far away places for months at a time.' ... Catechists don't need Padre Miguel's permission to conduct their affairs the way they see fit, but they would like his blessing. They recognize that he has sacrificed dearly on their behalf and that he loves their land and traditions" (Eber 2004: 227–28).

Summary

The work of the above priests is representative of many pastoral workers who evolved from their Tridentine formation to Action Catholicism— Germans in the Quetzaltenango area; Belgian pastoral workers among the Chortí in eastern Guatemala (Metz 2006); a New Ulm Minnesota group in San Lucas Tolimán; a Spokane, Washington, group in Nahualá; a Montana group in Santa María Visitación and Santo Tomás La Unión, and others. The priests' work involved a constant, ongoing dilemma. They agreed with the Catholics in arms that social justice demanded an overthrow of corrupt governments, but most disagreed that armed violence was the means to bring this about.

Appendix: Values and Interests in Social Change

In the 1960s a sociologist at Harvard, Sister Marie Augusta Neal, when confronted by the problems of social change within and outside of the church, developed a typology of the Catholic clergy (1965). Its predictive value was successfully tested among the clergy of the Boston archdiocese. Although it was impossible to replicate the study in the Maya area, the typology gives insight into clerical motivations when priests were confronted by the Maya crisis. As the noted sociologist Neil Smelser, of the University of California at Berkeley, pointed out in the study's introduction, that the typology has scientific validity far beyond its application to Catholic Church personnel. This writer has found it an invaluable tool for insights about workers in many kinds of public-service bureaucracies, that of their very nature have a worldview that looks beyond both self-interest and institutional interest.

Neal took the theoretical positions of Marx and Weber regarding religion, put them into the Parsonian framework of patterned variables, and developed the typology: some priests act from a sense of values, others from self-interest or institutional power; some priests are in favor of changes within Catholicism, and others are not. This gives a fourfold typology of clerical attitudes toward the problems of change facing Catholicism at the contentious time immediately preceding Vatican II: in Value-Change, a change should be made because it is the right thing to do as defined by the Catholic worldview; in Value-NonChange, a proposed change should not be made because it is wrong to do so as defined by the Catholic worldview; in Interest-Change, a change should be made because it will benefit either the person (in this case, the priest) or the institution (a variant is that the institution will look bad if the change is not made); and in Interest-NonChange, a proposed change should be rejected because it would be adverse to personal interests or to those of the institution.

Neal gives detailed descriptions of the four categories (1965: 15–16).

Value-Change priests are the biblical "prophets": individuals primarily oriented to the function of adaptation will identify with the ultimate values of Christianity but will have no deep binding commitment to any specific goals or norms defined and structured through time, especially if they seem no longer functional for the expression to the world of these ultimate values. Their major concern is articulation of the system with the larger community. They are more aware of the current needs of the world at large and the areas in which institutionalized patterns in the Church are not satisfying these needs. Their identification with the values is their major source of acceptance within the system; their orientation to change makes them more acceptable to members of the larger community who feel alienated from the Church in her historic role. Their emphasis is on the articulation of the system with society at large.

Interest-Change priests are the "cosmopolitan organizational men": individuals [who are] primarily oriented to the attainment of the specific goals defined in time by the Church's functionaries. . . . Their deep binding commitment is primarily to the goals which they strive to attain by traditional or change-oriented methods, whichever seem to promise the most effective realization of goals. Their ego involvement is identified with the interests of the Church with respect to its current goals. Even ultimate values at times are made subsidiary to the attainment of these goals. The action of role players may at times appear arbitrary, power oriented and non-normative.

Their emphasis on organization and power to achieve their ends is cause of concern for externs [non-Catholics], a security for the dependent, and a worry at times to the value-oriented members of the Church.

Value-NonChange priests are "the priest." Individuals [who are] oriented to latent pattern maintenance are primarily identified with the values traditionally stressed in the system. It is their concern to preserve whatever patterns have implemented the expression of these values in time. The fact that in the course of time the former implementation might have become less effective is rather difficult for them to perceive since their major experience is concentrated within the system. Deviations from expectations are defined as failure to follow the pattern on the part of individuals, rather than as failure of effectiveness of the system itself. This is very much reinforced through their consideration of the fact of God himself acting through the system in time. Emphasis is on the articulation of the system with God. They see norms and goals as inspired by God to inculcate values; hence, though they would never sacrifice a value in order to preserve either of these structures [norm, goal], they are not oriented to criticism of these aspects of the system, assuming they operate with God's approval.

Interest-NonChange priests are "the local organizational man"; individuals permanently oriented to the integrative functions of the Church tend to stress the primacy of the normative system. Ritual regularity is the major concern. Both goals and values can be made subservient to procedure. Any deviation from the traditional normative pattern appears to them a threat to the integrity of the total system. The mere suggestion of change in any aspect of the Church's structure is defined as heresy and tremendous affect can be released at the very suggestion of change. Their ego involvement is deeply imbedded in the normative structure. Their behavior is the target of much criticism currently [early 1960s]. Role players with this primacy can be so identified with the norms that both the goals and values are sacrificed to means. The rituals and the symbols come to be perceived as effective in their own right [ritualism]. The effective implementation of values is sacrificed in the interest of preserving the current pattern.

The typology gives insight into some classical types of motivation with which pastoral workers undertook their service among the Maya. While an altruism based on Christian love was the ideal, it was not always present, as the typology makes clear. Pastoral workers included many personality types.

Methodologically, it was impossible to make use of the typology in this study, because of its detailed data requirements. Neal's original study used a long questionnaire and follow-up in-depth interviews that discriminated the social and personality characteristics of the four types. This writer was involved in that study as a blind coder of the interviews.

25

Social Justice by Armed Rebellion

This chapter describes how some pastoral workers moved from Action Catholicism to joining the Marxist revolutionary EGP in Guatemala or the Marxist EZLN in Mexico, both organizations with agendas of armed rebellion.

Camilo Torres

Camilo Torres was a Colombian Catholic priest ordained in 1954. He died in Colombia as a guerrilla combatant in 1966 (Torres Restrepo: n.d.). He was never in the Maya area. He is included here because he became the Che Guevara, the inspiring model, for that sector of Liberation Catholicism that took up arms. He is mentioned in the dedication of the Melvilles' book (see chapter 22). Understanding his worldview gives insight into those who followed the same path in the Maya area.

During his career as a priest, Torres was also a university professor, a social science researcher, a social activist, and a combatant in the Colombian Marxist organization, the Army of National Liberation. Colombia at that time had a social structure and problems similar to those in Guatemala and Chiapas. Torres based all his activities on the teachings of Jesus. In a "Message to Christians" shortly before he joined the guerrillas, he expressed his view of liberation.

> The convulsions stemming from political, religious and social events in recent times, have possibly created a great deal of confusion for Colombian Christians. At this decisive moment in our history, it is necessary for Christians to be grounded in the basic essentials of our religion.
>
> The most basic essential in Catholicism is love of neighbor. "The one who loves his neighbor, fulfills the law" (Saint Paul in Romans 12:8). For this to be a true love, it must seek to be efficacious. If gifts, alms, a few free schools, a few housing projects, that which is called "charity" is unable to feed that majority of the hungry nor cloth most of the naked nor teach the majority of those lacking education, then we must look for effective means for the welfare of the majority.
>
> These means will not be found among the privileged minority who hold power because these effective means usually oblige the minority to sacrifice their privileges. . . . Therefore it is necessary to transfer power from the privileged minority to the majority who are the poor. To do this rapidly, a Revolution is necessary. It can be a peaceful Revolution if the minority does not put up a violent resistance. The Revolution is the means to obtain a government that provides food for the hungry, clothes the naked, teaches the uneducated, that fulfills its duties with works of charity, love of neighbor, not only occasionally and in a passing manner, not only for a few, but for the majority of our people. For this the Revolution is not only permitted but is obligatory for Christians who see in it the only efficacious and comprehensive means to realize *love for all*. (Torres Restrepo 1965: 3)

Torres's dedicated career and its message took place in the late 1950s and early 1960s before Vatican II and Medellín. While many did not fully agree with him, his spirit helped to give impetus to the Latin American liberation movement and Medellín's option for the poor.

Fernando Hoyos

Hoyos was a Jesuit for twenty years, a Catholic priest since 1973, a founder of the CUC in the late 1970s, a member of the EGP since 1980, and the leader of its combat forces in the Ixil area. He was killed by a civil patrol near the Huehuetenango-El Quiché border around July 13, 1982. Joining the EGP was the final step of Hoyos's long evolution from a middle-class Spanish family to Christian Marxism and, ultimately, to a Marxian faith in armed violence. Whether he saw this as a means to create the new man in a socialist state, or retained his Christianity and saw it as implementing the imperfect Kingdom on earth, is not clear. His sister believes the latter (personal communication, 2010). His personality has been sketched by Enrique Corral—a friend during their student days in Spain, later a Jesuit classmate both in Spain and Guatemala, and still later, like Hoyos, a member of the EGP. "Profound in his convictions, unaffected and austere, a devoted student of revolutionary theory and of the everyday conditions of the people for whom he was fighting, he was killed confronted by the most bloody military regime in Latin America. He was also killed due to his irrepressible impatience, his strong will, and in good measure to his underestimating the practical aspects, the fundamental elements of the life he ultimately chose" (Corral 1997: 6). Hoyos's letters to his family (Hoyos de Asig 1997, 2008) and the commentaries of fellow Jesuits (Falla 2008; Hernández Pico 2008) give some insight into the stages of his evolution and those who influenced him.

Jesuit Spirituality and the Search

The worldview of his twenty years in the Jesuits was the order's spirituality expressed in *The Spiritual Exercises of St. Ignatius of Loyola* (1990), a series of psychologically arranged meditations on sections of the Bible, especially on the life of Christ and its meaning for the life of the one meditating. Ignatius, an ex-soldier, emphasized a life of service to others based on love and imitation of Jesus.

There is a final meditation, "A Contemplation to Obtain Love," that summarizes the meditative journey. It begins with two notes about the essence of love. "The first is that love ought to be put more in deeds than in words. The second, love consists in an interchange between two parties; that is to say in the lover's communicating and giving to the beloved what he has or out of what he has or can; and so, on the contrary, the beloved to the lover.

So that if one has knowledge, he gives to the one who has it not. The same of honor, riches, and so one to the other" (Ignatius 1990: 28). The following meditations are interspersed with a repeated prayer, an offering of one's self for the service of others:

> The First Point [meditation] is, to bring to memory the benefits received, of Creation, Redemption and particular gifts, pondering with much feeling how much God our Lord has done for me, and how much he has given me of what He has, and then the same Lord desires to give me Himself as much as He can, according to His Divine ordination.
>
> And with this to reflect on myself, considering with much reason and justice, what I ought on my side to offer and give to His Divine Majesty, that is to say, everything that is mine, and myself with it, as one who makes an offering with much feeling.
>
> [Prayer] Take Lord and receive all my liberty, my memory, my intellect, and all my will—all that I have and possess. You gave it to me: to You, Lord, I return it! All is Yours, dispose of it according to all Your will. Give me Your love and grace, for this is enough for me. (Ignatius 1990: 29)

This prayer is repeated three times after reflecting on the wonders of the cosmos. For those Jesuits who take spirituality seriously, this is a powerful meditation that sets them on a quest to find where God's love and grace is incarnated for them, where they can use whatever talents they may have, no matter how great or humble.

We can see the resulting imprint of this Christology on Hoyos. As a Jesuit seminarian teaching in El Salvador, he wrote to his family about his young students preparing for the priesthood. "They are good people although perhaps too young to really understand Christianity. The important thing is to introduce them to Christ because this leaves a mark on them that is not erased for the rest of their lives . . . simply that they become acquainted with him. And to ask for this in a special way, these days I rise at four thirty in the morning to pray and insistently ask that they know him" (Hoyos de Asig 1997: 37).

In the *Spiritual Exercises*, the specification of the life of service to others is open-ended. It is to be determined by the circumstances of one's life and talents. Within the Jesuits, the dedication usually was to the various works the Jesuits had historically undertaken. But with the rapid rate of social change during the latter part of the twentieth century, the Jesuits were as confused as many other groups about what specifically they should be doing. Hoyos spent much of the 1970s trying to determine how to implement his love for the Maya, who had become the object of his Christ-based dedication.

Education

Hoyos studied philosophy in Madrid (Comillas) and Germany (Pullach). While still a seminarian, beginning in 1967 he taught for two years at the National Seminary in El Salvador. Returning to Europe for theological studies, he spent a year at Louvain. In addition to his studies, Hoyos became involved with Spanish migrants and their problems, an important step in his involvement with the poor. Hoyos left Louvain after a year. For him, the theology taught there was too abstract and distant from Latin American realities. Louvain was a center for the New Theology movement in Europe, concerned with expressing Christianity in modern philosophical terms in order to offset widespread European atheism. Hoyos next moved to Madrid, where he spent the next two years studying theology at Comillas University. He also took night classes about methods of teaching literacy and consciousness-raising. He and Corral spent their vacation periods among the rural poor.

Once again dissatisfied with the theology curriculum, Hoyos and a group of seminarians preparing to work in Central America formed a private study group under the direction of Father Ellacuría, a well-known Jesuit philosopher and theologian, and a theoretician of the liberation movement. This group delved into Christian Marxism. The direction given the group by Ellacuría is indicated by his subsequent career. He went to El Salvador and became a leading Central American intellectual and rector of the Catholic University (Universidad Centroamericano José Simeón Cañas). For their strong public stand against governmental corruption and against the torture and killing of peasants and workers by the Salvadorian military dictatorship (similar to their counterparts in Guatemala), Ellacuría and five other Jesuits at the university were murdered by the military in 1989.

In late 1972, Hoyos and Corral came to Guatemala for a final year of private theological study. By that time, he was a Christian Marxist, meaning that he used Marxism as a method of social analysis. It does not infer acceptance of Marxian metaphysics or tactics. He studied the history of the church in Guatemala, ecclesiology, the sacraments, pastoral methods, and the liturgy. An informal group studied Marxism (Hoyos de Asig 1997: 54). He also joined in discussions about liberation theology with priests and laity working in Christian base communities. In the evenings, he was taking additional courses about educational methods at the University Rafael Landívar. He wrote a paper, probably dating to this period, that makes reference to the work of Paulo Freire (Hoyos n.d.: 9). He spent his weekends visiting Maya communities in the Western Highlands, raising social consciousness and teaching literacy (chapter 17).

In August 1973, Hoyos wrote a paper about the violence in Guatemala (Hoyos 1973). In it, he discussed the Melvilles at length, and the challenging implications of their work for Guatemalan Christians amid the government's institutionalized violence. He also discussed the decision of Camillo Torres to take up arms. Hoyos himself had not yet made up his mind about the role of violence in the revolutionary movement. He concluded (12): "It is debatable the option of the expelled Melvilles to take up arms, but there can be no debate about their decision to combat the institutionalized violence that provoked an armed response in their case. We respect their decision that was not taken lightly, but was the result of their Christian commitment. We call upon those of us who have not made the same decision to ponder if we are not in reality collaborating with institutionalized violence? What is our denunciation of it?" Hoyos had been in the country for only ten months, but he was already confronting the question that would change his life.

Years as a Priest

Hoyos was ordained to the priesthood at the end of 1973. He describes his emotions: "at one moment fear, and in another moment great joy and hope, it feels a little strange at arrive at this point in my life and be fearful about it. What can I do for my fellow human beings? However I also feel a great tranquility and confidence in taking this step because I am fulfilling something to which I feel called by God, not that he has been talking to me about it in my dreams, but through all the poor people and their suffering that I have come into contact with this year" (Hoyos de Asig 1997: 57). Hoyos was beginning to find the object of his Christ-like dedication, the Maya poor of Guatemala and their subsistence crisis. But that was only part of the answer to his search. The unanswered part was the specific means to confront their most basic problem, the constant, ongoing grinding poverty, and the rampant suffering and death that accompanied it.

Hoyos and Corral joined a recently founded Jesuit community located in a poor section of Guatemala City, Zone 5. Here they joined a group of socially active Jesuits that included some who took important roles in the

Figure 25.1. Fernando Hoyos as a Jesuit visiting Lake Atitlán, 1977.

Nicaraguan and Salvadorian struggles. Ricardo Falla was also a member of this community. It became a gathering place for university students and Mayas on their trips to the city from the countryside. The Jesuits formed a research organization, the Centro de Investigación y Acción Social (CIAS). Its purpose was "to investigate, develop and broadly distribute models of just development for Central American societies." This involved "investigating the unjust conditions in which the majority of rural and urban workers live, and to put the results at the disposition of popular organizations and to accompany them in their justified struggles for a new and better society" (Hernández Pico 2008: 179). Hoyos became center's second director. "This work of research and social action is one of the most Christian things one can do because the more time one spends here and sees it going about its business, the more one realizes the real necessities and profound contradictions of the system that constitutes this society. We hope that our effort, however small, may be one more stone to change such injustices and create a society that may be more Christian" (Hoyos de Asig 1997: 52).

Hoyos continued visiting Maya communities in the Western Highlands, deepening his experience of the Maya subsistence crisis. There, he continued teaching literacy, raising Maya consciousness about their rights using the Book of Exodus (Hoyos de Asig 1997: 120), and laying the groundwork for the CUC, described in chapter 17. Hoyos abandoned clerical garb for rural dress and traveled with only a knapsack. He stayed in the homes of the Maya, where he celebrated the mass. Dispensing with the customary liturgical vestments and an ornate chalice for Eucharistic celebrations, he used a cup made of pottery, a tortilla, and any type of wine (González 2002: 339–42). He was determined to become one with the people.

When he taught at the seminary in El Salvador, a professor there was another Jesuit who became a friend, Father Rutilio Grande. In 1977 Grande was working in a rural parish in El Salvador organizing Word of God communities among his parishioners, mostly laborers on agricultural plantations. This work of empowerment was upsetting to plantation owners. Father Rutilio was murdered by a military or a paramilitary group. It was a frustrating event for many and was bound to have affected Hoyos. Also, Grande had been a friend of Archbishop Romero of San Salvador, and was the master of ceremonies for Romero's consecration as bishop. For his stand against the military, Archbishop Romero was assassinated in March, 1980. This was the period in which Hoyos was wavering between remaining in the Jesuits or joining the EGP. Hoyos kept track of the frustrating Salvadorian situation, similar to the one in Guatemala.

In 1979, the Jesuit house in Zone 5 was closed under pressure from the government. The community had strongly protested the stolen elections in 1974 and 1978. The critical tone of their publications brought them under government surveillance and harassment. Some went to another Jesuit house in the city, others to Nicaragua to help with the attempt to stabilize the country following the Sandinista victory, and others, including Hoyos, went to live with Guatemalan revolutionaries. His Jesuit ties were weakening and those with the EGP strengthening.

Joining the EGP

The year 1980 was a fateful one for Hoyos. On January 31, the Spanish Embassy was burned, with five members of the CUC among its victims. In the same year, the Jesuits called a meeting in Rome of delegates from their worldwide institutions to discuss the implementation of social justice in their works. As the director of CIAS, Hoyos attended. He returned deeply disillusioned. "He told us that he saw no possibility that the Jesuits as an institution would put forth any effort for political revolutions in Central America. At that point, I believe, he decided to deepen his relationship with the EGP and to replace cooperation with Jesuits for full-time cooperation with the revolutionary organization" (Hernández Pico 2008: 182). Around the middle of 1980 he made his decision to leave the Jesuits and become an active member of the EGP. His close friend Ricardo Falla recalls (1998: 9–10):

> In his September 9, 1980 letter, he told us he was joining the uprising in a place in the Guatemalan mountains . . . and, though he was not asking permission to leave the Society of Jesus [the Jesuits], he recognized that he could no longer fulfill his responsibilities to his congregation. He said, 'As always, I am taking this step filled with hope and based on the principle that my primary loyalty is to the people in whom God is present and for whom we are instruments.' Fernando, too, had fallen in love with the people. It was a love that called him to the armed struggle, and ultimately to sacrifice his own life.
>
> We could tell that in the dialectic between his two loyalties, Fernando's sense of belonging to the Jesuits was eroding. We tried to keep the links with him in the hopes that he would return to where he first learned to love the poor, but his guerrilla life . . . involved him in a way that was exclusive of other commitments. In March 1981, Fernando asked that we no longer consider him a member of our family and profession, a request we respected, though it pained us deeply.

He felt that if he continued as a Jesuit, he would be unable to do what he wanted to do; to live with the people (Hoyos de Asig 1997: 122). Shortly after he left the Jesuits, Hoyos wrote, "Never will I cease to be a Christian. . . . I think that even if I should cease to believe in God, He will never cease to believe in me. That is the source and the secret of my hope. A hope that does not avoid any sacrifice, any suffering nor any tears, but one that helps to transform everything in life, even if it means giving up one's life" (Hoyos de Asig 1997: 7–8).

Hoyos became a member of the EGP's National Directorate and soon was in charge of the Ho Chi Minh front in the Ixil area (see the introductory photograph). It was one of seven such EGP fronts in various parts of the country. In a rambling magazine interview, Hoyos outlined the place of the EGP in the socialist revolution. Asked about his view on the use of violence, he replied:

> Experiencing the great exploitation, the discrimination, the brutal repression by the army of the rich at the service of the rich, I was

getting to understand the root of the matter. I realized that there was only one road to obtain a true peace for the people, that there was no other possibility, and not for the least moment was there ever a problem for me about the moral prescriptions not to kill etc., no problem at all. Why? Because a gun is to give life, not primarily to kill, but to give life to the people which is our purpose. The gun is used to destroy those who are destroying the life of our people. Therefore not for one moment was this a problem for me. (Hoyos de Asig 1997: 191–92)

Fernando Hoyos had traveled far from his middle-class roots in Spain and his life as a Jesuit.

Hoyos's Influence on Others

The journey of Cristina shows a gradual evolution from Catholic Action to the early CUC and later to the EGP. Her journey was given a direction by Fernando Hoyos. She was from the village of Estancia, near Santa Cruz del Quiché. Its residents had traditionally relied on milpa agriculture and temporary work on the South Coast plantations (Samandú, Siebers, and Sierra 1990: 71–84). With the help of Catholic Action and the cooperatives organized by Padre Luis, the village had electricity, was using chemical fertilizer for agriculture, and had obtained sewing machines for the manufacture of hats. People no longer needed to work on the plantations. Padre José María Gran had organized monthly get-togethers for the young people of the aldeas. They discussed their common problems and related them to religious themes. Some became catechists. They were young, second-generation Catholic Actionists who were literate, had a developed social consciousness, and were determined to bring about reform. They were impatient with the older leadership of Catholic Action and became a dissenting group within its ranks. This problem arose in a number of Catholic Action communities (Samandú, Siebers, and Sierra 1990: 95, 107). Cristina tells how this background led her into the Ejército Guerrillero de los Pobres (EGP).

> I was born in La Estancia.... I am the oldest of four children. My parents, brothers and sisters, and I are poor Indian peasants. When I was three, I began going with my parents to the south coast plantations for the cotton and coffee harvests. The first child in my family to be killed, died there because of the poison sprayed on the coffee plants. After my brother died, my mother who was picking coffee, kept him

on her back the whole day. She waited until she had weighed the coffee before she put him down, and we buried him in a hole we dug behind the shelter where we slept with the rest of the workers from our village. None of them reported my brother's death, because the boss would have fired all of us on the next day. I never went to school because, being the oldest, I had to help my father plant and tend the corn. I learned all about farming the same as a man. At sixteen I went to church and learned my catechism. There I taught myself to read. (Carmack 1988: 48)

A catechist from the area describes the church services: "We used the Bible to preach at the masses and at the celebrations of the Word meetings; we always commented on it. In the beginning the activities were supervised by the priests, afterward the catechists did it themselves once they had completed the preparatory course. One had to think seriously about it because of the risks involved in teaching the word of God. The Bible was the main foundation of our formation. In it we saw the struggle of the poor" (Samandú, Siebers, and Sierra 1990: 78).

Cristina describes what she learned from the catechists:

Because of my outgoing personality, I joined the Catholic Action group when I was very young. I was the only woman who went to meetings. . . . I learned why there are rich and poor people, why we have to struggle, why we women can be involved, and that we have the same rights as men. I heard all these lectures in town [centers for Bible training and social-consciousness founded by Padre Luis] and when I'd go back to my village, I would seek out my women neighbors and relatives and teach them in the Quiche language what I had learned. In those days a priest began to come to the village with some Indian and ladino youths. They taught us how to read and write and how to fertilize the earth for planting; they spoke to us of civic-mindedness and human rights, and other things. With these talks many of us in my village began to "open our eyes and ears." (Carmack 1988: 49)

The priest was Fernando Hoyos (Samandú, Siebers, and Sierra 1990: 79–80). He began working in Estancia in 1972. A catechist recalled: "Fernando came for the sole purpose of uplifting the campesinos. He organized a work and study group with a vision of the future called NUKUJ (in K'iche', the preparation before the fiesta). There we analyzed campesino and other community organizations; for example the cooperatives, the political parties, the Ligas Campesinos and the church itself. We had an introductory study course called citizenship. This experience helped to wake up the people, to make our brothers realize that they had rights that had been denied them for years."

Cristina became the president of the women's section of Catholic Action. In 1974, the young leaders of Catholic Action began meeting with leaders of other K'iche' communities to discuss their mounting problems. Of serious concern was the perilous situation of the cooperatives. Guatemala was going through an economic crisis, described in chapter 17. Prices for fertilizer and other agricultural inputs had soared, so that members were unable to make enough to pay the interest on their loans. Burdened with these problems, and dissatisfied and impatient with the gradualism of diocesan Catholic Action and its older leadership (Samandú, Siebers, and Sierra 1990: 80–82), Cristina simply says that in 1976 the people of Estancia "shifted . . . membership to the CUC" (Carmack 1988: 51). These dissenters within Catholic Action had informally grouped around Hoyos, who took them into the CUC when it was formally constituted in 1978. The CUC did not turn to violence until 1980. The majority of Estancia's catechists and young men became guerrilla combatants. In 1980, Estancia was raided by the army during the "ashes" phase of counterinsurgency; many of its inhabitants fled or were murdered and the town reduced to ashes (Samandú, Siebers, and Sierra 1990: 84). Cristina lost her sisters, brother-in-law, and other relatives (Carmack 1988: 55).

Other Priests Who Join the Insurgencies

A priest from Northern Ireland, Donald McKenna, joined the Quiché Diocese and worked in Zacualpa. He had a reputation for being impatient and disliking pastoral duties. He was involved in community development projects before joining the EGP as a combatant. In the mountains, with a machine gun on his shoulder, he was asked by a Spanish reporter to explain how a priest became a guerrilla combatant. "For me the responsibility of a priest is to be on the side of his community. In Zacualpa . . . I celebrated Mass, baptized, and performed weddings. However the actual situation of the community obliged me to think about other things. I began to see my parish as a prison in which the community was held prisoner. I had two options: to bring food and money to the prisoners so as to mitigate their situation, or to help the poor to organize in order to break down once and for all the restraining bars of misery and oppression. I chose the second option and here I am" (Le Bot 1995: 144). Asked about other alternatives, he said that elections were a farce and that violence had been imposed by the military and their closest accomplices, the imperialistic Americans. There was no other alternative. "The violence of the poor is not violence, it is justice" (144).

Several other priests joined guerrilla organizations, but there is no information about their reasons. In Guatemala, Andrés Lanz of the Sacred Heart fathers was in charge of the EGP's clandestine radio network. He was captured in 1983 and is presumed dead (Stoll 1993: 335). A Jesuit and close friend of Hoyos, Enrique Corral, also joined the EGP. He survived the war and became an important spokesman during the peace accords for the coalition of the Left, the Unidad Revolucionaria Nacional Guatemalteca (URNG). In the Chimaltenango area, "not surprisingly, in the aftermath of the brutal repression some priests as well as congregants became convinced that the only way to redress the flagrant abuses was actively to join forces with the armed insurgency" (Green 1999: 152–53). In Chiapas, the priests in Simojovel, Palenque, Ocosingo, and Yajalón cooperated with the EZLN; another edited a clandestine newspaper; and several were Marxist liberationists in contact with key leftists (Pazos 1994: 154). The pastor of Sabanilla served as a chaplain in the EZLN army (Tello Díaz 2001: 154).

Summary

This chapter has described Liberation Catholicism in arms, a turning to violence seen as the only alternative to combat governmental oppression and those allied with it. Under the leadership of ladino revolutionaries, it was a desperate effort to implement social justice. This type of Catholicism had little patience, if not scorn, for Tridentine Catholicism. It looked at a Catholicism that stopped at empowerment of the Maya as well intentioned but unwilling to face what they considered the stark fact—that peaceful means did not work.

So far, Part VII has shown the diversity among Catholic priests in the Maya areas. Some Catholic priests retained the Tridentine outlook that emphasized a world to come and participation in sacramental rituals. They abstained from community involvement other than teaching catechism and administering the sacraments. The empowering Catholic priests, in addition to their catechetical and ritual duties, were personally involved in community development and liberation projects to overcome the subsistence crisis. In spite of their frustration with continual governmental corruption, they were willing to work within but against the system. They were revolutionaries, but peaceful revolutionaries seeking radical reforms.

A small number of priests, driven beyond frustration by the corruption and injustice they faced daily, came to the same conclusion as a number of their parishioners—that only by an armed rebellion could reform take place. They joined the insurgencies.

How many of each type of clergy were there in the Maya areas? The available information cannot answer the question, but Wilson's (1995: 215) description of Alta Verapaz may be helpful. "In several parishes, the teachings of the clergy tended toward liberation theology, predisposing the catechists to the revolutionary groups. In a few places, the clergy actively endorsed their congregations' decision to support the insurgents. These were rare occurrences, however, and the majority of the missionaries took a neutral stand in the war. Others . . . campaigned against the guerrillas, forbidding 'their' catechists to join them." In Chimaltenango, Green (1999: 152) observed: "Most church workers, however, were not revolutionaries. Although they had become convinced of the need for substantial social change to redress the institutionalized poverty and racism so evident in rural villages, their challenges were made through conventional means, such as leadership training, popular education, and small scale development projects, methods utilized by both base communities and USAID alike."

26

Liberation Catholicism

Its Relation to the Morality of Armed Rebellion

In bringing Maya to social consciousness and empowering them to take action to confront their crisis, the church played a role in the events leading up to the insurgencies. The preceding chapters have described the participation of some Catholic Action pastoral workers, catechists, and members of Word of God communities in the armed uprisings. As a result, it has been charged that the liberation movement and Catholic Action were responsible for the slaughter that took place during the government counterinsurgency in Guatemala and for the uprising in Chiapas (Meyer 2000:103).

Does Liberation Catholicism Necessarily Lead to Armed Insurrection?

That it does not is shown by the fact that in both Guatemala and Chiapas, there were many Liberation Catholics who did not join the insurrections.

The positions of Luis Gurriarán, Mardonio Morales, and many others show that there were pastoral workers who were active in community programs to empower the Maya, had often encountered and were frustrated by government corruption, but rejected the use of armed violence. A number of Action Catholics in the western regions of Guatemala, many in the Lacandón, and all in the Jesuit region of Chiapas refrained from armed insurrection. Within the Dominican region itself, there were pastoral workers and members of liberation communities who refused to join the EZLN.

Despite disagreements within the movement, the refusal to participate in a violent solution is reflected in the writings of liberation theologians. As McGovern (1989: 186) found: "Though I have tried to search out writings that deal with the question of violence, I have yet to find any statement by a liberation theologian calling for violent revolution or declaring it necessary for bringing social change in Latin America. One might expect to find some statements to this effect at least in the early years of liberation theology. With this in mind I read through a collection of testimonies (*Signos de liberación*) from Christian activist groups and liberation theologians covering the years 1969 to 1973. In 290 pages I found no statements that encouraged violence."

The intervening factor in the cases of those who did take up arms was the insertion of ladino Marxists from outside the region who were successful in turning the frustration and impatience of some Action Catholics, especially younger members, into armed rebellion. When nonviolent actions were frustrated by a corrupt system or were not given enough time to develop, the Marxist agenda implemented the "act" of the "observe-judge-act" of community discussions.

The Moral Question as Seen by the Liberationists in Arms

More than fifty men and women from the 116 families in the community of Santa María Tzejá in the Ixcán (see chapter 17) collaborated with the EGP in some capacity. They gave the following reasons for joining the insurrection.

1) . . . a pervasive and well-founded anxiety of losing their land to the army or the oligarchy or both, a concern fed by the uncertainty of land titles. . . .
2) . . ."paradox of success": the more the villagers prospered, the more they chafed against the weight of an unresponsive and brutal social system. . . .

3) ... the difficulty of peaceful change. Attempts at reform in Guatemala, no matter how modest, risked running afoul of the rigid social structure and a deeply authoritarian state with severe, frightening consequences. ...
4) The villager's new understanding of liberation from the Bible ... provided the moral underpinnings for their vision. These values transcended immediate self-interest and looked toward achieving a more equitable and democratic society.
5) Defense against the army and revenge for its atrocities fused into a powerful recruiting tool for the rebellion. ...
6) ... as the insurgency reached a climax, peer pressure played a role. Some particularly the very young joined because everyone else was getting involved, because of the initial excitement and adventure. (Manz 2004: 107–8)

These reasons were shared by many other Maya communities who took up arms. They had been empowered by the Bible, completely frustrated by the government inaction on land titles and the continuing threat of eviction, and further frustrated in their attempts to correct this and other governmental injustices. Added to this were the army's atrocities during the first phase of the counterinsurgency. Given all this, it appears that Maya communities almost automatically assumed that they were morally justified in opting for the violent solution when offered the opportunity by the EGP (Le Bot 1995: 140).

Was There Moral Justification for Those Action Catholics Who Participated in the Armed Uprisings?

The question of the just war has a long tradition in Catholic thinking. Taking off from Aristotle, it was reformulated by both Augustine and Aquinas, among others. There are many degrees of violence. The context here is armed violence in a war situation. To summarize, four conditions must be present to justify the use of offensive arms in these conditions:

1. There must be serious, lasting and certain damage inflicted on a community; 2. all other means to combat the source of the damage must have been shown to be ineffective (strategic consideration); 3. there must a reasonable possibility that the use of armed violence will accomplish its goal (military possibility); 4. there must be a reasonable basis for assuming that the use of armed violence will bring about social justice and not lead to a situation of greater evil and

disorder than that which it is attempting to replace (political possibility). (McCormick and Christenson 2003: 635–44; Catechism of the Catholic Church 1994, #2309)

The first condition addresses the moral issue in a narrow sense. The last three are pragmatic judgments that the use of arms is strategically the only possible option, that militarily and politically there are reasonable grounds to think that they may be able to attain the goal of social justice. All four conditions must be met to morally justify the use of armed violence.

The preceding chapters have briefly described the role of the Guatemalan and Chiapas elites and the local and national governments in causing and maintaining the Maya crisis. Most agree that it was a situation of tyranny, of systemic and prolonged injustice.

There were years of trying to trust in what the government said it was going to do about the land situation and public services, of trying to work within the system. But these efforts were constantly answered by governmental unconcern, corruption, and violence. Frustration had been building and increasing with each passing year. Many Maya who had been brought to consciousness about the systemic reasons for their crisis were now convinced of the ineffectiveness of trying to work within the system for social justice. Their frustration turned to anger and they seized the opportunity offered by the arrival of the radical Marxists, with their arms and promises. For these Action Catholics, the first two conditions of moral justification for taking up arms were sufficient. In the circumstances in which they were living, the conditions were so self-evident that there was little reflection about the moral dimension. This writer can find no discussion of the remaining two norms. Since all four conditions must be met, an examination of these possibilities is necessary to judge the objective morality of the armed rebellions.

Military Possibility of Achieving Its Goal

There must be a reasonable possibility that the use of arms will be successful in defeating the violence by which a system of social injustice maintains itself. This norm does not require certitude that the use of armed violence will be successful, nor does it require a very high probability that it will be. Reasonable means that the decision to take up arms is based on a military assessment of available information by competent persons. Ultimately, it is a question of risk, a prudential judgment. But before the vicious cycle of violence and counterviolence is unleashed, it is morally necessary to make a cold calculation of its worth, and not be swayed by the often understandable emotions of the moment. Sometimes, the question may not so much be whether violence is a military possibility, but rather when the use of violence is possible. In concrete circumstances, the question of military possibility can take many forms.

In Chiapas, it is clear there was little or no assessment of a successful military outcome. The bishop had initiated a campaign of dissuasion that had made significant inroads on EZLN military strength. The grossly inadequate arms of the EZLN used in the uprising were described in chapter 20. There were only two hundred automatic weapons among all the EZLN troops. Many combatants were armed with only wooden pikes and spears. The causalities would have been catastrophic if the government had pursued the EZLN and not called the unilateral cease fire. All these obvious military inadequacies raise the question of whether the "uprising" was really an uprising, really a military operation. The EZLN was so militarily unprepared that one wonders if their rebellion was intended to be a brief pseudo-military show for Marcos's and the FLN's political and propaganda purposes. As a propaganda effort, it was highly successful.

The Guatemalan insurrection was a drawn-out conflict. Significant Maya involvement, especially by the EGP, was largely confined to the late 1970s and early 1980s. After the army's "ashes" strategy of counterinsurgency, the EGP was essentially defeated. Two EGP combat insurgents, Mario Payeras and Yolanda Colom, along with Gustavo Porras, a staff member of the EGP's National Directorate, have written memoirs of their experiences, including critiques of the EGP's unrealistic strategy (Payeras 1983, 1987; Colom 1998; Porras Castejón 2009). Payeras (1991) synthesized the critiques in an unsuccessful call to keep the Guatemalan revolutionary movement alive. The question is: Could a prior assessment of the situation have raised serious doubts about the ability of the EGP's effort in its area of operations to defeat the Guatemalan army? If so, it means that the insurrection and the involvement of Action Catholics lacked objective morality. Although the EGP was only one of four Marxist groups involved, it appears that similar factors were at work in the other three groups, as they had similar outcomes.

Failure in the Rural and Urban Sectors

The immediate strategic failure was the EGP's inability to protect its support communities from the attacks and temporary occupations of the military. During the army incursions into the communities, it would resort to torture and murder to gain information about all aspects of EGP operations and kill all those identified as members of the group. The army's counteroffensive culminated with the "ashes" strategy. This meant that EGP was never able to establish any fully liberated zones and raise a popular army (Colom 1998: 237, 246, 310).

This lack of front-line capability arose from a series of other failures. In spite of many Maya joining the EGP, there was a constant shortage of guns and ammunition (Colom 1998: 147). Trainees used sticks in place of rifles (see the photos in Simon 1987: 131–33). New combatants were forced to fashion weapons made of wood (see the photos in Kobrak 2003: 44). Even if guns were available, EGP commanders could not devote sufficient time and resources for Marxist indoctrination of the flood of Maya recruits. Workers from the city were needed to instruct the Maya about the economic and political structure of the country, and of the EGP's political strategy, which required patience and a long-term view. All military considerations were supposed to be subordinate to the larger political picture. Some Maya were politically sophisticated, had experience of the national scene, and were well equipped for EGP leadership and understanding its worldview. But as Colom (1998: 228, 247, 310) points out, many others from outlying areas only identified with their local community, were not conscious of a larger Maya identity, much less of the political entity called Guatemala in which they supposedly had rights as citizens. Many were illiterate. These characteristics were the result of their social and geographical isolation, not any inability on their part. These Maya were provoked, frustrated, and wanted to immediately fight the local figures who were oppressing them. In EGP units, disputes arose over short-term military objectives versus longer-term political goals. At times, the military perspective prevailed, which created long-term problems.

The EGP strategy was "to incorporate as the principal moving force of the revolution, the poor indigenous campesino" (Payeras 1991: 13). Without any experience among the Maya, the ladino Marxist theoreticians saw them only as a poor exploited class who would respond to the EGP by accepting their political philosophy and forming a popular army. The dependence of the Guatemalan revolution on the Maya, a people from not only a different class than that of the Marxist theoreticians, but one from a different culture, made that struggle uniquely different from the Cuban, Nicaraguan, Vietnamese, and other Marxist rural revolutions. Yet the EGP model was primarily taken from these insurrections. While some Maya attained high positions within the EPG, no Maya were ever promoted to its all-ladino National Directorate.

The problems of the rural sector, especially lack of arms and insufficient training of recruits for guerrilla warfare, were largely the result of the failure of the EGP's unit in Guatemala City to attain its objectives and supply the rural sector. Politically, the EGP's recruitment was largely restricted to university students. It was never able to establish a popular base among the working class and forge the class consciousness needed for development of both the urban and rural sectors (Payeras 1987: 19, 25). It never grew strong enough to draw government troops away from the rural sector.

An important reason for the inability of the EGP to expand was the strength of army intelligence, which continually restricted the EGP's capabilities in Guatemala City and finally destroyed it. The Guatemalan army was held in low esteem by many Guatemalan intellectuals, university students, and others, including some pastoral workers—an attitude inherited by the EGP (Kobrak 2003: 63; Porras 2009: 71). Its officers were seen as a confused group of military bureaucrats more interested in their personal finances, mistresses, and military advancement than in running an army. But Guatemalan intelligence officers, in addition to their American advisors, sought instruction from foreign intelligence agencies with experience in defeating urban insurrections, especially Israel and Argentina (Porras 2009: 66, 78). The Guatemalan intelligence operatives learned their lessons well and developed a highly efficient operation that constantly hobbled EGP strength in the city so that it was unable to provide the necessary logistical and trained manpower support for the rural units (Colom 1998: 147–48; Kobrak 2003: 45, 47).

Ultimate Source of the Chain of Problems

Many of the EGP's problems can be traced back to its very beginnings. The EGP had been formed by a small group of urbanized ladino intellectuals after the defeat in eastern Guatemala. Most were middle-class, well-educated patriots dedicated to a Guatemala that embodied social justice. They were also radical Marxists who saw warfare as the only means to accomplish this goal. None of them had a military background. They were a small clandestine group that formed a revolutionary subculture of their own, one defined by a 1960s Marxist worldview that had been encouraged by the Cuban uprising. As events proved, they suffered from both class and culture ethnocentrism. They were never able to sufficiently insert their movement among the Maya in the countryside, nor among the working class in Guatemala City. Educated, dedicated, and highly capable Maya, such as Emeterio Toj Medrano, were refused positions on the National Directorate, the highest level of EGP authority (Konefal 2010: 134, 215n6).

The ethnocentrism of the Marxist subculture had some similarities to the clerical subculture of the Catholic Church. Experience among the Maya changed some members of both subcultures. Mario Payeras played an important role in the Marxist subculture—as a member of the National Directorate of the EGP, as a member of the original invading force in 1972, and as the commander of the EGP front in Guatemala City. In 1984, he broke with Comandante Rolando Morán, the leader of the EGP, and resigned based on his firm conviction of the impossibility of military success. His collected essays, *Los fusiles de Octobre*, sum up his critique, showing that the EGP's basic assumptions were wrong, and that they were responsible for its devastating defeat.

An underlying problem was the EGP's strategy: form small guerrilla units in the rural areas that would enlist frustrated and oppressed peasants to form a popular army because they would readily understand class warfare; defeat the military; and take over the government. This strategy gives primacy to military over political considerations. Based on the history of successful revolutionary movements, Payeras maintains that these first two steps were inverted. The oppressed classes must first be brought to consciousness of their place within the class structure of the country, won over politically to seek change, and then, only when conditions were favorable, would the military phase be launched. But the EGP attempted the opposite, especially after the Sandinista victory in Nicaragua and the successes in El Salvador. They never politically educated the working class in the city, nor the Maya in countryside. They rigidly hung on to the class-versus-culture dichotomy, even when Maya such as Toj Medrano saw it as a false dichotomy that was hindering the recruitment of middle-class Maya and those in their sphere of influence.

It is interesting to see Payeras, a man of many talents, coming to consciousness about the Maya of his own country. His evolution is similar to that of some pastoral workers seen in the preceding chapters. In 1982, he published an article in the EGP journal. It reads like the introduction to a Maya ethnography, as he begins to realize the implications of the fact that the Maya comprise approximately 50 percent of his country's population, that Guatemala is a bi-ethnic culture (1997 [1984]: 71–89). This is at least twelve years after planning the EGP insurgency. Later, in 1992, he wrote: "Twenty five years after the hopes of comandante Turcios and his companions in the northeast of the country, our understanding of the national ethnicity question still finds itself as nothing more than babbling, rudimentary, if not ignorant. For the average Guatemalan ["average" here seems to exclude the Maya], the new vigor of the indigenous struggle (all through colonial times they rebelled many times and were put down without mercy) may allow all of us to look into the mirror of history, because without doubt, the future of Guatemala is intrinsically linked to the destiny of the indigenous communities" (1997: 50). In his later years, Payeras (1997 [1984]) proposed a nonviolent restructuring of Guatemalan departments into autonomous zones based on the differing Maya cultures of the country and employing the social sciences in this effort.

All of this raises a question for the Action Catholics who turned to Marxian violence. Was there an assessment of the military and political possibility for the success of an armed uprising as a means of seeking social justice? From the available accounts, this writer could find no record where the four conditions were discussed by pastoral workers in their meetings with catechists, much less in meetings of catechists with their Word of God communities. The communities that took up arms seem to have been so intent on overcoming their immediate crisis that the larger questions of the future, the military, and political possibilities were not raised. The morality of the first condition seemed self-evident, and frustration with using peaceful means appeared equally self-evident. These two conditions alone were the basis for the moral justification of the acts of many Action Catholics who undertook armed rebellion.

One Who Did Reflect on the Morality of the Situation

Father Ronald Hennessey (see introductory photograph) questioned the military and, especially, the political possibility that the EGP would bring about social justice. He was from an Iowa farm family, was a veteran of the Korean War, became a Maryknoll priest, and was assigned to Guatemala in 1964 (Melville 2005). The following year, he was sent to Cabricán as an apprentice to Thomas Melville, who was frequently absent, organizing the Petén colonization project (chapter 22). Hennessey ministered to the parish and worked with the cooperatives. He also became the town's only mechanic for its few vehicles, stemming from his experience in maintaining farm equipment and running a motor pool in the army. After Melville's expulsion, Hennessey went to the Petén in an effort to hold the colonization project together. In 1975, he became an administrator, serving for two terms as the elected regional superior of the Maryknollers in Central America. In 1980, he returned to parish work in San Mateo Ixtatán, high in Kuchumatán Mountains in northern Huehuetenango, but close to the lowland Ixcán areas of both Huehuetenango and El Quiché. During this period, the EGP was clandestinely recruiting in his parish and the army was carrying out the first phase of its counterinsurgency plan—kidnappings, tortures, and indiscriminate murders.

A Catechist's Question about the Morality of the War

In one of the outlying communities of his parish, the head catechist, the son of the first Maya in the village to make the break from traditionalism, became a member of the EGP. When Hennessey visited the community, the catechist asked him about the morality of the EGP agenda. Here is a summary of the exchange (Melville 2005: 111–12):

> The catechist asked, "It be a sin for us to . . . fight . . . protect ourselves, our trees . . . take back our stolen land . . . save our children, our siembras [crops], animals, our lives?" Hennessey was not sure how he should answer. In a sympathetic tone, he said that God's authority rested with those who had been chosen by the people to rule over them. This was a rejection of the stolen elections by the Guatemalan military. He continued "But you must first think of how much it will cost you in blood and destruction to defeat a military government [military possibility]. . . . And how can you be sure that an EGP government will serve your interests more than the present regime [political possibility]. It could be worse! . . . The Catholic church doesn't have moral answers necessary to take sides in every political or social conflict. . . . The Lucas García government is patently unjust, even insanely murderous [first condition]. . . . But that doesn't necessarily mean that the EGP is to be trusted to come up with something better [political possibility]. . . . You have to decide how you should do this [strategic consideration]. I can't give clear answers to every problem that comes up! . . . You know God's laws and you've always respected them. Just try to continue that way."[1]

Here, in this exchange, Hennessey posed the four criteria for the catechist to ponder.

An Attempted Recruitment

Given his background, the EGP thought that Hennessey might be willing to join them. Late one night, an EGP officer and four Maya soldiers stole into town and came to the priest's house. They were admitted by the padre and, after some pleasantries, a long discussion ensued (Melville 2005: 148–58). The officer began to present his case, "Look, padre, we're EGP guerrillas! We're going to defeat the fucking Lucas García dictatorship [referring to the president, an army general] and we're going to make a true agrarian reform. The land should belong to those who work it, not to thieves or the children of thieves! I'm sure you approve of that."

Hennessey, while having no respect for the government, was also wary of the EGP. He calmly expressed his reservations. "You folks have hurt a lot of innocent people around here.... If the tactics you've been using are an indication of the kind of government you'd set up ... then, it seems that Guatemala won't have any more justice than it already has with Lucas García [political possibility]."

The officer resented the comparison. He rattled off the names of the departments where the army had carried out massacres and burnings of entire villages. He cited the kidnappings and murders of priests and catechists. "We thought that you, after having worked at a colonization program in El Petén, would understand our struggle and be on our side."

Hennessey replied, "I was not equating the EGP with the army. I was in Yacá and saw what you people did to Juan Silvestre. They suspended him by the neck for four hours because he was a *contratista* [labor recruiter] for the finca owners! ... And another incident! The EGP killed Andrés Sebastián in Ixbajau [the leader of a town of staunch traditionalists] simply because the people there taunted your troops.... And your compañeros killed several elders in Yolcultac without the semblance of a fair trial [political possibility]. Who's next?"

The guerrilla replied that he could not defend everything the EGP had done, that recently some had been allowed to join the organization without the required period of training. He pointed out the problem of organizing the Maya. "Some of our people think this fight can be won with the same kind of popular uprising that occurred in Nicaragua. But we have a centuries-old ethnic problem here, something the Nicas didn't have. Most of us believe that it will take years of slowly escalating armed struggle, training people as we go. We've got to be prepared to take on your Pentagon.... We have not forgotten 1954, even if you have! ... Your country ended our Ten Years of Spring [the Arévalo-Arbenz reform period]. The only sincere attempt at democracy we ever had! ... And your fucking government wouldn't let us have it. The fucking United Fruit Company meant more to ... your fucking government, than justice and democracy ever did, even now, for the Guatemalan people."

Hennessey, waiting for the guerrilla to calm down, replied, "This is not

1954.... Maryknoll has come a long way since then, as has the Church. We have Pope John XXIII to thank for that!... Maybe it's true that President Reagan and the current pope, John Paul II, see a communist conspiracy behind every cry for social justice, but that doesn't mean that all of us do. ... But our sympathy for the Guatemalan people doesn't mean that we give carte blanch approval for all and any kind of revolutionary activity." The guerrilla was not listening carefully. He continued to complain about the lack of preparation and training for the swelling numbers wishing to join the EGP in the Maya areas, and about some of them acting on their own without EGP supervision. Hennessey continued, "I guess the things that bother me most are those incidents where not even a trace of the victim's guilt is present." Hennessey cited a case were the EGP had used a Claymore mine to blow up a truck carrying Maya workers under its tarp whom the EGP thought were soldiers. When they discovered their error, the EGP had simply fled the scene rather than summon help [political possibility].

The guerrilla tried to use the accusation to show how the revolution needed the padre's support. "Look, Ronaldo, I like you! I want to come to an understanding with you. This revolution can't be successful unless we have all the exploited pulling together. But if the people think that we're cowards, that we're capable of setting them up so that the survivors join us, then we've already lost." This refers to the accusation of provoked repression, that the EGP would bait the army to commit acts of terrorism in order to increase recruitment to its own ranks. The accusation was that the EGP would enter a community, receive help from them, and then purposely depart, leaving the community defenseless. In retaliation, the army would enter the community and carry out its tactics of tortures and murders. This, in turn, would infuriate the community, and they would retaliate by joining the EGP (Morrissey 1978: 83). The guerrilla had anticipated this objection and answered it before it had been raised.

After some more back and forth, Hennessey addressed the main purpose of the EGP's visit. "What do you want me to say? That I'm 100 percent on your side? ... I've never made a blanket statement that revolutionary war is always wrong. The last pope, Paul VI, gave the [four] conditions under which revolutionary war can be considered moral. ... But he also said there's danger that such movements will bring new disasters [political possibility].... You must know the history of the Mexican Revolution! How many dead? Five million? Ten million? And for what? And the French Revolution, with Robespierre's Terror? ... Russia had Stalin and China, Mao."

The guerrilla made one last effort to overcome Hennessey's reservations. "Let's face it, padre! You're a North American and a priest! As a person, as an individual, is your primary loyalty to the U.S. government and its hypocritical support of fascism masquerading as emerging democracy, and with the bishops who dress in silk robes, drive around in BMWs, and tell us misery and oppression are God's will? Or are you really with the poor and oppressed of Guatemala?"

Hennessey knew their dialogue would eventually come down to this. He replied, "My recognition of the role played by the United States, the Church and Maryknoll in 1954 [the overthrow of the Arbenz government], if anything, has made me more receptive to your perspective.... They promoted a revolution in favor of Guatemala's wealthy ... and I'm not about to challenge your right to promote one to benefit Guatemala's poor.... But that doesn't mean I'll overlook the injustices your people commit in the name of the poor ... nor am I convinced that you will inevitably form a government that will bring a better life for most Guatemalans [political possibility]."

Based on the past conduct of the EGP in the local area, Hennessey had confronted the officer with serious doubts about the EGP's military and political capability to achieve social justice for Guatemala. The officer realized there was little more that could be said. He replied, "To put all I've said in very blunt words, we will kill all those who would kill us or would have us killed. The Church has always permitted self-defense. Please remember that when you hear about some of our so-called injustices or crimes.... Anyway we leave this discussion for another day.... And despite the way our talk has gone tonight, I appreciate your frankness."

The Second Condition: Was Armed Violence the Only Alternative?

Those who chose armed rebellion to obtain social justice usually did so on the basis that it was the only alternative after all other means had failed. But the preceding chapters have shown that the use of armed violence was also a failure—that it was not a viable alternative at that time in those circumstances. But given a longer time frame, a period of continued suffering, given the necessity of patience for a long, drawn-out effort, a question can be raised about the second condition—that all peaceful means for reform were doomed to failure. This was certainly true for the immediate future. But as one of the better sections of the Vatican document on the liberation movement suggested, "That which today is termed 'passive re-

Figure 26.1. A delegation from Tenejapa arrives in San Cristóbal for the Guadalupe procession.

sistance' shows a way more conformable to moral principles and having no fewer prospects for success" (Hennelly 1990: 487, #79).

In Guatemala, the CUC had been successful in some of its nonviolent efforts before merging with the EPG. A strike on the South Coast had led to increased wages for agricultural workers. The work of Father Luis Gurriarán in organizing cooperatives became a threat to the corrupt ways of the civil and military authorities in Quiché. They complained to the national government, which issued an order for Gurriarán's second expulsion from the country. When the Maya of the department learned of it, a crowd estimated at five thousand flowed into the central square of Santa Cruz and protested in front of the governor's palace. They were adamant in their demand that Father Luis remain and continue his work. The governor, furious and stymied, gave in to their demand, and Luis continued his work of empowerment. In Chiapas, the protest marches of Pueblo Creyente, Abejas, and Xi' Nich' show that these nonviolent efforts were beginning to make some progress by forcing the government to release human rights advocates unjustly accused of crimes.

Figures 26.1 to 26.6 illustrate the way in which a number of human rights organizations with biblical roots organized religious processions, a nonviolent strategy for social justice. These pictures were taken in San

Figure 26.2. A section of the delegation with its placard, "The Christian faith demands justice and charity," with the name of the marchers' community.

Figure 26.3. The Guadalupe image superimposed on the Mexican flag.

Figure 26.4. "Most holy Virgin of Guadalupe, let ladinos and indigenous peoples treat each other as brothers."

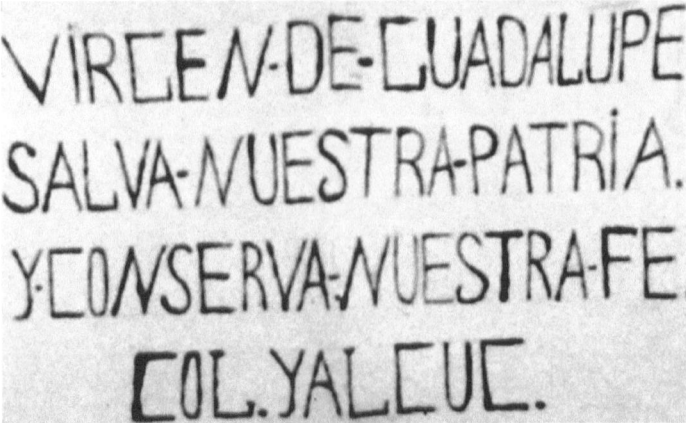

Figure 26.5. "Virgin of Guadalupe save our country and preserve our faith."

Figure 26.6. "Irreverence and injustice destroy nations."

Cristóbal de Las Cases on December 12, 1964, the feast of Our Lady of Guadalupe, the patroness of Mexico and the symbol of national unity. On this day, Maya groups in Chiapas converged on San Cristóbal for a procession to her local shrine. In that year, there was a bitter land dispute in the Tenejapa area arising from ladino efforts to seize Maya land. Figure 26.1 shows a large delegation from Tenejapa arriving in San Cristóbal with their Guadalupe flags and protest signs. They are led by two priests in the foreground, and another one on the extreme right, about mid-picture. Figure 26.2 shows a section of the delegation. The message of their placard places the land dispute in a moral context. Figure 26.3 continues the moral context—the unity of all Mexican people symbolized by the Guadalupe image on the Mexican flag—while figure 26.4 says it explicitly. Figures 26.5 and 26.6 imply what will happen if social justice does not govern group relations. There is no mention of the land dispute itself. But all present knew the specific context behind these generalities.

If these strategies had been patiently pursued and allowed to mature, and if human rights groups had been allowed to build a massive nonviolent movement, could such a movement have brought about the reforms that would have enabled the Maya to escape from their crisis conditions without large-scale violence? How familiar were the pastoral workers with the history of other liberation movements and the successful nonviolent strategies those movements had developed? Nonviolence does not mean that violence will not be used against the nonviolent. All such movements have their martyrs. But in such movements there is no continual cycle of large-scale violence and death. Eventually, the defenders of injustice may be worn down, when they finally realize that violence alone will not prevail in the long term. If the nonviolent sector of the liberation movement had organized, it might have restrained the impatience and frustration of those who resorted to violence. The Maya situation might not have been perceived by the military as similar to the wars in Nicaragua and El Salvador. While the military undoubtedly would have reacted with brutality against a nonviolent movement, it probably would not have reached the depths it did, because it would not have been as threatened. Could the small but incipient successes of the Pueblo Creyente, Abejas, and Xi' Nich' in Chiapas in freeing political prisoners, and, in Guatemala, could the early nonviolent CUC successes, along with the growing political power of the Christian Democrats in local communities, have been the basis for the development of a mass nonviolent movement that would have enabled the Maya in the long run to overcome those who were primarily responsible for their crisis? Yvon Le Bot (1995: 26, 86n11) and David Stoll (2008: 203n2) raise similar questions. The Marxist's precipitous push to armed violence precludes an answer. The Maya situation was far from unique. Protest marches and civil disobedience were successful in the long run for many liberation movements, including the black liberation movement in the United States, Gandhi's movement against British rule in India, and effort to replace the regime of Ferdinand Marcos in the Philippines. Nonviolence as a strategic possibility for confronting the Maya crisis was seldom, if ever, discussed.

Summary

In the worldviews of the EGP and the EZLN, the strategy to obtain social justice was formulated by urbanized, middle-class, ladino intellectuals. In their frustrations with corrupt national governments, they built a grandiose model of leftist reform by violence. They went to the remote, relatively isolated Maya areas of Guatemala and Mexico. In spite of the rampant injustice of governmental corruption and exploitation by landed elites, the Maya, inspired by the biblical reflections and development programs of Action Catholicism, were beginning to make some progress in confronting their problems. But the Radical Left were able to sell their vision of quick social justice by insurrection to frustrated, mostly young Maya suffering through a grueling crisis of lack of land, food shortages, sickness, and early death. There, they were aided by some frustrated, militarily naive Catholic pastoral agents who bought into the radical leftist vision. Earlier chapters noted that close collaborators of both Hoyos and Bishop Ruíz described them as being naive about aspects of insurgency. Few Maya had the ability to evaluate the larger picture into which the Marxists were trying to insert them. The Maya had experiences of violence with regional police and paramilitaries. But most of them had little or no idea of the power of a modern army specifically trained for counterinsurgency.

In brief, the Maya were subjected to the effects of the systemic injustice of the political-economic structures of Guatemala and Chiapas, which were enhanced by Maya's own internal problems. This was the root cause of all that was to follow. Discussions about the failure of attempts to remove the injustice should not distract from the root cause itself. In their efforts to overcome this injustice, in this writer's opinion, some Maya impatiently seized upon insurrection as the solution to their problems. They were misled by theoretical ladino Marxists, and helped by some idealistic but militarily naive Liberation Catholics, both pastoral workers and Maya catechists. Therefore, they participated in an undertaking that lacked objective morality, as it lacked military possibility and perhaps political possibility. This is an easy statement to make, but, understandably, it was not so clear to those living in the fiery cauldron fed by the flames of social injustice. There appears to have been little or no serious consideration of nonviolence as a strategic consideration. The Maya of Guatemala paid a very high price.

VIII

The Search for a Revitalized Maya Worldview

During the last half of the twentieth century, Maya communities came into contact with a variety of theological worldviews, all seeking to convert them. As seen in previous chapters, some clung to the Maya tradition as it had been expressed for much of the colonial and national periods. Others embraced forms of evolving Catholicism. At the same time, Protestant churches and other religious groups were growing in the Maya areas. In becoming aware of these differing theological paradigms, the Maya also realized their ability to choose among them. They added the Spanish loanword, *religión*, to their languages to describe this new experience. There had been no need for such a concept/word in Maya languages based on a worldview whose main outlines had been universally accepted without challenge. But in the minds of numerous Maya, cycles, covenants, emanations, and communication with the internal sky-earth remained important concepts, even as they accepted Western theological paradigms.

27

Choices Faced by Catholic Maya in a Turbulent Society

The previous volume described the Maya absorption of Tridentine Catholicism into their own traditional system during the early colonial evangelization. In the mid-twentieth century, after the liberal suppression and the reentry of the Catholic Church into Maya communities, as described in previous chapters, there was the same initial reaction to the same catechesis, until a breaking point was reached. Tridentine Catholicism then evolved into Action Catholicism, as either Catholic Action in the restricted sense, or as the liberation movement, or both. Catholicism became more involved in Maya communities, helping them to meet their subsistence crisis and become aware of systemic social injustice. Using this as springboard, urbanized ladino Marxists encouraged some Maya to join national insurgencies. The racist and genocidal aspects of counterinsurgency strengthened Maya identity and led some to seek a worldview in which Maya tradition and its moral sense played a more prominent role. At the same time, some sought to escape their Maya background by taking on ladino ways. This chapter describes some paths chosen by Maya Catholics as they sought to establish their identity in the confusion of the last quarter of the twentieth century.

Ladinization

Action Catholicism had emphasized the importance of education for the Maya. It established programs of adolescent and adult literacy, founded parochial schools in numerous communities, and in others encouraged attendance at government schools. A number of Catholic students received scholarships that enabled them to continue their studies at higher levels, some eventually becoming university graduates. Most of these experiences led to inculturation in the national ladino culture. The introductory photograph shows marching Atiteco students from the government school celebrating Guatemala's independence from Spain. These experiences raised a question for Catholic Maya. As Action Catholics seeking to overcome the crisis of their culture, would their educated children still identify with their Maya background and the interests of their own people? Chapter 12 quoted the fear of Catholic Action parents about the younger generation forgetting their roots. An unknown number of Catholic Maya did leave their roots and identify themselves as ladinos, although some may have retained their Catholicism as ladinos. Such retention could be ambiguous, however. Many ladinos were "cultural Catholics," because they were born into the religion but never took it seriously.

Another option was to join a traditional Protestant denomination, where conversion frequently meant taking on many ladino characteristics. Scotchmer (1991: 46, after Emery) lists eleven characteristics shared by Protestants: "1) high moral standard, 2) emphasis on family stability and marriage sanctity, 3) increased education, 4) political liberalism and anti-communism, 5) positive values attached to Ladinization, 6) financial independence and the value of work, 7) egalitarianism, 8) democratic church government, 9) intentional separation from the world (breaking down), 10) Protestantism more valued than community ties, 11) mutual recognition of '*hermanos*' (brothers)." Obedience to civil authority was also stressed. There was total rejection of traditional Maya culture, especially the saint cult in any form.

Charismaticism

For many, this option, with its Life of the Spirit experience, was a revitalization of dream visions and shamanic experiences, core elements of traditional Maya worldview. As seen in chapter 2, and as the Maya ethnographer Jacinto Arias (1991) repeatedly points out, communication with the inner sky-earth, the world of emanated gods, was an essential element of the tradition and the role it played in the self-definition of individuals. Several Q'eqchi' Baptist charismatics communicated with the mountain spirits of their community, who in turn urged them to retain some traditional customs (Adams 2001: 210). "The Pentecostal and Neo-Pentecostal movements are attractive because they sanction within an institutional framework a form of spirit possession which permits the Guatemala believer to continue with his previous worldview and understanding of the holy. . . . Pentecostalism is only living up to its own understanding of its purpose; it is being revivalistic" (Evans 1990: 286; see also Cook 2001: 148, 161).

Both Tridentine and Action Catholicism had moved away from the Maya experience of communication with the internal sky-earth (Samandú, Siebers, and Sierra 1990: 120). But as seen in chapter 8, many Catholic Maya did not reject all the traditional worldview. It lived on in the background, and was still called upon to help explain Maya experience in the confusion and anomie of a world so different from the traditional one to which they had been accustomed. These inward experiences were a type of religious activity many Maya preferred during the tense environment of militarization. The charismatic sessions involved Maya providing the music, singing, preaching, and interpreting the experience for other Maya. Everyone could undergo the charismatic experience anywhere. It was independent of any hierarchy: Catholic priests, Protestant pastors, or traditional Maya shamans. While Action Catholicism also demanded personal renewal and asceticism, in the turbulent conditions of the 1970s and 1980s, it sometimes paid inadequate attention to personal religious experience and its role in both personal and social reform.

The emphasis on material prosperity was another attractive aspect of charismaticism for the Maya. The neo-Pentecostals "place great emphasis on 'healing in the atonement' which roughly translates into empowerment on earth through proper faith in God. Closely related is the belief that material prosperity is the entitlement of the faithful; money, good health, security are all tangible evidence of God's benediction. The believer is thus right to demand such things from God, for personal prosperity is witness to His power and grace. This message is underscored by the preaching of televangelists broadcast from within Guatemala and by satellite from the United States. This strong emphasis on prosperity in Guatemalan neo-Pentecostal churches has been dubbed . . . 'health and wealth' theology" (Garrard-Burnett 1998: 164). This worldview appears to be an updated version of the Maya covenant's emphasis on material security in return for observance of the moral order and the required rituals. This was a revitalization quite different from that of Action Catholicism.

Charismaticism emphasizes individual religious experience. When disputes arise among kin groups, persons claiming such an experience may found their own charismatic church. This splintering effect contributes to the appearance of so many different charismatic churches. It is also attractive because leadership depends solely on a personal experience of the Spirit instead of the education and probation periods required for priests and ministers. The charismatic withdrawal from social action was attractive to some Maya Catholics. They disliked Action Catholicism, with its criticisms of the political and economic system. They thought it was a mixing of the sacred with the secular. For others, it interfered with their goal of bettering themselves materially within the existing system. The charismatic demand for a disciplined life also fitted in with the attainment of this goal. Scotchmer's (2001: 249) description of a certain type of Protestant leadership also fitted some Catholics: "No attention is given to promoting health care, agricultural innovation, literacy or education, as these will be resolved somehow in God's time without human intervention.... Literacy efforts, development projects, and local as well as national political topics are not part of the Indian subordination model for these are seen as worldly and problematic."

Diffusion of Charismaticism

Chapter 11 described the beginnings of charismaticism in Guatemala during the early 1970s. The 1976 earthquake brought a significant increase of personnel and resources from Pentecostal organizations in the United States that greatly assisted the growth of the local charismatic groups.[1] This was followed a few years later by the tumult of insurgency and counterinsurgency. During this period, significant numbers of Catholic and Protestant Mayas became Catholic or Pentecostal charismatics. "The uncertainty of the mid 1980s, it seems, produced a fertile environment for Protestantism—with its emphasis on eternal truths and verities—to flourish among the poor whether they resided in the damaged communities of the countryside or the miserable *barrancas* (slums) of the city" (Garrard-Burnett 1998: 164). This was especially true of Pentecostalism. In San Juan Ostuncalco, 57 percent of all Protestants were Pentecostals. Scotchmer (1991: 128, 167) estimates that one-third to one-half came to

Pentecostal churches from other Protestant congregations, most of which were ladino-oriented. "The energy and excitement are nearly irresistible to persons looking for a more inward expression of conversion when compared to the sedate and rationalized Protestantism typical of some non-Pentecostal churches."

Much the same was true of Tridentine Catholicism, with its formal system of belief and standardized rituals oriented toward reform of one's personal life. A number of Action Catholics joined Pentecostal congregations. Frequently, this was a defense against the counterinsurgency, with its stereotypical thinking that all members of Catholic Action were subversives in some capacity. But after undergoing the charismatic experience, a number of Maya Catholics embraced it, and either remained in the Pentecostal congregations or formed Catholic charismatic groups.

Chapter 11 described the beginnings of charismaticism in Guatemala. Following the murder of Father Woods and the expulsion of Father Stetter, the five Maryknoll colonization communities in the western Ixcán were without the regular services of a priest. In 1979, the Maryknoll priest in Barillas, Father Banazcek, began to visit the area a few times during the year (Falla 1985: 12). He was a charismatic and had spread the movement to many communities in the Barillas area. During one of his visits, he asked the Ixcán communities if they would like to have a charismatic retreat. They agreed and sent a group of fifteen singers to Barillas to learn the songs and music. Upon their return, they taught the songs to a group of about sixty-five catechists. Then, in each of the five communities, Maya charismatics from Barillas conducted three-day retreats, as described in chapter 11. Father Banaszek made a brief appearance at each to give a general absolution in place of the personal sacrament of penance. The retreats were well received, and many Catholics in these Ixcán communities became charismatics. They began to hold weekly sessions in the community church without a priest, and frequently during the week in private homes. Some from Catholic Action opposed the movement because of its inward-turning philosophy.

In the Ixil sector of Quiché, in 1980, some Catholic Action groups became charismatics following the murders of three Sacred Heart priests. For safety, some of the catechists from Chajul fled to the Maryknoll colonization settlements in the Ixcán. Through their influence, the Chajul catechists became charismatics, returned to Chajul, and spread the movement, so that it became the dominant form of Catholicism in the town. One of the Catholic charismatics preached in the aldeas throughout the area and in Sacapulas (Stoll 1993: 174). These Catholics adopted a charismatic interpretation of the counterinsurgency as a punishment from God, therefore believing that Catholics should seek his pardon. During militarization, the military appointed a Catholic charismatic as mayor and another as the head of the civil patrols (104–5).

Catholic Reactions to Catholic Charismaticism

Chapter 11 mentioned Ignacio's observation of charismaticism in San Juan La Laguna and the negative reaction of the priest. This was typical of many Catholic priests. A Chajul catechist, after working with Protestant charismatics in Guatemala City, returned to Chajul and began preaching a version of it among Catholics. He was rebuked by the local priest. As a result, the catechist founded his own Pentecostal church and took about half of the Catholics with him. In 1989, more Catholics left for the Pentecostal church (Stoll 1993: 189, 192).

In Nebaj, the leader of Catholic Action founded a Catholic charismatic group after the local priest was murdered. He went from house to house with its message of repentance and reconciliation. In 1981, he and his small group of Catholic charismatics were expelled by the local Catholic Action from their group (Stoll 1993: 174). Ten months later, his successor founded a Pentecostal church that included most of the town's Catholic charismatics (174, 189). But in 1989 this group made their peace with the Catholic Church, due to the efforts of a new K'iche' Catholic priest. With his approval, they met in a separate chapel and conducted a busy schedule of prayer meetings in their homes (192).

A charismatic Maryknoll priest introduced the movement in Jacaltenango, and demanded that it be accepted as the only recognized form of Catholic worship (Delgado 1998: 189–97). There was a strong negative reaction, especially by Catholic Action. The priest was forced to leave the parish. He was succeeded by another Maryknoll priest who attempted reconciliation but was vehemently attacked by the charismatics. Evaluating the situation as impossible, he also left the parish. A diocesan priest, Arnulfo Delgado, was then assigned to the parish. He sought to reconcile the two groups. For doing so, he also was attacked by the charismatics, who accused him of reporting them to the military. Delgado sees the Maryknoll history of having introduced evolutionary Catholic changes as helping to strengthen charismaticism, which was seen as yet another change. If catechists became charismatics, their communities usually followed them.

Personal versus Social Reform

In concentrating on individual reform, charismaticism was more agreeable to the Tridentine mentality prevalent among many in the higher levels of the Catholic bureaucracy. It did away with many of the problems created by Action Catholicism, with its emphasis on social reform. The Vatican clearly prioritized personal renewal over social reform in a discussion of liberation theology. "The church is . . . aware of the complexity of the problems confronting society and of the difficulties in finding adequate solutions to them. Nevertheless it considers that the first thing to be done is to appeal to the spiritual and moral capacities of the individual and to the permanent need for inner conversion, if one is to achieve economic and social change that will truly be at the service of humanity. The priority given to structures and technical organization over the person and the requirements of human dignity is the expression of materialistic anthropology, contrary to the construction of a just social order" (Hennelly 1990: 485, #75).

By taking this position, the Vatican was downgrading Action Catholicism and strengthening the position of Tridentine and Charismatic Catholicism. The document ignores the dialectical interaction between the experience of working for a just social order and the inner conversion of individuals. It creates a false dichotomy, as most liberation theologians plainly saw. Giving primacy to individual conversion could easily be used as an excuse for social inaction. Numerous cases of inner conversion have come about as a result of working for a just social order. More recent developments within the Pentecostal movement have recognized its past shortcomings regarding communal problems and have undertaken programs of assistance and campaigns for human rights in what is called "Progressive Charismaticism" (Miller 2009).

Catholic charismaticism in Guatemala was accepted by Catholic bishops, but always with warnings about what they saw as its risks. The emphasis on the inner experience threatened to supplant or relegate the sacraments and the clergy to minor roles. In 1986, the Guatemalan bishops attempted to delineate for Guatemalan Catholics the acceptable and unacceptable aspects of the charismatic movement (Cleary 1989: 67–71). Pope Benedict XVI, in a 2008 address to the bishops of Guatemala, gave this warning. "You know very well that firmness of faith and participation in the sacraments strengthen your faithful against the risk of the sects or of groups that claim to be charismatic but that create a sense of confusion and can endanger ecclesial communion" (Benedict XVI 2008: 1). Cleary

(2007: 156), summarizing pronouncements of Latin American bishops, lists five characteristics that should distinguish Catholic from Pentecostal charismatics: (1) they are more aware of the Catholic mystical tradition and do not rely exclusively on more recent theories of conversion; (2) they see spirit baptism as a new work of grace that does not take the place of traditional Catholic baptism and confirmation; (3) they tend to see a wider range of gifts as evidence of spirit baptism and deemphasize speaking in tongues; (4) they hold the mass with the real presence of Christ in the Eucharist as a pivotal practice; and (5) they see Marian devotions as an integral part of their spiritual life.

There are no studies of the worldviews of Maya Catholic charismatics. But as seen previously, the Maya usually adopt practices of other cultures on their own terms. It is doubtful that all the above characteristics describe Maya Catholic charismatics. The importance of the Life of the Spirit, the inner experience in the charismatic movement, tended to downgrade the Catholic sacraments. As a result, tensions arose with local Catholic priests in the communities of Nebaj, Chajul, San Juan La Laguna, and Nahualá, as described in chapter 18. Similar tensions also arose with Catholic Action organizations that retained their focus on community action. Not infrequently, these tensions resulted in Catholic charismatics simply joining Pentecostal congregations or founding their own Pentecostal churches, as happened in Chajul and Nebaj.

The Pan-Maya Movement

Another option for Maya seeking meaning amidst confusion was the pan-Maya movement. Chapters 8 and 12 described Tridentine Catholicism and its evolution to Catholic Action in San Andrés Semetabaj. In the 1990s, Warren (1998: 177–93) returned to San Andrés to find out what had happened to members of a leading Catholic Action family since that time. Her account provides a three-generation picture of the evolution of this family's worldview, from traditional colonial Maya, to Tridentine Catholicism, to Action Catholicism, to the pan-Maya movement.

Late in the nineteenth century, a traveling merchant from Totonicapán decided to immigrate to San Andrés. He sought acceptance in the community by performing service in the cofradías and by sponsoring marimba groups for the community's festivals. He had four sons who assisted him in performing the rituals, thereby learning the traditional worldview. The sons had three years of elementary schooling where they learned Spanish.

One of these sons, Gustavo, wanted to obtain the skills of the national society for both personal and community advancement. As a young man, he and his brothers worked on the coastal plantations. During the Arbenz era, Gustavo served on a local agrarian reform committee whose task was to identify plantations with large amounts of unused land for potential redistribution. This effort was cut short with the fall of Arbenz in 1954.

Stemming from his interest in improving his community, Gustavo began to question those aspects of traditionalism that hindered this goal. He and his brothers helped found the local Tridentine Catholic group in the 1950s and became catechists (chapter 8). The group transitioned to Catholic Action in the 1960s. Gustavo adopted a bicultural strategy to confront the subsistence crisis. Maya should master ladino ways and the Spanish language in order to make their way in the world. He participated in many Maya-led community activities and, because of them, is proud of his achievements toward ending ladino domination of his community (chapter 12). He held a number of public offices, but developed a drinking problem as a result of the required social obligations. With the help of Alcoholics Anonymous, he fought to control it. "He presents his life story as a witness to the struggle against alcohol abuse and the constancy of religious faith that informs his ceaseless drive for institution building in the community" (Warren 1978: 181). He saw Catholic Action as a reform of what he considered degenerate aspects of Maya rituals and of those parts of the traditional worldview that legitimated them.

The Third Generation

Gustavo's brother, Luis, was also one of the first catechists. Luis had two sons who were brought up in Catholic Action. The eldest, Alfonso, became bilingual due to six years of schooling locally, and later in Quetzaltenango. He left Catholic Action as a young adult, unwilling to submit to the moral discipline required of its leaders, but retaining his disdain for shamanism. In the late 1960s, he lived as a ladino in Guatemala City, working as a mechanic. Returning to San Andrés, for six months he assisted Warren with her research, learning skills that he later found useful when he worked organizing agrarian cooperatives as a bicultural or ladinoized Maya.

His work brought him into contact with many people throughout the region who were helpful in the late 1970s when he successfully ran for the national congress. But the election was nullified by the military government. Many of his supporters were traditional Maya who had benefited from his networking on agrarian issues. They introduced him to the traditional culture that had been ignored or disrespected by his years of schooling, his socialization in Catholic Action, and the years he had spent living in the capital as a ladino. In the 1980s, the pan-Maya movement cultivated rural supporters and encouraged the revitalization of shamanism as a pan-community activity. Alfonso immersed himself in its study and became a practicing Maya shaman. He represented Maya spirituality at major pan-Maya and regional meetings. Alfonso came to see institutionalized Christianity as an instrument of oppression of his people.

The youngest son of Luis, No'j, had a similar experience. He was brought up in Catholic Action, had six years of schooling, and became fluent in Spanish. His and other local families emphasized a bilingual strategy for education and advancement. As a result, he initially was more fluent in Spanish than Kaqchikel. He received a scholarship to assist in a linguistics project working on Maya languages. He was quickly recognized as a gifted analyst and became a well-known member of OKMA, a national group of professionally trained Maya linguists. This group was interested in Maya-language revitalization and the publication of technical linguistic research. No'j became a nationally recognized pan-Maya linguist. He began observing the traditional Maya rituals and speaking a standardized Kaqchikel that rejected Spanish loanwords commonly used in contemporary Kaqchikel. He rejected all religious authority tied to any form of Christianity. After two generations, these brothers had cyclically spiraled back to the pan-Mayanist adaptation of their grandfather's cultural world, now as a project of self-conscious revitalization.

Reactions of the Second Generation

Meanwhile, Gustavo, the uncle of Alfonso and No'j, was greatly upset by the actions of his nephews and others of their generation. He remained devoted to Catholic Action. He wanted to show that Catholic Action, in spite of its rejection of the manner in which the cofrades and shamans celebrated the traditional rituals, still respected the traditional culture and wanted the community to retain its cultural identity. Gustavo had come to see himself as a *k'amöl b'ey*, a specialist in traditional ritual knowledge learned from his father. He organized *actos culturales*, dramatic presentations with explanations of traditional rituals by members of Catholic Action. The catechists were careful to explain that they were only plays. They lasted about an hour and omitted some of the objectionable parts, such as the extended drinking, dancing, worship of saint images, and their processions. Local Catholic priests attended the presentations. They approved and encouraged Gustavo and the catechists. Both Gustavo and the priests saw the traditional values expressed in the rituals as similar to those of Catholicism. As in many other cases, Gustavo's original conversion to Catholic Action was never a total rejection of the traditional worldview, although at that time, with the polemics of its introduction, it may have appeared so. For Gustavo and the priests, the two systems were not in complete contradiction. But the local bishop did not think so, and the actos were discontinued.

While Gustavo still respected the traditional culture, he did not support the pan-Maya Movement. He disputed that it was a full-circle return to traditional Maya culture. He remarked, "I've talked to those who head this [revitalization] movement. To really recover Maya culture, we would have to return to the ways of our ancestors. To recover everything we would have to speak our indigenous language.... None of our children speak the language.... My view is; how are we going to recover our culture, how are our children to recover it if we are the ones who teach modern things, new things? To recover this, we would have to go back to the kitchen and cook on the floor over the three stones, the *tenemastes*. At the great university and grand meetings of our indigenous people, the indigenous intellectual gives his talk on indigenous culture. But when he goes to give the talk, he should be wearing indigenous dress. But no. Why? To recover our culture, he should wear indigenous dress. I tell you with pride this is what I wear. Some may say we look like clowns. But they should put on indigenous dress" (Warren 1978: 180).

To overcome the subsistence aspect of the Maya crisis, Gustavo saw that it was necessary to adopt the economic and technological ways of contemporary society in spite of their non-Maya origin. For the theological crisis of the traditional worldview, he found the needed reforms expressed by Catholic Action. The reforms largely centered on what he considered the degenerate ritual practices of the cofrades and shamans that had taken the place of those originally received from the Spanish priests, as described in chapter 3.

Luis, Gustavo's brother and the father of Alfonso and No'j, had also been a catechist and a founder of Catholic Action in San Andrés. He reacted to the pan-Maya revitalization efforts of his sons in a more philosophical manner. From the traditional Maya worldview, he invoked the metaphysical principle of the cyclical spiral of all entities. His observations were briefly recorded in chapter 12. Here is an extended account of his remarks.

> Don Luis speaks with equanimity of the cosmic Maya life cycle in which an older generation dies so new generations might be born.

> He sees the inevitable end of what his generation has been able to achieve through their efforts in the church, schools, commercial crop production and local government. Given the historical circumstances, they have advanced the cause, especially from the days when indigenous people were expected to step off the sidewalk with hats in hand so Ladinos could pass unimpeded. Now the next generation will assume the struggle with different tactics, only to be replaced by still other generations. Both elders [Luis and Gustavo] deeply value the dedication of their entire adult lives to faith in Christ and religious activism. No doubt they are pleased that one of the family's sons [not previously mentioned] has combined leadership in neighborhood improvement projects with religious devotion as a catechist offering religious instruction at a nearby Catholic girls school. The idea that the God of their people and their children might be another divinity . . . is far beyond these elders' capacity for personal transformation. (Warren 1978: 190–91)

These reflections in a moment of intergenerational tensions show how the traditional worldview lived on in two ardent members of Catholic Action.

Gustavo's reaction represented a trend among a number of Maya Catholics. He was devoted to their vision of Action Catholicism, a revitalization of the Maya religion in the epoch of Jesus Christ, and how it had helped him to overcome his personal problem and ladino oppression in his local community. He saw the pan-Maya movement as regression back to the cycle of another epoch that had been surpassed. Luis also places his Catholic experiences into the larger Maya metaphysical framework of cycles that are not unlike the biblical cycles seen in chapter 13. He does not see the pan-Maya movement as regression, rather the beginning of a new cycle following the death of the Catholic cycle. Behind both cycles abides the pantheistic One seen also as the Judeo-Christian God.

Summary

This chapter has examined several important trends of the continuing Maya search for identity and a worldview. This search has taken place during a half century during which their communities lost their isolation and became involved in a market economy and the national society dominated by ladino culture. These experiences have made the Maya aware of a multiplicity of religious and other worldviews, as well as of their ability to choose among them. Some have joined congregations of traditional Protestant denominations or other religious groups. But many more have chosen to join Pentecostal churches or the pan-Maya movement.

As a result, there has been a substantial decrease in the influence of Catholicism among the Maya. In 1980, Pentecostal churches and Protestant denominations comprised about 20 percent of the Guatemalan population; by 2008, the represented 30 percent (Holland 2008). The percentages for the Maya area are probably higher, owing to the strength of Pentecostalism in the rural areas. Three-fourths of Protestants are Pentecostals (Scotchmer 1991: 8). A significant number of Maya Catholics also became Pentecostals. In 1970, the municipio of San Pedro Sacatepéquez, in San Marcos, had only two evangelical churches, both in the municipal center. By 1990, there were twenty-five in the municipal center and at least one in each of the seventeen outlying aldeas. "The Catholic Church, seriously dispirited by the political assaults of the 1980s, has been further demoralized by the new competition" (Ehlers 2000: xxiv). In 1989, there were twenty-two evangelical pastors in Nebaj, 46 percent of whom came directly or indirectly from Catholic Action (Stoll 1993: 186). From scattered accounts, this appears to be typical of what was happening in the Western Highlands. In Chiapas, factional divisions within Maya communities resulted in substantial numbers forming Protestant congregations. Chamula is typical of this trend. Of all the Mexican states, Chiapas has the largest percentage of non-Catholics.

Catholic charismatics, while fewer in number than Pentecostal charismatics, are a significant group that have an ambivalent relationship with Catholic authorities. In the future, this may lead to more of them joining Pentecostal churches. A common denominator among Pentecostal charismatics, Catholic charismatics, and pan-Mayanists is the retention or revitalization of aspects of the traditional Maya culture. This was also true of some members of Catholic Action, as they placed their Catholicism within the Maya metaphysic of epochs and cycles of births, deaths, and rebirth. This may be interpreted as covenant strength, degeneration, or covenant revitalization through Action Catholicism. The traditional worldview was still percolating in the background of many Maya minds, helping them to interpret new experiences.

28

A Bishop's Evolving Worldview

Samuel Ruíz García was consecrated the bishop of the San Cristóbal Diocese on January 25, 1960, and remained until 1999, when he reached the mandatory retirement age for bishops. He died in January, 2011. In forty years of wrestling with the problems of Catholicism and the Maya, the bishop pioneered significant changes in the relationship between the non-Western Maya tradition and a Western-oriented Catholicism. This chapter sketches his evolution from a conservative Tridentine Catholic to a promoter of an inculturated Maya Christianity with a theology based on the wisdom of the ancestors and synthesized in the life and teachings of Jesus. His evolution gives further insight into the evolution of religious worldviews and of how they can break out of their ethnocentrisms and degenerations stimulated by cross-cultural experiences. The introductory photograph shows the bishop in 1963, a few years after being consecrated a bishop.

Chapters 2 and 6 pointed out the necessary link between Maya subsistence and worldview. Although the Maya began questioning the covenant, chapters 8, 12, and 27 indicated that aspects of the traditional worldview continued to function in the background, even as the Maya accepted Tridentine Catholicism and later Action Catholicism. In addition to liberation from the subsistence crisis by social justice, liberation for some Maya and for some pastoral agents also meant cultural liberation, freeing the traditional Maya worldview from the depreciation and ignorance of it by the Catholic Church and others. The traditional worldview included the wisdom of the ancestors, the values and spirituality learned in the family circle, which had guided the Maya moral sense for centuries. Vatican II began to recognize this problem, which carries the label "inculturation." It saw the essential Christian message as either already expressed or as capable of being expressed in other cultural forms. This chapter traces Don Samuel's personal evolution to becoming an ardent advocate of a Maya Christianity.

The Starting Points of His Evolution

Don Samuel's parents were from families of agricultural workers in an area of intense conservative Catholicism, the state of Guanajuato in central Mexico (Fazio 1994: 9–11, 23–55). As young adults, they traveled separately to the United States to labor in the fields of Utah and California as "wetbacks." They met while attending church services, and were married in California. They returned to Mexico, to the town of Irapuato in Guanajuato, where Samuel Ruíz García was born in 1924.

Mexico during Don Samuel's youth of the 1930s witnessed numerous political insurrections and government attacks on the Catholic Church. Maclovio Ruíz, Samuel's father, took an active part in the local church as a director of Catholic Action and as president of the Knights of Columbus. In response to Calles's violent persecution, Catholics formed defense organizations that included an armed guerrilla band, the Cristeros. Maclovio was about to join them in the mountains but was dissuaded at the last moment by the insistence of his wife. Since the government had closed all Catholic schools, Samuel was schooled at home by his parents. His father had a fifth-grade education, and his mother's schooling had stopped after the third grade. But both were avid readers and continued to educate themselves. Don Samuel's sister recalls the family's religious formation. "The atmosphere in which we grew up in Irapuato was very Christian. Every day we went to mass together. My father read the Bible to us; my mother was a catechist in our neighborhood even during the times of persecution. My father was a director of Catholic Action and my mother was head of the women's section. Samuel also was a catechist. I believe that it was the religious life lived in our home that led to his decision to enter the seminary" (Ruíz García 2003: 146). At the age of fifteen, Samuel enrolled in the seminary in nearby León. He excelled in his studies, and in 1947 was selected to pursue his theological studies in Rome at the Gregorian University and the Biblical Institute.

He returned to the seminary in León as a professor of theology and sacred scripture. Soon, he was made the Dean of Studies, and in his thirtieth year, 1954, he became the rector (president) of the seminary. Up to this point, Tridentine Catholicism had dominated every stage of his life: his socialization as a child, his duties as a catechist, his seminary studies in León and Rome, and the content of his teaching at the seminary.

In 1959, he was designated the bishop of San Cristóbal. He was consecrated on January 25, 1960, in a ceremony fitting for a "prince of the church." The ornate liturgical vestments symbolized the princely conception of the role. They included a fifteen-yard flowing train held up by two acolytes. This was a display of European medieval Catholicism for a diocese of predominantly poor Maya. The San Cristóbal Diocese had been weakened under the liberal Mexican governments. At the beginning of 1960, there were only thirteen priests in the diocese. This number was reduced in 1965 when most of the predominantly ladino area was made the separate diocese of Tuxtla Gutiérrez. Under the previous bishop, some Franciscans had initiated a catechetical program in the Chol region (Meyer 2000: 60). A diocesan priest in the Bachajón area (Maurer 1996: 61) and a few Maya on their own initiative had gathered acquaintances together in community chapels to teach them songs, prayers, and the elements of a question-and-answer catechism (Iribarren 1985: 48). At that time, many Maya lived in serflike conditions on large plantations created out of lands they had formerly owned. Many others lived in isolated communities that lacked electricity, potable water systems, health services, sanitation systems, schools, paved or dirt roads, and public transportation. They were linked together by a series of spider-like foot trails, some of which led to the few dirt roads that connected to the main market towns. The Catholic Church, through the infrequent visits of its priests for Maya festivals, had a real but slim connection to these communities (Early 2006).

Given his background, the new bishop had met few indigenous people in his life. He began to visit the Maya communities. He saw short, thin men and women, along with their dirty, ragged, or naked children. Most of the women were barefoot, and the men wore sandals made of discarded automobile tires. Bishop Ruíz was overwhelmed by the poverty that he mistook as a sign of their being uncivilized. His reaction: "We have to teach them Spanish. We have to put shoes on the indigenous people and help them to eat better." In other words, we have to civilize them by teaching them Spanish and turning them into ladinos. Later, when he better understood Maya culture and the systemic injustice behind the poverty, he retracted these statements. "Shoes and eating better" was his naive, on-the-spot formulation of the subsistence crisis (Ruíz García 2003: 18, 23). The teaching of Spanish to Maya groups was also one of the first instructions of the Spanish king in the sixteenth century after the initial Maya-Spanish encounters (Early 2006: 169). Just as the friars reacted negatively to the king's proposal, the diocesan vicar convinced the bishop of the impracticality of any such effort. Neither the bishop nor any of the priests understood the Maya perception of the infrequent rituals the church enacted among them (211–15). The founding of the catechetical schools, described in chapter 7, was Ruíz's first effort to initiate Tridentine Christianity in the Maya communities.

Vatican Council II

Given the bishop's Tridentine background, Vatican Council II was a threshold for the expansion of his worldview. He attended all four sessions. Womack (1999: 27) interprets the bishop's experiences:

> Given the reality of his diocese and the ever clearer strength of his character, Ruíz quickly knew who he was—a cultural [theological] conservative and an economic and social reformer. The debates of every draft interested him. . . . Ruíz's interest rose on the reality he was witnessing of the great differences in the Church, and caught on the continual references to the matters of most concern to him at home, God's word incarnate, "social action," Protestants, catechists, missions. For his feelings on these matters he was learning words and ideas to think about them and argue them. He met French and Italian intellectual priests who introduced him to the social science on "countries in development." The less philosophical, the more

historical, Biblical, pastoral and practical the debates, the more they engaged him. . . . He made many friends among the young churchmen there. And he cultivated clerical sociologists and anthropologists of development.

The Council recognized the existence of widespread hunger and poverty stemming from unjust social systems. The economic section of the Council's document, *The Church in the Modern World* (*Gaudium et Spes*), focused on social justice. After describing the large differential between the wealth of the haves and the widespread poverty of the have-nots of the world's population, the Council insisted on "the right to have a share of earthly goods sufficient for oneself and one's family belongs to everyone. . . . Men are obliged to come to the relief [of] the poor, and to do so not merely out of their superfluous goods. If a person is in extreme necessity, he has the right to take from the riches of others what he himself needs. Since there are so many people in the world afflicted with hunger, this sacred Council urges all, both individuals and governments, to remember the sayings of the Fathers: 'Feed the man dying of hunger, because if you have not fed him, you have killed him'" (Gaudium et Spes, #69). Earlier, the Council had said, "This social order requires constant improvement. It must be founded on truth, built on justice, animated by love; in freedom it should grow every day toward a more humane balance. . . . Profound and rapid changes make it particularly urgent that no one, ignoring the trend of events or drugged by laziness, content himself with merely a individualistic morality. It grows increasingly true that the obligations of justice and love are fulfilled only if each person, contributing to the common good, according to his own abilities and the needs of others, also promotes and assists the public and private institutions dedicated to bettering the conditions of human life" (Gaudium et Spes #26, #30). Given the social and economic conditions underlying the Maya crisis, this insistence on social justice was particularly pertinent. The Council's decrees awakened the bishop's resolve to confront the unjust social structures of his diocese. It helped him understand that the problems were not restricted to Maya communities, but involved the social, economic, and political structure of the whole state of Chiapas, the Mexican nation, and, ultimately, the world economic system.

The Council and its decrees also initiated the bishop's thinking about the Maya worldview and the problem of differing theological formulations, but left him confused, given his narrow Tridentine formation. As his thinking evolved, he would later refer back to the decrees of the Council as substantiating the worldview to which his later evolution would lead him. Upon his return from the last session of the Council in 1965, he formed an advisory group, the Consejo Diocesano de Pastoral, to study the problem of expressing the faith within a cultural framework. On the advice of a French priest-sociologist, the bishop divided the diocese into language zones and formed teams of pastoral workers for each zone (see map 1). This permitted greater concentration on the specific needs of each zone.

The Melgar Awakening: A Lesson in Anthropology

As seen in chapter 10, the Second General Conference of Latin American Bishops (CELAM) was held in Medellín, Colombia, in August 1968 to implement the recommendations of Vatican II for Latin America. In preparation for it, Don Samuel attended a meeting of bishops from dioceses with predominantly indigenous populations in nearby Melgar. Specialists were invited to give talks on topics of interest to these bishops. This meeting was probably the most important event in the evolution of the bishop's worldview.

One of the invited experts was the distinguished Colombian anthropologist, Professor Gerardo Reichel-Dolmatoff, who discussed indigenous cultures. Don Samuel describes the experience and how it broke through the confining walls of his Tridentine worldview:

> The meeting was sheer torture for me—the content was way over my head, and so were the questions they kept asking me. But as I sat there next to several bishops from Guatemala, I began to hear for the first time about anthropology, and its bearing on our missionary activity. One of the speakers was Professor Reichel-Dolmatoff, a prominent anthropologist. His ideas shook our very foundations like an earthquake. What is culture? Culture, he told us, is the set of responses that a group of human beings make to questions of transcendence and of living together, and passes along from one generation to the next. And he distinguished culture from civilization, which is the complex of technical means whereby a human group exercises dominion over its environment. Civilization is whatever a culture, or the people living in a culture, need to survive. Some human communities may have very little civilization, very few technical resources—subsisting on the fruits of the field or climbing the trees for coconuts—yet have a very rich culture with many values. Conversely, there can be human communities with enormous technical resources, but very meager cultural values . . .
>
> Through these ideas, Reichel-Dolmatoff began to tell us how the evangelizing activity of the Church was guilty of destroying cultures.

Missionaries have been ethnocentric, acting like their own culture was the only true one in the world, and all other agents and all other factors must submit to it. I must confess that I have not been blameless in this regard. When I first came to Chiapas, I too was guilty of destroying the native culture.

As I listened to all this, I turned to my Guatemalan bishop friends and said, "Here we go again. First they told us we had to be good theologians, in order to hand on the faith. Then they told us we had to be canonists, in order to establish discipline in our dioceses. Now we must be anthropologists as well!"

I was angry. I stood up and addressed the speaker. "Mr. Reichel-Dolmatoff, I have two questions. First, in studying various cultures do you find that they have a central point around which all other factors are organized?" He said, "We wouldn't put it quite that way, but yes, every culture has a nucleus, a center." Then I asked, "In the cultures you have studied, would you say that religion is more central to its content, or more marginal and peripheral?" And Reichel-Dolmatoff replied without a moment's hesitation: "Religion is always central. In any culture of the world, and certainly in all the cultures of North and South America which I have studied, religion is the central point of the culture."

I turned to my Guatemalan bishop friends—Bishop Flores, and also Bishop Gerardi . . . and said, "So what do we do now? What is there to evangelize?" And I thought to myself: I am supposed to preach the Gospel. I thought that meant converting the pagan Indians to be Christians, even if we destroyed the Indian culture in the process. But why then has God permitted the existence of so many diverse cultures in the world? Just so missionary activity can come along later and destroy them? Or should I be sitting down and studying the culture, and waiting for the proper moment to preach? And since human beings ultimate destiny is not the here and now, but beyond history, why would we work so hard to destroy cultures and replace them with one common culture? Why did God choose to have Jesus born into a particular culture, that of being a Jew in that time and place? (Ruíz García 1999b: 2)

This was a moment of great awakening for the bishop. How could God wish to destroy cultures when there were so many in the world? The bishop was confused and his worldview would never be the same. A quick lesson in introductory anthropology had been a mental earthquake. The anthropologist's explanations had broken through his narrow Tridentine worldview and its catechesis used in the catechetical schools of his diocese. In his evolution, he still had to work out the implications and answers to the many questions he now had regarding the inculturation of Maya culture.

Another speaker at the meeting was a Peruvian priest, a theologian with experience in working among the poor and discovering their squalor, misery, sickness, and death. This had led him to explore the theme of liberation from these miserable conditions as a theological subject. This talk, coming after the questions raised by the anthropologist, gave the bishop a theological direction for his ongoing evolution. Don Samuel continues his commentary:

> Gustavo Gutiérrez came to discuss with us the Vatican II document on the missionary activity of the Church [*Ad Gentes*]. Now I had been present at all four sessions of Vatican II, and I was familiar with that document. But now through Gutiérrez I began to understand what it really meant. God loves all humankind, he said, and therefore God reveals Godself to all the ethnic and cultural groups of the world. So there is a process of salvation going on in every culture throughout the world. Before any missionary arrives to speak of Christ, a process of salvation and a revelation of God is already there!
>
> This changes the whole meaning of the Church's missionary efforts. Vatican II talked about the seeds of the Word already planted in every culture. The missioner's first task is to see what God has been doing in that culture—to face the people and say face to face, not with our backs turned: Dominus vobiscum, The Lord IS with you. [The reference here is to the change of the priest's position at the altar when celebrating mass, dictated by Vatican II.]
>
> But to do that, we must have an instrument to help us discover what God is doing. Our evangelizing activity must recover what God has been doing in that culture, and take it forward in an evangelizing process....
>
> If God is present in all these cultures, missionary work has a whole different meaning: announcing the presence of God already there. I remember talking about this once with a deacon in Canada, a Native American. He told me, "Bishop, what you say is so true. Before I became a Christian, I worshipped the Great Spirit. And when I did so, I was not worshipping the devil. Now that I am a Christian, I know that the Great Spirit is a family—Father, Son and Holy Spirit. But this is the same God I formerly worshipped as the Great Spirit." With this new understanding, pastoral work in our diocese changed profoundly. (Ruíz García 1999b: 3)

From Gutiérrez's talk the bishop found a theological interpretation for the concept of culture and its values that he had heard in Reichel-Dolmatoff's talk. This theologically conservative bishop was undergoing an evolution in his understanding of Christianity.

In the same year that the Melgar meeting uprooted Don Samuel's Tridentine worldview, he met with the Tzeltal catechists. After asking why Tridentine Catholicism was not doing anything for suffering Maya bodies, only their souls, they asked another question: "If the word of God is like a seed to be found everywhere, can we not assume that the seeds are to be found where we live in the mountains and forests? Why should we have to come to your centers, to your schools to seek these seeds and harvest them? Why can't we do it in our own communities?" (MacEoin 1996: 153). This was a question about the Maya worldview and its theology—the wisdom of the ancestors—which already existed within Maya communities, and how it expressed or could express a Christian theology. The Maya catechists joined the Colombian anthropologist in expanding the bishop's worldview

The Bishop at Medellín

Bishop Ruíz was selected to present one of the seven opening position papers to focus the discussions of the conference. His topic was "Evangelization in Latin America" (Ruíz García 1968: 153–77). The paper stressed that Latin America could no longer be considered a group of Catholic nations whose citizens identified with Catholic sacramental practice. He stressed the great need for evangelization and missionary work among all sectors of Latin American society. Only in two paragraphs of the final section (176) did he explicitly mention the indigenous cultures and the necessity of evangelization among them.

From a bishop of a predominantly Maya diocese, and in many ways a proxy for indigenous populations at the conference, this lack of attention to the indigenous cultures of Latin America may appear odd. At Melgar he had undergone a revelatory theological experience, a conversion. It had upset his prior view of evangelization among indigenous cultures, and had left him confused and unprepared to discuss it at the conference itself. This confusion concerned the content of a Catholic evangelization that was to be communicated to a different culture that already had its own religious worldview. To complicate matters, at a late hour, he was assigned the talk on evangelization that had previously been assigned to another bishop.

I started shaking in my boots. There wasn't much time. I went to Caracas and asked the other speaker for the documentation he had assembled for my topic. But what he had, I could not deliver as my own. My anxiety mounted: I had no idea what to say. I went to a meeting with other conference speakers to coordinate our presentations so we would not invade each other's territory. I told the Archbishop of Panama I hadn't time to prepare anything. He told me he had some ideas, and gave me some of his own notes on the subject. . . . A Peruvian bishop friend of mine invited me to go downtown with him to a bookstore. While he was shopping, I was browsing through the shelves. Suddenly I came upon a book by a Spaniard named Beristáin about evangelization. It was my salvation! I started to read it. I was embarrassed to just lift quotes from the book, so I would change the words around a little. I practically transcribed the whole book.
When I gave my talk at the CELAM conference, no one realized that I was like Jesus when he said: "The words I speak to you are not mine!" So I became a specialist in evangelization! (Ruíz García 1999b: 4)

Under these circumstances, the bishop was elected president of the Department of Missions of the Latin American Bishop's Conference, and a short time later, the first president of the Commission for Indigenous Affairs of the Mexican Bishop's Conference.

Anthropologists Strike Again

In 1969, the Commission for Indigenous Affairs of the Mexican Bishop's Conference faced the question of what kind of pastoral plan should be formulated for Mexican indigenous communities. The commission decided it needed the assistance of those with knowledge about these communities. Over a four-day period, January 25–28, at Xicotepec de Juárez, the commission sponsored meetings with representatives of four such groups: the government's Instituto Nacional Indigenista (INI); six distinguished Mexican anthropologists; Catholic and evangelical missionaries; and a discussion group composed of nineteen indigenous catechists, including six Maya from Chiapas (Ruíz García 2003: 26–30).

For the meeting with the anthropologists, the commission intended to pose a question about the survival of the indigenous cultures. If assimilation to the national culture, the goal of INI and most Catholic and evangelical groups, was taking place at such a rapid rate, the commission wondered if a distinctive pastoral plan was needed. At first, the anthropologists flatly refused the invitation to attend the conference, and requested that there be no publicity about the invitation. They did not want to be associated with such a meeting in any manner. This was a blunt repudiation of the bishops and all missionaries for their past policies that attempted to destroy indigenous cultures by conversion to Christianity. However, the commission, through some intermediaries, continued their efforts to have them attend. Finally, the anthropologists agreed, on one condition—that all the missionaries and the INI representatives sit as if they were in a courtroom accused of a crime by the anthropologists. The commission replied, "Agreed, the price to be paid is of no concern if it gives an answer to the question about the survival of indigenous communities" (Ruíz García 2003: 28).

In three separate meetings, the anthropologists presented their accusations against the assimilation policies of INI, the conversion efforts of evangelical missionaries, and, finally, those of Catholic missionaries. The gist of their case against the missionaries was best summed up in a section of the Declaration of Barbados. Although issued a year later, one of the anthropologists at Xicotepec was the Mexican signatory of the declaration.

Responsibility of the Religious Missions

Evangelization, the work of the religious missions in Latin America also reflects and complements the reigning colonial situation with the values of which it is imbued. The missionary presence has always implied the imposition of criteria and patterns of thought and behavior alien to the colonized Indian societies. A religious pretext has too often justified the economic and human exploitation of the aboriginal population. The inherent ethnocentric aspect of the evangelization process is also a component of the colonialist ideology and is based on the following characteristics: 1. its essentially discriminatory nature implicit in the hostile relationship to Indian culture conceived as pagan and heretical; 2. its substitution effect implying the conversion of the Indian and his consequent submission in exchange for future supernatural compensations; 3. its spurious quality given the common situation of missionaries seeking only some form of personal salvation, material or spiritual; 4. the fact that the missions have become a great land and labor enterprise, in conjunction with the dominant imperial interests.

As a result of this analysis we conclude that the suspension of all missionary activity is the most appropriate policy on behalf of both Indian society as well as the moral integrity of the churches involved. Until this objective can be realized the missions must support and contribute to Indian liberation in the following manner: 1. overcome the intrinsic Herodianism of the evangelical process, itself a mechanism of colonialization, Europeanization and alienation of Indian society; 2. assume a position of true respect for Indian culture, ending the long and shameful history of despotism and intolerance characteristic of missionary work, which rarely manifests sensitivity to aboriginal religious sentiments and values; 3. halt both the theft of Indian property by religious missionaries who appropriate labor, lands and natural resources as their own, and the indifference in the face of Indian expropriation by third parties; 4. extinguish the sumptuous and lavish spirit of the missions themselves, expressed in various forms but all too often based on exploitation of Indian labor; 5. stop the competition among religious groups and confessions for Indian souls—a common occurrence leading to the buying and selling of believers and internal strife provoked by conflicting religious loyalties; 6. suppress the secular practice of removing Indian children from their families for long periods in boarding schools where they are imbued with values not their own, converting them in this way into marginal individuals, incapable of living either in the larger national society or their native communities; 7. break with the pseudo-moralist isolation which imposes a false puritanical ethic, incapacitating the Indian for coping with the national society—an ethic which the churches have been unable to impose on that same national society; 8. abandon those blackmail procedures implicit in the offering of goods and services to Indian society in return for total submission; 9. suspend immediately all practices of population displacement or concentration in order to evangelize and assimilate more effectively, a process that often provokes an increase in morbidity, mortality and family disorganization among Indian communities; 10. end the criminal practice of serving as intermediaries for the exploitation of Indian labor.

To the degree that the religious missions do not assume these minimal obligations they, too, must be held responsible by default for crimes of ethnocide and connivance with genocide.

Finally, we recognize that, recently, dissident elements within the churches are engaging in a conscious and radical self-evaluation of the evangelical process. The denunciation of the historical failure of the missionary task is now a common conclusion of such critical analyses.

This was a scorching indictment. When the Catholic representatives responded, they admitted that their past policies were destructive of indigenous cultures, but based on Vatican II, said that they now were engaged in seeking "the seeds of the Word" within Indian cultures. The anthropologists were taken back, a bit skeptical about what all this really meant, but could see a definite change in Catholic assumptions about pastoral planning for indigenous communities. This was the reason for the final paragraph in the missionary section of the declaration. (The remaining sections of the declaration indicting national governments and academic anthropologists for their treatment of indigenous populations appear as an appendix to this chapter.)

For four days, the nineteen members of the indigenous group lived together at Xicotepec, sharing meals, recreation, and informal talks among themselves in which they compared the experiences of their various communities (Centro Nacional de Pastoral Indigena 1970: 9–11). On January 27, they held two formal discussion sessions with the commission on the subject "The situation of the church and its priests with regard to indigenous communities." During the second session, the catechists presented a brief summary of their biting critique of the way ladino priests interacted with indigenous communities (Centro Nacional de Pastoral Indigena 1970: 63–64). This was previously seen in chapter 8. Their presentation reinforced the points previously made by the anthropologists. Coming after the experience at Melgar and the questions of the Tzeltal catechists, the bishop realized that he was attempting to plan for something he knew little about. Don Samuel began sending diocesan pastoral workers to take courses in anthropology at the Universidad Iberoamericana in Mexico City, and attended some sessions himself.

Overcoming Resistance to Inculturation

Early in the discussions of inculturation, in 1973, Don Samuel gave a talk, "Christians and Justice in Latin America" (Ruíz García 1973). He sharply attacked Tridentine theology and its catechesis as one of, if not the main obstacle to acceptance of the ideas of inculturation by the clergy. He clearly showed the continued evolution of his theological thinking, influenced by his experiences with anthropologists at Melgar and Xicotepec.

There persists in Latin America a conceptual, deductive, abstract theology that claims to be all embracing. It arrived from Europe at the beginning of colonization. It has retained its grip on our seminaries thanks to our neocolonial imitation of European universities. That theology scanned through Scripture, the Fathers of the Church, the teachings of the Councils and the Popes to prove a catalog of pre-established [orthodox] theses and *had little concern for the real and complex lives of human beings*. The majority of the bishops and priests ordained before Vatican Council II, today [1973] the majority of all bishops and priests, studied this type of theology that used scholastic philosophy as a fundamental auxiliary science in the search for an understanding of the faith. It is well known, however, that even in Europe before Vatican Council II, there began an attempt to develop a new theology based on a modern philosophical reflection about human beings and their existential experiences. In Latin America, meanwhile, theology oriented itself toward a concrete theology of event that sought enlightenment from faith for the real and every day life of Latin American people and an interpretation of the Word of God based on our present and past reality. (Ruíz García 1973: 33; emphasis added)

Here Don Samuel is addressing a group of missionaries, many of whom were likely to see any theological change as morally wrong because of their Tridentine education. In the terms of Neal's typology, described in the appendix to chapter 24, some were Value-NonChange and others were Interest-NonChange. Yet for theology to help people lead a Christian life, to show "concern for the real and complex lives of human beings," to deal with "the real and every day life of Latin American people and an interpretation of the Word of God based on our present and past reality," such a change was necessary. By 1973, this conservative Catholic prelate had moved away from, and was now attacking, the Tridentine theology that had been behind the catechesis he had both studied and had taught as a child, and that had constituted a significant part of his studies for the priesthood in León and Rome, and of his teaching career upon his return to León. At the time of Vatican II in the mid 1960s the bishop had been a theological conservative and social progressive. By 1973, he had evolved to being both a theological and social progressive.

The Vision: A Fully Inculturated Maya Church

Throughout the history of the Maya-Catholic relationship, there has been a Maya desire for a Christianity that respected their culture, as many saw no contradiction between it and the wisdom of the ancestors. This is shown by the continuation of elements of the traditional Maya worldview among Maya Catholics, the key role of the fiscales and the catechists in teaching Christianity, and the examples of fiscales becoming priests during Maya rebellions (Early 2006: 231–37). Given his own experiences in Maya communities, given the new perspectives of Vatican II, given the anthropological revelation at Melgar, and given the anthropological critiques of Xicotepec, Bishop Ruíz had come to the same realization and now undertook to implement it. He now understood that the Maya ignorance of orthodox Catholicism did not mean that they were without any religion—uncivilized, as he originally thought. In later years, he wrote: "We have committed many mistakes on our pastoral pilgrimage. Our first actions, years before Vatican II, were destructive of the culture. We had only our own criterion with which to judge the traditional customs, shaping our judgment with ethnocentrism and moralism, attitudes that regrettably were very common at that time" (Ruíz García 1994: 599).

Figure 28.1. Bishop Ruiz in 1998.

In his visits to Maya communities and in his dealings with Maya catechists, the bishop had come to greatly admire the virtues he saw in the Maya and in their culture: "the wisdom of the ancestors." Probably this was reinforced by his negative dealings with many Chiapas ladinos, who were always conscious of their superior class status and often devoted to their own self-promotion at the expense of the Maya. The principal value Don Samuel found in the Maya communities was their sense of community, of working together, which contrasted with the individualism, egotistical outlook, and factionalism of the non-Indian world (Meyer 2000: 129). He grew to love the Maya, even to the point of exaggerating their good qualities. Maya communities have a long history of presenting to outsiders a strong community solidarity that covers over the internal dissensions, enmities, and jealousies that are present in all human communities. The bishop's idyllic view of Maya communities was not unlike that of a number of nineteenth- and twentieth-century predecessors, such as Tolstoy and the anthropologist Robert Redfield (1941), with his folk-urban continuum based on the changes of Yukatekan Maya life. They had idealized the small rural community.

The bishop also drew his inspiration from the Bible, especially the description of the "Kingdom of God" as the redemptive moment of the current epoch (phase 7 in Figure 13.1). Writing to the Jesuits of Bachajón, he sees the goals of their inculturating work as:

1. To discover the religious significance of the symbols and rituals proper to the Indian culture to adopt them in the evangelization process;
2. To incarnate the revealed message in the religious experiences of the culture, through the communal celebration and weekly reflection of the Word of God;
3. To incarnate the meaning of the Kingdom of God in concrete and gradual commitments that promote solidarity and fraternity present in the Indian cultures. (Zatyrka 2003: 260)

"In order to understand the efforts of Don Samuel, it is necessary to emphasize the light that guided him and sheds an understanding on what he did: that light was the construction of the Kingdom of God on earth. He repeated it on many occasions: seek a utopia, construct the new man who will counteract the domination of the strong over the weak, and who has as his mission to fight a homogenized and ethnocidal world" (Meyer 2000: 129). The following chapter will examine the development of the Teología India movement that he encouraged in his diocese.

Summary: A Life Devoted to Change

When the bishop first arrived in the diocese, he saw civilizing the Maya as his main problem. This was to be done by teaching them Spanish and having them dress with shoes on their feet and eat a proper diet. The end result would be turning them into ladinos and orthodox Tridentine Catholics. As the bishop deepened his knowledge of the Maya and the situation in Chiapas, his focus changed. As he came to know and admire the Maya, he realized that what he had understood as their uncivilized condition was actually a condition of their extreme poverty and sickness, largely imposed by the unjust social structure in which they were enmeshed. The system was maintained by Chiapas oligarchs for their own benefit and reinforced by the national government. The option for the poor was his commonsense response; no theology needed. But Catholics needed a theology of liberation for their conversion to embrace the option for the poor. This theology helped him inaugurate a liberation program of empowerment in the diocese with the support of personnel and resources from outside Chiapas.

In his forty years as the bishop of a predominantly Maya diocese, Don Samuel focused on two main problems: social justice and the recognition of a Maya Christian theology. Due to the urgency of the subsistence crisis, the problem of social justice took priority. The shortage of diocesan clergy and resources hindered his initial efforts. In 1966, he attempted a pilot project in Chamula that ended in failure because of problems with its priest-director and external opposition by traditionalists bought off by the government's political party (PRI). Most of the social programs in the diocese were carried out by religious groups that drew their support from outside the diocese: the Dominicans working out of Ocosingo; the Jesuits in Bachajón; a Mexican congregation in Tenejapa; the Sisters of Divine Shepherd in a number of parishes. All received strong encouragement from the bishop.

It took time for the conversion experience about worldviews at Melgar to be expressed in diocesan policy. By allowing his name to be used for the 1975 treatise on the biblical theology of liberation (see chapter 14), the bishop attempted to shift the focus of theology from using Western-oriented theological paradigms to that of an already existing revelation in contemporary Maya communities. This was to become the basis of a Teología India that began to develop in the last decade of his episcopate, because it had to wait for the ordination of Maya Catholic priests. The bishop also had to take into account how the Vatican and other bishops were considering inculturation. Efforts on his part had to be put into the framework of the church's thinking at that time. Premature implementation could bring condemnation, when he was already being harassed by the Apostolic Nuncio for his position on liberation as social justice.

In 1994, after the cease-fire and suspension of military action, the bishop served as the mediator between the government and the Zapatistas. This mediation occupied most of the final years of his service to the diocese. In 1998, he resigned as mediator when he became convinced of the insincerity of the Mexican government. In 1999, at the age of seventy-five, the mandatory retirement age for bishops, he submitted his resignation to the Vatican.

Appendix: Rest of the Text of the Declaration of Barbados

The declaration was the outcome of a meeting of anthropologists sponsored by the World Council of Churches and the anthropology program of the University of Berne (Switzerland). It was drawn up by anthropologists from countries in Latin America with significant indigenous populations. Because of the continuing relevance of this document, below are its other sections: the preamble and the critiques of national states and of professional anthropologists for their past treatment of indigenous cultures.

The Preamble: For the Liberation of the Indians

The anthropologists participating in the *Symposium on Inter-Ethnic Conflict* in South America, meeting in Barbados, January 25–30, 1971, after analyzing the formal reports of the tribal populations' situation in several countries, drafted and agreed to make public the following statement. In this manner, we hope to define and clarify this critical problem of the American continent and to contribute to the Indian struggle for liberation.

The Indians of America remain dominated by a colonial situation which originated with the conquest and which persists today within many Latin American nations. The result of this colonial structure is that lands inhabited by Indians are judged to be free and unoccupied territory open to conquest and colonization. Colonial domination of the aboriginal groups, however, is only a reflection of the more generalized system of the Latin American states' external dependence upon the imperialist metropolitan powers. The internal order of our dependent countries leads them to act as colonizing powers in their relations with the indigenous peoples. This places the several nations in the dual role of the exploited and the exploiters, and this in turn projects not only a false image of Indian society and its historical development, but also a distorted vision of what constitutes the present national society.

We have seen that this situation manifests itself in repeated acts of aggression directed against the aboriginal groups and cultures. There occur both active interventions to "protect" Indian society as well as massacres and forced migrations from the homelands. These acts and policies are not unknown to the armed forces and other governmental agencies in several countries. Even the official "Indian policies" of the Latin American states are explicitly directed towards the destruction of aboriginal culture. These policies are employed to manipulate and control Indian populations in order to consolidate the status of existing social groups and classes, and only diminish the possibility that Indian society may free itself from colonial domination and settle its own future.

As a consequence, we feel the several States, the religious missions and social scientists, primarily anthropologists, must assume the unavoidable responsibilities for immediate action to halt this aggression and contribute significantly to the process of Indian liberation.

The Responsibility of the State

Irrelevant are those Indian policy proposals that do not seek a radical break with the existing social situation; namely, the termination of colonial relationships, internal and external; breaking down of the class system of human exploitation and ethnic domination; a displacement of economic and political power from a limited group or an oligarchic minority to the popular majority; the creations of a truly multi-ethnic state in which each ethnic group possesses the right to self-determination and the free selection of available social and cultural alternatives.

Our analysis of the Indian policy of the several Latin American nation states reveals a common failure of this policy by its omissions and by its actions. The several states avoid granting protection to the Indian groups' rights to land and to be left alone, and fail to apply the law strictly with regard to areas of national expansion. Similarly, the states sanction policies which have been and continue to be colonial and class oriented.

This failure implicates the State in direct responsibility for and connivance with the many crimes of genocide and ethnocide that we have been able to verify. These crimes tend to be repeated and responsibility must rest with the State which remains reluctant to take the following essential measures: 1. guaranteeing to all the Indian populations by virtue of their ethnic distinction, the right to be and to remain themselves, living according to their own customs and moral order free to develop their own culture. 2. recognition that Indian groups possess rights prior to those of other national constituencies. The State must recognize and guarantee each Indian society's territory in land, legalizing it as perpetual, inalienable collective property, sufficiently extensive to provide for population growth. 3. sanctioning of Indian groups' right to organize and to govern in accordance with their own traditions. Such a policy would not exclude members of Indian society from exercising full citizenship, but would in turn exempt them from compliance with those obligations that jeopardize their cultural integrity. 4. extending to Indian society the same economic, social, educational and health assistance as the rest of the national population receives. Moreover, the State has an obligation to attend to those many deficiencies and needs that stem from Indians' submission of the colonial situation. Above all the State must impede their further exploitation by other sectors of the national society, including the official agents of their protection. 5. establishing contacts with still isolated tribal groups is the States' responsibility, given the dangers—biological, social and ecological—that their first contact with agents of the national society represents. 6. protection from the crimes and outrages intrinsic to the expansion process of the national frontier that are not always the direct responsibility of civil or military personnel. 7. establishing a national public authority responsible for relations with Indian groups inhabiting its territory; this obligation cannot be transferred or delegated at any time or under any circumstances.

The Responsibility of the Religious Missions

[This section was quoted in full in the chapter.]

The Responsibility of Anthropology

Anthropology took form within and became an instrument of colonial domination, openly or surreptitiously; it has often rationalized and justified in scientific language the domination of some people by others. The discipline has continued to supply information and methods of action useful for maintaining, reaffirming and disguising social relations of a colonial nature. Latin America has been and is no exception, and with growing frequency we note nefarious Indian action programs and the dissemination of stereotypes and myths distorting and masking the Indian situation—all pretending to have their basis in alleged scientific anthropological research.

A false awareness of this situation has led many anthropologists to adopt equivocal positions. These might be classified in the following way: 1. a scientism which negates any relationship between academic research and the future of those peoples who form the object of such investigation, thus eschewing political responsibility which the relation contains and implies; 2. a hypocrisy manifest in the rhetorical protestation based on first principles which skillfully avoids any commitment in a concrete situation; 3. an opportunism that although it may recognize the present painful situation of the Indian, at the same time rejects any possibility of transforming action by proposing the need "to do something" within the established order. This latter position, of course only reaffirms and continues the system.

The anthropology now required in Latin America is not that which relates to Indians as objects of study, but rather that which perceives the colonial situation and commits itself to the struggle for liberation. In this context we see anthropology providing on the one hand, the colonized peoples those data and interpretations both about themselves and their colonizers useful for their own fight for freedom and on the other hand, a redefinition of the distorted image of Indian communities extant in the national society, thereby unmasking its colonial nature with its supportive ideology.

In order to realize the above objectives, anthropologists have an obligation to take advantage of all junctures within the present order to take action on behalf of the Indian communities. Anthropologists must denounce systematically by any and all means cases of genocide and those practices conducive to ethnocide. At the same time, it is imperative to generate new concepts and explanatory categories from the local and national social reality in order to overcome the subordinate situation of the anthropologist regarded as the mere "verifier" of alien theories.

Conclusion: The Indian as an Agent of His Own Destiny

That Indians organize and lead their own liberation movement is essential, or it ceases to be liberating. When, non-Indians pretend to represent Indians, even on occasion assuming the leadership of the latter's groups, a new colonial situation is established. This is yet another expropriation of the Indian populations' inalienable right to determine their future.

Within this perspective, it is important to emphasize in all its historical significance, the growing ethnic consciousness observable at present among Indian societies throughout the continent. More peoples are assuming direct control over their defense against the ethnocidal and genocidal policies of the national society. In this conflict, by no means novel, we can perceive the beginnings of a pan Latin American movement and some cases too, of explicit solidarity with still other oppressed social groups.

We wish to reaffirm here the right of Indian populations to experiment with and adopt their own self-governing development and defense programs. These policies should not be forced to correspond with national economic and socio-political exigencies of the government.

Rather, the transformation of national society is not possible if there remain groups, such as Indians, who do not feel free to command their destiny. Then, too, the maintenance of Indian society's cultural and social integrity, regardless of its relative numerical insignificance, offers alternative approaches to the traditional well-trodden paths of the national society.

Barbados 30 January 1971

Miguel Alberto Bartolomé
Guillermo Bonfil Batalla [one of the Xicotepec prosecutors]
Héctor Daniel Bonilla
Gonzalo Castillo Cárdenas
Miguel Chase Sardi
Georg Grünberg
Nelly Arvelo de Jiménez
Esteban Emilio Mosonyi
Darcy Ribeiro
Scott S. Robinson
Stefano Varese

29

The Movement for a Maya Christianity

The bishop's vision for a Maya Christian theology is the goal of a group led by Maya Catholic priests. They have joined indigenous groups throughout Latin America in a theological movement, Teología India. The members of this movement are developing Christian indigenous theologies based on the values expressed by and in their indigenous worldviews. They ask Catholic authorities to recognize that the professed universality of Christianity demands that the religious expressions of their indigenous cultures be respected, and that they be seen as capable of expressing the Christian message as summed up in the life and teachings of Jesus.

The Need for a Maya Christian Theology

As growing numbers of Maya became Catholics, either following the Tridentine form of Catholicism, Catholic Action, or the liberation movement, there was a growing distancing away from Maya culture. Catholicism among the Maya was more and more coming to mean ladinization. Maya Catholic priests, seminarians, and catechists noted the serious shortcomings of the Catholic liturgy and theology for their people. The dissatisfaction of catechists was clearly expressed by their question to Bishop Ruíz: "If the word of God is like a seed to be found everywhere, can we not assume that the seeds are to be found where we live in the mountains and forests? Why should we have to come to your centers, to your schools to seek these seeds and harvest them? Why can't we do it in our own communities?" (MacEoin 1996: 153). In the introductory letter to the Tzeltal catechism, the pastoral workers and catechists had complained: "The dynamic of the previous catechesis prescinded from and even destroyed the culture of [Maya] communities, seeing many of its elements as incompatible with the Gospel because of confusing the content of the Good News with the western cultural patterns" (Los Tzeltales 1972–74: ii). Chapter 16 described how this group composed their own catechism by reflecting on the contemporary reality of their communities with the help of the Bible.

The seminary experiences of Maya Catholic priests forced them to form split personalities that cried out for an indigenous theology. A Maya priest in Guatemala, Padre Tomás García, reflected on his experience.

> The formation in the seminary is a deformation of the Maya person. Personally I did not experience rejection, but our world was rejected. This brought on doubts, will I be good? Will I be bad? The Maya education was in the family. But sometimes one heard that we are backward, that we do not know Spanish, that we do not eat well, etc.
> In the seminary, life changes, one begins a life a little more comfortable, there are more things available. After five years in the seminary I felt that I had become a ladino. But after leaving the seminary, I realized what I had lost.
>
> My mother was the one who educated me; but upon entering the seminary I lost those characteristics. I did not like being an Indian. I left the seminary. I studied, I saw reality in another way. Once again I wished to become a priest, my work would be with my own people, and I asked the Bishop, "Monsignor, if you think I am worthy of ordination, I want to go to an Indian parish." He said to me that the priesthood was universal. But I replied that the indigenous peoples have been marginalized. Alright, he said, we will see what the Holy Spirit tells me.
>
> After my ordination the Bishop sent me to the Cathedral in Quetzaltenango. I worked like a Spaniard, and I thought I had taken a

wrong turn. When I recovered the love of my community, my culture and my background, I recovered those characteristics that my mother had taught me. (García 1996: 90)

Don Samuel recalls his attempt to help some Maya seminarians struggling to maintain their Maya identity in the ladino-oriented culture of a Catholic seminary.

> I have the crucial experience of two indigenous seminarians...."You have to work your way here locked into two cultures, and I'd like to help you to take that step." They understood what I was saying. One of them, who had a good grasp of theology, said, "Excuse me, my bishop, but I can't continue here. I want to be a priest, but not a priest from the outside. I have come to understand my culture. Before I did not understand it. Now I feel and am an indigenous person. And I feel that I will be accepted by my indigenous brothers and sisters—but tolerated rather than accepted. They will accept me because I come from Western culture as a mestizo priest. But that's not what I want. I want to be a priest of my culture, of my own culture." And so he left. He is working now as a member of our [pastoral] team.... [The other] left too....
>
> There are other situations where they swallow their pride and go on to ordination. But the pull of their culture is strong. I remember the trauma of one. "I came to the seminary," he told me. "In a matter of three or four days, the rector asked me: 'Do you have a spade somewhere round here?' Then he took a spade and began to dig as if to plant a tree. I took the spade and with my soutane [cassock] hitched up to my belt, I asked if I should continue to dig. 'That's enough,' he said. 'We're not going to plant a tree. Just toss your Indian complexes into that hole, cover them up and be the same as the rest of us.' I felt like a fish that had to live out of water. And I learned to live out of water. But after I was ordained, my bishop sent me to my place of origin. My parents were still there, and I had forgotten the language they spoke. And the people rejected me because I was a traitor to the community. I had abandoned my culture for a different one. Then I forced myself to relearn the language and made every effort to assimilate the culture again." (Ruíz García 2000: 4)

It is significant that a number of leaders of the pan-Maya movement were educated in Catholic seminaries. The founding of pan-Maya is in part a reaction to their seminary experiences. Among them was Antonio Pop Cal, regarded by many as the patriarch of the movement (Pop Caal 1992). A Maryknoll priest recounted that a former seminarian from Santa Eulalia had become a member of a pan-Maya group. "Something happened in the seminary. . . . He suddenly turned on the church and came back and was preaching atheism in the parish, and his father was horrified" (Konefal 2010: 199).

Padre Eleazar López Hernández, a Zapotec Indian, a Catholic priest, a Catholic theologian, a prominent spokesman for the Teología India movement, and a consultant to the Maya section of the movement, has described his seminary experience (López Hernández 2000:18–22). By the end of his philosophical studies, which he passed summa cum laude, he had acquired a Western mentality. In 1969, during his theological studies, he attended a meeting of priests and theologians working among Mexico's poor. This had a deep impact on him and he became highly critical of the traditional theological curriculum of his seminary.

The following year, he attended a meeting devoted to developing a Mexican pastoral plan for indigenous groups. The agenda was formulated by drawing on the conclusions of the earlier Xicotepec meeting, described in the preceding chapter. The questions raised about indigenous identity launched him on a theological career of developing a Teología India as an accepted expression of Christianity. It has not been an easy task.

> The encounter of indigenous people with the church has not always been characterized by respect for our identity. Completely the contrary, the prevalent attitude has been to do away with indigenous identity as a sine qua non prerequisite to enter into the service of the church, especially as a member of its hierarchy. . . . For indigenous people there has not been a dignified place in ecclesiastical structures. Indians converted to Christianity have been forced to renounce our identity or to hide it behind many masks in order to be accepted. Indigenous priests, pastors and nuns are the most effected by this type of schizophrenia brought on by the malformation received in convents and seminaries. (López Hernández 2000: 188)

Father López describes the psychological pain felt by Catholic priests from indigenous backgrounds. "We are divided in our hearts by a double love—we love our community and believe in its way of life, but we also love the church and believe in its way of Salvation. . . . We are convinced that it is possible and worth while to reconcile the two loves, because we know that there is no insuperable contradiction between the two fundamental truths, those of the church which are the same as those of Christ, and the theological truths of our communities. The deepest desires of our people are the same as those of Christ, the differences are of appearance, not content. Further many of these "truths" are better stated in our communities because of the cleanness of heart of the poor, and, because of this, we believe the theological dialog will be beneficial not only for the indigenous communities, but also enrich the church so that by means of the indigenous peoples, it will recover with greater purity the Christian tradition itself" (Flores Reyes 1996: 259).

In recent years, Vatican officials have emphasized the necessity of having Catholic priests from indigenous backgrounds. Don Samuel reacted against this concern taken in isolation from inculturation: "An autochthonous church is not the result of simply ordaining indigenous priests. Even though an indigenous person is ordained a priest and even though he speaks an indigenous language and considers himself a member of his culture, for these reasons alone he will not be an indigenous priest, but an indigenous person who has been ordained a priest. There must be an evangelization incarnated in its own culture and not a westernized version that continues identifying the evangelizing message with western culture" (Meyer 2000: 140). In his report to Rome for the years 1988–93, the bishop wrote, "The goal is this: to work that there may arise a church with an indigenous face, with a truly inculturated evangelization, a liturgy that manifests its faith by its own symbols, a priesthood from its own people" (141).

Vatican II's Recognition of the Problem

An indigenous liturgy and theology were anathema to the European-dominated Tridentine church. With its emphasis on the one true belief, it made every effort to stamp out such liturgies and theologies, including the sometime use of torture resulting in death. But in the developing global society, Catholicism was encountering a panorama of different cultures along with their religions and concomitant problems of ethnocentrism, both the church's own and those of the other religions. The recognition of this problem gave birth to the inculturation movement. It aims to formulate principles for expressing the essence of Christianity in terms of the many cultural worldviews now coming into ever closer contact with each other and often becoming a source of conflict rather than an underpinning of peace.

In the 1950s, the New Theology movement in Europe was spearheaded by Catholic theologians, including the Rahner brothers, deLubac, Congar,

Chenu, and others. They realized that Catholicism had to move beyond its expression for a European-based culture in which many of its medieval assumptions and expressions were no longer pertinent. In its Tridentine form, Catholicism could not cope with the contemporary quest for meaning and offer anything significant in place of the dominant secular atheism. Immersion in the history of Christian theology led to the realization that Christianity had been inculturated in differing cultures and at different periods of time, and that it employed diverse philosophical systems for its expression. The New Theology opposed the narrow perspective of biblical fundamentalism, with its basis in an epistemology of exaggerated realism. This had been the methodological assumption of the Council of Trent and much of the Christian tradition.

The advocates of the New Theology contributed to the formulation of the documents of Vatican II. "The Church," "Revelation," "The Church Today," and "The Missions" all stressed the importance of taking into consideration the diversity of and respect for the various cultures of the world. The African bishops with their experience of tribal cultures were especially forceful in bringing up the question of inculturation. The Council's document on the church's missionary activity, "Ad Gentes," contains several key sections—the equivalent of the significant "two sentences" mentioned in chapter 9 about the interpretation of ecclesiastical documents and how change takes place within Catholicism. "But whatever truth and grace are to be found among the nations, as a sort of secret presence of God. . . . Let them [missionaries] be familiar with their national and religious traditions, gladly and reverently laying bare the *seeds of the word* which lie hidden in them" (Ad Gentes #9–#11; emphasis added). "Seeds of the Word" became the often-repeated key phrase. The document on "The Church in the Modern World' slightly clarifies the phrase.

> For God revealing himself to His people to the extent of a full manifestation of Himself in His Incarnate Son, has spoken according to the culture proper to different ages. Living in various circumstance during the course of time, the Church, too, has used in her preaching the discoveries of different cultures to spread and explain the message of Christ to all nations, to probe it and more deeply understand it, and to give better expression in liturgical celebrations and in the life of the diversified community of the faithful. But at the same time, the Church, sent to all peoples of every time and place, *is not bound exclusively and indissolubly to any race or nation, nor to any particular way of life or customary pattern of living,* ancient or recent. Faithful to

her own tradition and at the same time conscious of her universal mission, she can enter into communion with various cultural modes, to her own enrichment and theirs too. (Gaudium et Spes, #58; emphasis added)

This was the ideal. But the document continues with a distinction that both keeps these statements within the bounds of traditional Catholic orthodoxy and at the same time has prevented significant inculturation beyond orthodox Catholicism's historical areas of influence. "For the deposit of faith, or revealed truths are one thing, the manner in which they are formulated without violence to their meaning and significance is another" (Gaudium et Spes, #62). Guarding the deposit of faith is seen as the primary occupation of ecclesiastical authorities. The deposit of faith is understood to be the divinely revealed truths derived directly from a literal interpretation of the Bible, or indirectly by later ecclesiastical interpretation.

The understanding of worldviews by various anthropological, literary, and philosophical disciplines has shown that the Catholic conception of the deposit of faith is based on two false assumptions. The first is that the deposit of faith itself is free of any cultural expression. This is an impossibility. Any statement, theological or otherwise, must be made in terms of some cultural context. It cannot float about in some disembodied space, free of any cultural expression, and thereby become the template to which all inculturated theological expressions must conform. Inculturation of Christianity viewed in this manner is an impossibility. The deposit of faith is a collection of theological constructs directly or indirectly derived in large measure from a literal interpretation of the Bible, whose writings are drawn primarily from Hebraic culture. They too wear cultural clothes. But many of them have been interpreted by philosophical systems not familiar with Hebraic culture.

The other false assumption sees the Bible primarily as a book of revealed truths to be believed, as demanding, above all, intellectual assent. But chapter 13 has shown that the Bible is a manual of moral teachings. Its focus is on the way people should act. Belief in biblical persons is assent to the moral message of action that they represent. This is the primary function of theology—to help people live a meaningful life. As seen in the theological polemics of the sixteenth century, this purpose was secondary to belief as intellectual assent to theological constructs. This undermines the primary purpose of theology and the Bible. As seen in chapter 10, Tridentine theology was frequently turned into an ideology that supported an unjust social order. The reasons behind this would lead to a discussion of political and economic power that is beyond the scope of this work.

In 1990, twenty-five years after Vatican II, Pope John Paul II's encyclical, "Redemptoris Missio," returned to the question of inculturation.

> As she carries out missionary activity among the nations, the Church encounters different cultures and becomes involved in the process of inculturation.... It is not a matter of purely external adaptation, for inculturation means the intimate transformation of authentic cultural values through their integration in Christianity and the insertion of Christianity in the various human cultures. The process is thus a profound and all embracing one, which involves the Christian message and also the Church's reflection and practice. But at the same time it is a difficult process, for it must in no way compromise the distinctiveness and integrity of the Christian faith [deposit of faith]. Through inculturation the Church makes the Gospel incarnate in different cultures and at the same time introduces peoples, together with their cultures, into her own community. She transmits to them her own values, at the same time taking the good elements that already exist in them and renewing them from within. Through inculturation the Church, for her part, becomes a more intelligible sign of what she is, and a more effective instrument of mission. (John Paul II 1990, #52)

This statement shows the continuing Catholic dilemma about inculturation. The Pope is still trying to harmonize inculturation with the deposit of faith, as seen in the section beginning with "But at the same time."

Inculturation and Latin American Bishops

At Medellín, in 1968, the emphasis was on the problem of poverty, which included the indigenous populations, but there was little specific mention of them or of their distinctive cultural needs. The problem of poverty demanded immediate action and took priority. At Puebla, in 1979, the emphasis was still on indigenous populations as poor people. But as some Mexican bishops observed: "Compared to Medellín, Puebla made significant advances.... The high value it placed on the indigenous world did not completely displace the ethnocentric model inherited from a society that measured the Indian from the viewpoint of the Church and not that of the indigenous peoples themselves" (Voces Episcopales de Mexico 1993: 321).

In 1988, the Mexican Bishop's Commission for Indigenous Affairs issued a farsighted document.

> But revelation is not only contained in Sacred Scripture and [Church] Tradition, it is the personal communication of God through history, and therefore has used and ought to use in all the world, the paradigms specific to the cultures it encounters there. . . . When the gospel came into contact with Greek, Latin and other cultures around the Mediterranean, its message was Hellenized, Latinized and inculturated in all of them. But this has not happened with Chinese, Hindu, Africanized, many Asiatic cultures nor in the cultures of Latin America. The inculturation and evangelization of the Gospel in these cultures is a task that has yet to be undertaken, and is perhaps the most important challenge and problem facing contemporary evangelization. (Comisión Episcopal Para Indigenas 1988: 71–73)

The document was signed by ten Mexican bishops representing heavily indigenous dioceses of Mexico, including Don Samuel.

Early Catholic discussions of inculturation among the Maya often concerned ritual, such as the use of liturgical vestments made from Maya weavings. This was described as "adaptation to the culture" (Angrosino 1994). The above statement sees this approach as superficial. Liturgical inculturation should be related to the more-difficult question of theological inculturation.

In 1992, the Fourth Council of Latin American Bishops was held in Santo Domingo, in the Dominican Republic, in conjunction with the 500-year anniversary of Columbus's chance encounter with the long-existing New World and its inhabitants. Given the occasion of the conference, there had to be recognition of indigenous and Afro-American peoples as distinctive cultures, and discussion about their relationship to Christianity (Hennelly 1993b: 138–41, #243–#251). Historically, the relationship had been one of disdain, oppression, and slavery. The Bishops asked forgiveness for the Church's role in these social sins (#248).

The conference was the scene of an open clash between the Vatican and the Latin American bishops, to be discussed in the following chapter. The result of this clash can be seen in the council's still-ethnocentric view of inculturation: "The aim of inculturation is to bring a society to discover the Christian nature of those values [already inherent in it], to esteem them, and to maintain them as values. It also seeks to incorporate gospel values that are not present in the culture, either because they have been obscured or have even disappeared. . . . By becoming incarnate in these cultures, faith seeks to correct their errors and prevent syncretism" (Hennelly 1993b: 136, #230). The ethnocentric bias is again seen in the Catholic assumption that its formulations are disembodied absolutes lacking any cultural expression.

First Steps toward a Maya Christian Theology

Around 1990, Maya Catholic priests initiated a movement for the inculturation of the "wisdom of the ancestors" as a Maya Christian theology. A theological paradigm starts from a series of religious experiences that become its assumptions. A paradigm builds on these assumptions and can take various forms.

"The point of departure of the Maya-Christian theology is the life of our community, a life constantly threatened by dominating powers. The point of departure is not the world of clear ideas, of eternal truths considered in themselves. The point of departure is life itself; a life that is maintained and constantly unfolding. To think about God is seeking life" (Anonymous 1993: 249). "God, for us, is not the inaccessible, the absolute, the abstract, the distant one, nor the Teotl of the Nahuatls without face or form, not the one who lives in the high heavens; but is and always has been the one who embraces us and the one we embrace, who is around and together with us, the one by whom we live; the heart of the sky and the heart of the earth; our Father, our Mother" (Teología India Mayense 1996a: xi). It is the "understanding we have of our entire life always guided by the hand of God. It is the reflective reasoning that accompanies, explains and guides the journey of our indigenous communities throughout all their history. For this reason it has existed as long as our communities have been in existence" (López Hernández 1993: 355). Maya theology insists that it is nothing new, that it existed long before the arrival of the Spanish and later missionaries.

Indigenous theology continues as a tradition. "It is elaborated and founded in the tradition of the community. . . . What the contemporary community lives and believes is compared with the sources of its faith: this theology drinks from the present wisdom rising in the knowledgeable men and women of the community. It is the theology that enriches itself with the symbols and myths that tells us about God's plan that the community lives" (Siller 1993: 368). Father López Hernández sees the Quetzalcoatl model as providing a framework for an indigenous theology, similar to the Old Testament preparing for the Christ event (López Hernández 2000: 37–39, 117, 140, 166). The Guadalupe story can do the same for the role of the Virgin (171–84). He has written extended discussions on indigenous Christology and Mariology (154–70). The Xólotl story has some characteristics of the Christ event (42–47). The San Cristóbal theological group sees the cycle of Manuel stories as closely resembling the life of Christ (Ruíz García 2003: 78).

From the reflective level, Teología India passes into practice. "It is like consciousness, accompanies action, becomes the motor that drives the transformation of reality. It is prophecy, engagement, praxis" (López Hernández 2000: 94).

Relation to Traditional Christian Elements

This methodology raises a question about the place of the Bible in Teología India, given its privileged and, in many cases, supreme role in traditional Christian theologies. The Bible has an important place, but it is no longer privileged or supreme (Richard 1993: 385–403; Teología India Mayense 1993: 49; López Hernández 1996: 111–28). It is used to find common elements between the two theologies, a comparative perspective. This is seen as helping to better understand the theological import of the Bible itself as well as enriching the message for the Maya community (Anonymous 1993: 255). One of the founders of Abejas, the human rights organization discussed in chapter 21, sees Teología India as the Maya equivalent of the Old Testament. This is also discussed in the South American sectors of the movement (Ruíz García 2003: 26; Shapiro 1981: 143).

A question raised by the reflections of Teología India is its Christology, the place of Christ in it. An indigenous theology would clarify the diffuseness of the wisdom of the ancestors, leading to seeing Jesus in terms meaningful to indigenous peoples. It requires a deeper understanding of Christianity itself.

> If our grandparents have placed Jesus Christ in the center of our communities, it is because He speaks our language and gives life to our communities in the same way that Jesus of Nazareth gave it to the Apostles and his disciples. But it cannot be forgotten that the revelation of the Son of God, the Word Incarnate, was not exhausted by the historical, cultural and religious experience of Jesus of Nazareth and the Jewish community from which he came. The dynamic of the incarnation of the eternal Son of the Father has become a part of the life of all communities in the world. It has been inculturated in every human reality. . . . The Son of God is not a stranger to our history and our cultures. He has also been made incarnate in our reality by

many "epiphanies" or irruptions of his saving presence in the form of events and persons that have been for our communities instruments of the saving action of God. It is for this reason that we undertake to make a synthesis of what is ours and what is proposed from outside. (Teología India Mayense 1993: 49)

The Task: Recovery of Meaning

The reflections on the presence of God that constitute Maya theology were developed by the ancestors of the contemporary Maya. Beginning with the Spaniards and continuing until recent times, this theology was suppressed and attempts made to destroy any trace of it. But it was encoded in the oral myths passed on from generation to generation and in the traditional community rituals, many of which were practiced in secret because of their being viewed as "pagan," "heretical," "devil worship." The tradition became diffuse, as it was no longer passed on in schools by a priesthood who became full-time corn farmers in order to survive. Therefore, one of the tasks of Teología India is to recover those meanings that became obscured. This is a communal task done in workshops that have been held at local, regional, and international levels beginning in the 1990s. Maya Catholic priests, deacons, catechists, shamans, interested bishops, and laymen discuss traditional myths and rituals to draw out their theological meaning (see, e.g., Teología India Mayense 1996a: 41–146; Teología India Mayense 1993: 143–56). A methodology for these reflections about symbols and myths has been codified by Siller (1993: 365–84; Siller 1996: 267–86).

The diocese of San Cristóbal formed a department of Teología India around 1986 (Ruíz García 2003: 78–90). Four workshops were held between 1991 and 2000. Prior to a workshop, local community elders are asked to recount traditional myths and the values contained in them. In San Pedro Chenalhó a Maya Catholic remarked: "Today, the culture of our ancestors is revived. It's as if our ancestors are resuscitated. Before, wherever people went, they had God in their hearts. Wherever they went walking, they prayed. They crossed themselves when the Sun rose, they joined ... their hands to make the sign wherever they went walking. It was very good, the culture of our ancestors, but we lost it, the traditions of our ancestors. We lost how their culture is, it was lost because of us. But today they say that the traditions of our ancestors are being revived, that's what they say" (Moksnes 2003: 154). An example of what the recovery of meaning is trying to accomplish can be seen in Montejo's (2001; 2005: 139–57) account of the story of Q'anil: The Man of Lightning, and the values it teaches that were important for the Jakaltec community during the war years.

A Synopsis and Some Cautions

The recovery of this theology is an ongoing task still at an early stage. From these efforts, its characteristics are beginning to emerge (Flores Reyes 1996: 259–60).

1. Its expression is concrete, not abstract, since it is drawn from community life.
2. It is holistic, integrated with all aspects of community life. It is not religion in the sense of a specialized part of community life.
3. It has its own language that is characteristically religious, a language of myth and symbol.
4. Its subject is the Maya community and is developed by the community, not by individuals who want to impose their own ideas for their own interests.
5. This theology is not something to be developed in the future; it already exists in the daily life of the community.
6. The singular is used for the title of this theology because it refers to all those designated as "Indians" since 1492.
7. While this theology has been mentioned in the documents of Medellín, Puebla, and Santo Domingo, it should be stripped of all stereotypes and other categorizations. There are a plurality of indigenous theologies, siblings to each other and to other theologies of the world.
8. The question arises as to whether Teología India seeks to recover the religious thinking of indigenous communities prior to their encounter with Christianity or whether it seeks to express the Christian experience of these communities. It seeks both as they are saying the same thing.
9. The effort to formulate a Teología India is based on the conviction that there is no single Christian theology. It wishes to develop its own distinctive theology to take its place among "the concert of theological voices." The movement emphasizes its different concept of theology than that prevailing in Western societies.
10. In becoming Christian, the Teología India should not renounce its own distinctive indigenous characteristics, its mythical and symbolic content, its holistic method, its collective subject.

The ninth characteristic is further explained by Padre Lopez: "'Theology' in the West defines itself as necessarily tied to the intellectual work of the churches, and for the purpose of the discussion here, the Catholic church. But theology is not the exclusive property of anyone. All human groups who have an experience of God, give birth to theologies that seek to express this transcendental experience. Such is the case of those original communities of the so-called 'New World,' or the American continent, who for thousands of years have been in constant communion with the Creator and Designer of all. But the theological voice of these communities has had no recognition within the Catholic church" (Teología India Mayense 1996a: xiv).

Bishop Gerardo Flores of Verapaz sees some dangers in the movement and advises caution (Flores Reyes 1996: 263).

1. The desire to advance too quickly that outpaces the slow and sure rhythm of the community.
2. Due to flattery and congratulations the movement has received, some may lose their sense of reality and fall into error. . . .
3. The superficiality of some who think that inculturation is obtained by some expressions and rituals that are, in reality, ridiculous folklore.
4. Manipulation by local and international advisors who use the movement for their own political purposes and self gain. . . . [Interest-Change in terms of the typology in the appendix to chapter 24]
5. Some pastors, due to a fear of the unknown or a closed conservatism, attempt to restrain the movement using the same criteria as the Spaniards did five hundred years ago by labeling it diabolical and perverse. [Value-NonChange and Interest-NonChange]
6. Some may use the movement as an outlet for their own personal resentments and frustrations. [Interest-Change]

The first point, the pace of the movement, states a problem some Maya may have with a Maya Christian theology. "The people ask themselves, 'If the church formed me as a Catholic, and now it comes and says the opposite. Who is right?' I see a problem in this" (Teología India Mayense 1996a: 92). While at first glance it may appear that a Maya theology would be welcomed by Maya Catholic communities, it cannot be forgotten that in the past there were sharp clashes between Catholics and traditionalists in many Maya communities (see chapter 8). Also, the split between Catholics and traditionalists was not limited to religious affairs, but often spilled over into community politics. A sudden about-face would not be understood by a number of Maya Catholics. In Yibeljoj, in the mid-1990s, "people—including catechists—were still quite hesitant about what to think. Many appreciated that the Teología India catechists pointed out the value of the traditions of their parents and ancestors, but were more questioning when it came to what traditions and ceremonies they should actively take part of themselves. During the catechist course in 1996, there was a lot of very heated discussion on this matter, where the head of the Teología India group tried to convince upset catechists that there were good traditional ceremonies that one should respect etc., that not all beliefs and ceremonies were bad and idolatrous, etc. This re-evaluation process has continued, and now [2008] it appears that people even more have integrated the Maya cultural legacy—pride in their discourse, at least the young and politicized" (Moksnes, personal communication).

Building a Social Structure for a Maya Catholicism

In 1958, the Jesuits assumed responsibility for the Catholic parishes in the municipios of Chilón and Sitalá with residence in Bachajón (Maurer 1984). Originally, the Jesuits concentrated on translation work. One of their first efforts was to translate a seventeenth-century Tridentine catechism by P. Ripalda from Spanish into Tzeltal (440). This writer found this three-hundred-year old Tridentine catechism in wide use in the San Cristóbal Diocese in the early 1960s. In 1970, there was a translation of another catechism, along with passages from the Bible. The priest would explain the biblical passages to the catechists for use with their students (446). Later, they used the tijwane method of biblical reflection, as described in chapter 14. In 1965, they translated into Tzeltal the mass prayers, and in the early 1970s the book of Exodus. By 1984, they had finished the Gospels and the Acts of the Apostles, by 1987 the various letters of the New Testament and the Book of Revelation (171, 175). To meet the subsistence crisis, they initiated the community-action programs described in chapter 12.

An Inculturated Church Structure

The Jesuits organized a number of Catholic Action ministries into what they hoped would be a fully inculturated Maya church (Zatyrka 2003: 254–313). Maya communities began building chapels where the visiting priest would administer the sacraments and a local catechist would conduct biblical reflection sessions. Initially, the catechist was also a health promoter, conflict mediator, agricultural agent, and a community organizer, as seen in the curriculum of original catechetical schools (chapter 7). These functions evolved into specialized roles as ministries within a small bureaucracy to attend to the needs of about 32,000 people. There were four levels of authority: the 532 local communities, 56 zones of several adjoining communities, 15 interzones joining together 3 zones, and, finally, the entire Jesuit-administered area. The levels beyond the local community were created to take care of specialized roles that did not warrant having someone in every community, and for purposes of coordination, training, and supervision.

At the local community level, there were catechists, one of whom might be a deacon; a presidente and his wife, who was in charge of the chapel; a health promoter; and an elder (principal) and his wife, who were community advisors and admonitors of the catechist or deacon and their wives. In 2003, there were 1,614 active catechists, or one for every twenty persons; 202 deacons; and 80 health promoters.

At the zone level, there was a coordinator of the catechists who managed their training and advised on difficult matters; a community visitor who was a mediator of conflicts to restore harmony both within individual communities and between local communities; and a community promoter in charge of various community development projects, such as building roads, latrines, and potable water systems. At the interzone level, there was a supervisor-coordinator of each of the three officials of the zone levels.

The next level embraced all the zones and communities. It was composed of committees which were ultimately responsible for the training and performance of the specialized roles at the two lower levels. Several of these committees were concerned with conflict mediation in disputes between extended kin. Others dealt with women's issues. The Maya representatives to the governing body of the whole area composed another committee. Another group, spiritual assistants, were deacons whom the Jesuits hoped would eventually become ordained priests. They were experienced, mature catechists well-grounded in the Bible and Catholic doctrine. They took courses leading to a degree in theology sponsored by the Iberoamericana University. In 2003, there were twenty-four spiritual assistants.

At the highest level was the Regional Pastoral Council, composed of the Jesuit pastoral workers, lay pastoral workers, and Maya representatives—for the most part, older deacons. This was the highest policymaking body. It was still dominated by non-Maya, but the goal was eventually to turn the entire structure over to the Maya as an independent (autochthonous) inculturated church. For each of the ministries there is a religious ceremony of induction attended by all those from any level concerned with the training and supervision of the individual. Prayers are recited asking for help in the proper performance of the role, words of encouragement are given by various attendees, and an object symbolic of the role is presented.

Maya Diaconate

At Vatican II, African and other bishops argued strongly for changing the celibacy requirement for the priesthood in the Latin Rite, but they were unsuccessful. As a substitute measure, Vatican II restored the ancient office of the permanent deacon for any diocese that wished to make use of it (Ad Gentes, #16). The role allows married laymen to become ordained ministers of the Catholic Church for preaching and administering baptism, witnessing marriages, and to assist at funerals. Celebration of the mass, hearing confessions, and administering the last rites with confessions are the only priestly roles they do not perform. Ordination as a permanent deacon is a consecration like that of a priest. It cannot be withdrawn, although, as in the case of priests, a bishop can forbid its exercise in his diocese or the Vatican can defrock a priest that excludes any exercise of priestly functions under any conditions. Before consecration, the selected candidates serve three trial years as temporary deacons (predeacons). Having Mayas as consecrated deacons was seen as an important step toward the formation of an inculturated Maya Catholic Church. It would provide personnel with intimate knowledge of their culture to take on the tasks of theological and liturgical inculturation. In 1985, Bishop Ruíz ordained the first group of 20 permanent deacons. By 1996, there were 137. The diaconate program was extended to other parishes of the diocese, including San Andrés Larraínzar, where Rudolfo (chapters 6, 15, and 21) became a deacon. The introductory photograph shows the ordination of a deacon by Bishop Felipe Arizmendi, who succeeded Bishop Ruíz and has supported the program. The deacon's wife is at his side, symbolizing the sharing of the role. Maya administering some of the sacraments was a significant step toward a Maya identification with the Catholic Church.

The Maya Seek a Married Catholic Priesthood

From the earliest days of the Spanish colony, some Maya have desired to become Catholic priests. The Mexican synods of 1555 and 1565, that at that time included Guatemala, forbade Indians to recite the canonical hours of the Divine Office, to say mass, to hold funeral services, or to organize processions without the presence of a Spanish priest (Ricard 1966 [1933]: 67). These prohibitions imply that the catechists of that period, the fiscales, were acting as unauthorized priests. During the Maya rebellions—in 1712 in Cancuc, in 1867–70 in Chamula—and during the Yukatecan caste wars in the nineteenth century, fiscales were ordained as Catholic priests by self-declared Maya bishops (Early 2006: 231–37).

While the 1974 Indian Congress had stressed four problem areas of oppression currently suffered by Maya communities, and had strongly attacked the state government, there was no mention of the church in the final presentations of the congress. Bishop Ruíz wondered how the Maya viewed the contemporary church and their role in it. In 1975, the Jesuits organized a meeting of catechists to discuss this theme with the bishop in attendance. To help stimulate reflection about Maya taking more responsibility for their own church, the bishop presented them with a hypothetical situation. "Suppose that tomorrow all the missioners and I leave here by plane. And that our plane has an accident and we all die. That would mean no more missioners to accompany you. You would be left alone. Do you think you are ready for that? What do you have to show for more than 15 years of missionary presence among you? Please think about it, discuss it among yourselves and give me your answer." A principal, speaking for the catechists, gave an unexpected reply.

> Tatik Bishop, we think that perhaps you and the missioners are not doing your job correctly. You see, we know through Scripture that our Lord Jesus Christ formed his followers during three years. Then he was killed on the cross, resurrected and lived among his disciples for another 40 days, at the end of which he left them to go to heaven. But his work is still present among us two thousand years later. However, you have been here for 16 years and we still need your presence to continue in our faith life. What did our Lord do that you are not doing now? Before he left, our Lord Jesus Christ left his Holy Spirit to continue his work. But you have not given us this Spirit. You have not shared it with us Tzeltal. It is true that we receive the Holy Spirit in our Baptism, in our Confirmation, in other Sacraments. But you have not given us the Spirit that builds the community and keeps it united. If you give this Spirit to us, then you will see how your work continues, even if you have to leave us. (Zatyrka 2003: 177)

They were requesting a Maya Catholic priesthood, the Spirit that builds community and keeps it united. There was an underlying assumption to this reply: "Even though the Maya do not consider the missionaries as real ladinos (because they do not exploit, they are good ladinos), nevertheless they are not real Tzeltales. Some Maya say that the padres think like ladinos and never will be able to perfectly understand the Tzeltales" (Maurer 1984: 462). Consequently, if there is to be an inculturation of Catholicism in Tzeltal Maya culture, it can only be done by a Maya clergy.

The obvious candidates were mature catechists and deacons, almost all married men. In 1997, and again in 1998, the San Cristóbal Diocese sent petitions to Rome for a priesthood that allowed marriage, as another step toward an inculturated church (Vos 2002: 241–43, Zatyrka 2003: 195). Such a church would be united with Rome in much the same way as the Uniate churches. A number of these are Eastern Rite churches with their own theology, liturgy, and a married clergy that remained united to Rome instead of becoming Greek or Russian Orthodox churches (Sable 2005: 9462–66). Therefore, the Maya request has precedents, as there already is a married priesthood within the Roman Catholic Church. The Uniates kept pointing this out during the celibacy discussions at Vatican Council II. Celibacy is an issue only in the Latin rite of the Roman Catholic church, but by far the numerically dominant one. It became the law of the Latin rite in the fourth century when part of the monastic discipline was imposed on all its clergy (Delhaye 2003: 324).

Summary

This chapter has described the need for a Catholic theology and church based upon Maya culture instead of the Western-imposed forms that presently exist. Bureaucratic Catholicism slowly began to wrestle with the problem of inculturation at Vatican Council II and at the conferences of Latin American bishops. In the Maya area, groups composed predominantly of Maya Catholic priests met to make explicit an already existing Maya theology. Some Maya parishes have established the permanent diaconate as a hopeful step toward a married Maya priesthood. The question is: Will these efforts be fully recognized and allowed to advance by bureaucratic Catholicism so that a distinctively Maya Catholicism will be fully integrated and accepted as orthodox Catholicism?

IX

Conclusion

30

A Look Backward and Forward

This volume, and the preceding one, have examined the interactions of the Maya and Catholic worldviews that began in the sixteenth century and have continued to the beginning of the twenty-first. The previous volume showed that at the time of the Spanish Conquest, the Maya interpreted their military defeat as having been due to the stronger power of the Spanish gods. Consequently, they incorporated the Spanish Catholic gods, especially the saints, into the Maya covenant and defined these gods according to the covenant's concepts and logic. The previous volume and Part II of this volume showed that this slightly modified Maya worldview continued to define Maya life into the twentieth century. At that time, it was called into question, due to the Maya subsistence crisis primarily resulting from loss of their lands. This loss impacted all parts of their culture.

By the mid-twentieth century, following the liberal suppression, the Catholic Church reentered the Maya communities in an attempt to establish orthodox Catholic communities as defined by the sixteenth-century Council of Trent. Amid much misunderstanding, orthodox Tridentine Catholicism gradually distinguished itself from the Maya worldview and, in some areas, attacked it in an effort to suppress it. But as Maya catechists forcefully pointed out, Tridentine Catholicism was inadequate to relieve their sufferings due to the subsistence crisis. This lack of concern for contemporary social problems was a systemic Catholic problem, as Tridentine theology placed primary emphasis on correct belief and ritual as preparation for a life to come, to the neglect of social justice in contemporary communal life.

Vatican Council II and the Catholic Action movement, with its concern for contemporary human communities, attempted to reform what had become an archaic church. The implementation of these reforms initiated the second phase of Catholicism's interaction with Maya communities in the current era. It was based on a biblical theology that stressed the moral message of the Gospels, especially the prophetic theme of social justice. As a result, numerous Catholic programs of health, education, and cooperatives were established in Maya communities. They increased life expectancy and began to give the Maya the institutional tools to defend themselves and take their place in contemporary society.

This evolved into the third phase of interaction, the liberation movement that brought into critical focus the systemic nature of the social injustices against Maya communities and how they were the source of many of the problems encountered by Catholic Action at the local level. As the liberation movement developed, it divided over the strategy to confront systemic injustice. One sector allied itself with radical Marxists and chose violence as the only feasible solution to the problems of sustained injustice, itself protected by violence. They joined in armed uprisings against the national governments of Guatemala and Mexico. Others renounced violence as the answer and attempted systemic reform by nonviolent methods. The investigation has shown that the traditional Catholic position about the morality of taking up arms was not carefully considered.

In the fourth phase, the post-insurgency period, systemic injustices continue with non-implementation of the peace accords, poverty, land seizures, and the murder of human rights advocates. The subsistence crisis has been somewhat alleviated by some increase in government efforts, programs funded by non-governmental organizations, and monies sent back from large-scale migration to the United States. In the fourth phase, another aspect of Catholic liberation became important—liberation from Western theological categories that Catholicism had imposed in the past to express universal values. This liberation became the agenda of the indigenous theology movement. It sought an inculturated Christianity that expressed the message of Christ using the theological forms of Maya culture, the wisdom of the ancestors that was readily found to be the same as the moral message of Jesus. Both worldviews see time as cyclical—a process of birth, death, and regeneration, in Maya terms; one of election, sin, and redemption, in Judeo-Christian terms; and cultural vibrancy, degeneration, and revitalization, in anthropological terms. The death-sin-degeneration segment of the cycle took place when both Maya and Catholic cultures dimmed their moral sense and became ritualistic. Ritual assumed supreme importance by closing in on itself and severing itself from the primary function of religious systems: helping people to live meaningful lives in the human community. Regeneration-redemption-revitalization came about when both cultures once again confronted the moral dimension of human community in another cycle of their existence. The regenerative aspect of religious worldviews gives them great persistence through time in spite of degenerative crises that superficially appear to foretell their permanent deaths.

The preceding chapters have shown the persistence of the Maya worldview into the last half of the twentieth century—through all four phases of the Catholic interactions, and outside of Catholicism, in other Maya responses to their crisis. In the last half of the twentieth century, Maya communities were increasingly penetrated by a Western missionary presence that rejected the Maya worldview; by national governments with assimilationist intentions; and by the dynamics of a market economy. In analyzing these interactions, Tridentine and Action Catholic Maya were seen to retain the Maya metaphysic of epochs and cycles. Some interpreted their Catholicism as a cyclical revitalization of the malfunctioning Maya worldview. The Catholic inculturation movement is an attempt to revitalize previous cycles of the Maya worldview to express a Maya Christian theology and to implement a liturgy based on it, to be administered by a Maya Catholic priesthood. Charismatic Catholicism reconnects to the importance of inner experience as practiced in traditional Maya dreamvisions and shamanism. Outside of Catholicism, Pentecostal charismaticism attempts to do much the same thing, along with other traditional characteristics (De Leon Ceto 2006: 120–24). The pan-Maya movement is an explicit call to a revitalization of the pre-Columbian phase. Traditionalists continue to retain the Maya worldview of its colonial phase. The only trend that completely rejected all aspects of the Maya worldview is ladinization, in any of its forms. Some Maya Catholics have made this choice.

Conversion to traditional Protestant denominations and other religions usually involves adoption of ladino characteristics, including the rejection of Maya culture.

This persistence of the Maya worldview is primarily due to its metaphysical structure. It enables it to endure in spite of changing. It posits reality as dynamic, as constantly going through emanations and cycles of birth, death, regeneration. This gives it great elasticity to see changes as cycles within its own structure. In this sense, it lacks firm systemic boundaries. In contrast, a number of religious worldviews are static, based on a one-time revelation recorded in a sacred book and presented as essentially unchanged since their inception.

The Future of the Maya-Catholic Relationship

In the interactions of the persisting Maya and reforming Catholic worldviews, are they evolving toward a mutual understanding or are they distancing themselves from each other?

The effort for an inculturated Maya Catholic Church with its own theology, liturgy, and clergy represents a significant effort of mutual understanding. Bishop Ruíz and some other bishops of indigenous dioceses see it as part of the liberation movement, liberation from imposed Western theological forms.

In spite of these efforts, the Vatican has moved against various facets of liberation on behalf of the Maya and other indigenous groups. Nonfunctional seminary training, whose curricula abstracted from the needs of indigenous and other communities, was a continual problem for many poor Mexican regions. To overcome this obstacle, Mexican liberationists founded seminaries at Tula in the state of Hidalgo and Papantla in Veracruz with curricula oriented to the option for the poor. In southeast Mexico, Chiapas was one of several predominantly indigenous dioceses that formed a pastoral region, Pacifico Sur (Norget 2004: 154–86). It founded its own Regional Seminary of the Southeast (SERESURE) with a realistic curriculum for future work in indigenous communities, the only one in Mexico. A former student recalled: "There were intense months of study and then other months in equal number of intense work with the people. It was fantastic. Most of the students at SERESURE were from indigenous communities; Chiapas, Oaxaca, Puebla" (161). But the Vatican moved against these efforts. The two liberation seminaries were forced to restructure themselves along more traditional lines. SERESURE was forced to close in 1990. It reopened in 1992 as a ladino-oriented seminary.

The Vatican disliked the liberation and inculturation policies of some Mexican bishops. They did not remove them from office, as this would generate criticism and bad publicity. Instead, in these dioceses the Vatican appointed auxiliary bishops known for their dislike of liberation and inculturation, with the right of succession upon the death or retirement of the current bishop. In addition, the Vatican gave these auxiliary bishops veto power over the policies of the bishops to whom they were supposedly assistants. This made the bishops figureheads. The bishops of the liberation dioceses of San Cristóbal, Oaxaca, and Tehuantepec were treated in this manner. But the assistant bishop assigned to Bishop Ruíz, upon understanding the crisis conditions of the Maya, was converted to the liberation programs of Don Samuel. Consequently, he was not allowed to succeed Bishop Ruíz, but was shipped off to the northern Mexican diocese of Saltillo.

For an extended period, there was an attempt by Mexican government officials, in collusion with the Vatican's Apostolic Delegate, to remove Bishop Ruíz from office. This is an important story in itself (Fazio 1994). It has been passed over here because the Maya were not directly involved. Ironically, Bishop Ruíz would probably have been removed from his diocese if the EZLN uprising had not occurred. Since he was the only acceptable mediator to both sides in their later negotiations, the Vatican was forced to leave him in place.

In Guatemala, much of Liberation Catholicism had been initiated by foreign priests. But toward the end of the century, many of them had been replaced by ladino priests trained in conservative seminaries. This type of personnel has had little concern for inculturation and, coming after the ravages of the civil war, little interest in Maya action programs. There has been an increase in the number of Maya who have become Catholic priests. Some are important members of the Teología India movement, while others have become ladinized as a result of their seminary training.

Both liberation reflection on the Bible and the Teología India movement have a common element in their methodologies. They begin with reflection on the contemporary reality of their communities and only then go to theological sources for assistance in assessing this reality without imposing the sources on reality. The Vatican was threatened by this methodology since it diluted the authority of its pronouncements. It attempted to take control of the Santo Domingo conference away from the Latin American bishops and to role back the advances of Medellín and Puebla (Hennelly 1993b: 23–36). The Vatican set the agenda instead of the bishops themselves. It placed in key positions bishops willing to act as Vatican puppets, and used disruptive parliamentary tactics to thwart the established procedures of episcopal conferences. All of this was a continuation of many Vatican efforts since 1965 to deter the advances of Vatican Council II. They were not entirely successful. The resulting document of the conference, the "Message of the Fourth General Conference to the Peoples of Latin America and the Caribbean," is an uneasy attempt to reconcile irreducible positions. Hennelly (29) has suggested an order in which the documents may be read to make some sense of them. A liberation theologian has assessed the Vatican effort. "Theologically there have been steps backward, fundamentally in terms of theological procedure or method. Historical reality is no longer seen as a sign of the times in a strictly theological sense—that is, as the place where God may speak and be present as God. This is an important regression from Vatican II, which enunciated the reality and importance of the signs of the times for the mission of the church, and from Medellín and Puebla, where the signs of the times were scrutinized *in actu* (in themselves) and became the basis of theological reflection.... The presidency [a puppet bishop] imposed the principle of judging, seeing, acting [a reversal of the customary first two steps]: theology comes first followed by observation of the world, and then by the application of theology to the world. This means 'judging' from God's viewpoint what has not yet been 'seen.' With regard to biblical texts, it means seeing God in those texts from the past without having seen God in present reality" (Sobrino 1993: 177).

The Vatican has moved against the previously quoted Father López Hernández and the Boff brothers, as well as other liberation theologians. Father López worked for over twenty years at the Mexican center of pastoral planning for indigenous peoples (CENAMI). There he did his research and writing, as well as being a consultant to many indigenous groups in Latin America, including the Maya. The Vatican sent letters to Mexican bishops questioning his orthodoxy. Upon his bishop's return from Rome in 2006, no charges were filed against him, but the bishop requested Father López give up his position at CENAMI, return to his diocese (Tehuantepec), and become the pastor of a local church. This was an effort to diminish his influence on the movement. Father López, with his divided heart, complied (López Hernández 2006).

The importance and reliance of Catholicism in the Maya areas on the fiscales (Early 2006: 226–37) and their successors, the catechists, has been a constant for almost five centuries. In Maya communities, many of these men are tireless workers implementing the spirit of Christ in their communities, men of maturity and self-giving, some of extraordinary intelligence and abilities. They should be the priests in their communities. The San Cristóbal Diocese has worked toward building an inculturated church with such priests. As a first step, it ordained married catechists as permanent deacons. Some have received degrees in theology. Then, in 1997, and again in 1998, the diocese sent petitions to Rome for a priesthood that allowed marriage, as another step toward an inculturated church (Vos 2002: 241–43, Zatyrka 2003: 195). Thirteen years later, in 2010, the Vatican still had not responded to the Maya request.

In addition to the Vatican's moves against what had been a growing bond between the Maya and Catholic worldviews, orthodox Catholicism has a strained relationship with a growing sector of Maya Catholics, the charismatics. These Catholics emphasize a personal experience, the Life of the Spirit, that connects with the dream-visions and shamanic experiences of the traditional worldview. There is a tendency to substitute this experience for participation in the sacraments administered by priests or Maya deacons. Its withdrawal from community empowerment activities has alienated Catholic Action. These strained relationships have led significant numbers to become Pentecostal charismatics.

Additional loses have been attributed to the pan-Maya movement. Second-generation members of Tridentine and Action Catholicism were brought up in isolation from and disrespect for the traditional Maya culture. As they matured and realized what had happened, some resented it, went back and absorbed Maya culture. They initiated or became part of the pan-Maya movement, which rejects Catholicism as a Spanish imposition and does not forget Catholicism's four-hundred-year oppressive legacy.

Therefore, by the end of the twentieth century, in spite of the efforts of Maya and non-Maya Catholics, it appears that bureaucratic Catholicism is gradually separating itself from nearly five hundred years of involvement with Maya culture. The divided hearts of Maya Catholics can only take so much pain for so long. There is the possibility of a Maya Catholic Church with its own hierarchy independent of the Vatican. Or, as happened to many Tridentine Catholics, they may cease to find meaning in a Western ethnocentric Catholicism and drift away to other alternatives. The introductory photograph shows a young Atiteco Montessori student perusing pictures in the *National Geographic* magazine. The question for him and other Maya Catholics is: Where will we seek meaning for our lives in the twenty-first century?

Some Implications of the Research

For Catholicism, the tension between contemporary Maya Catholics and the Catholic bureaucracy is not a phenomenon limited to Guatemala and

Chiapas. Nor does its outcome have only local significance. The Catholic-Maya encounter is one among many, a relatively small example of what is happening in a global society. A supposedly universal Catholicism is encountering non-Western cultures and is attempting to overcome its own ethnocentrism, as well as the ethnocentrism of other cultures and their religious systems. Previous Catholic efforts of a similar nature by the Jesuits—Ricci in China in the sixteenth century, and de Nobili in India in the seventeenth—failed due to Vatican rejection. The Maya inculturation movement appears headed for the same ethnocentric rejection. But Maya culture, in spite of centuries of domination by others, is alive and demanding respect in a world given to ignoring the cultures of indigenous groups. If Catholicism's problems with the Maya are any indication of its reactions to similar problems worldwide, it appears that the Catholic theological crisis of the twentieth century will intensify during the twenty-first century. Action Catholicism may come to be seen as a brief interlude of attempted reform that failed.

For the Maya, this study may help clarify the problems they face in the current epoch. Like all religious systems, theirs has to face its own ethnocentrism, its continuing problem of unity among its diverse groups, and the danger of using religious simply as a political tool. The problems faced by the Maya are common to many indigenous groups, just as the problems faced by Catholics are common, not only to all Christian groups but to all religious paradigms in a global world in constant, twenty-four-hour communication. As religious systems struggle to reform themselves due to internal degeneration, and to break out of their no-longer-isolated worlds, they are challenged by ethnocentrism in many forms. Sometimes religions, rather than being instruments of peace, become a tool or the basis of intranational or international conflict. Perhaps the interactions of the Maya and Catholic worldviews seen in this research may offer some insights about the problems stemming from the encounters of religious worldviews in a globalizing world, and about the efforts to move beyond these problems.

Anthropology and the Encounters of Religious Systems

In a globalizing world in which communities with different cultures frequently come into contact with each other, anthropology can play an important role. Most cultures have posited a religious worldview to help individuals and their communities live meaningful lives. Cultural anthropology is based upon "field work," the experience anthropologists gain by living in different cultures and undergoing the cross-cultural experience. Such experience is not limited to anthropologists, as seen in previously described examples of pastoral workers and Marxian revolutionaries. Nor do all anthropologists have this experience. But for all who do, the experience allows anthropologists to record the results of formal theological and philosophical paradigms when inserted into the everyday lives of human communities. There, the formal paradigms undergo mutations and unintended consequences because of the complexities involved in everyday life. In so doing, their strengths and weaknesses are revealed. Anthropology can contribute to cultural self-understanding and reform, and in so doing, it can foster mutual acceptance rather than the conflict that the encounters of theological paradigms often engender.

Anthropology played a helpful role for Catholicism during several pivotal events already discussed. It provided the moment of insight for Bishop Ruíz at Melgar as he broke through the walls of his Tridentine background to an understanding of culture and religion's place in it. This resulted in his becoming one of the leaders of the liberation-inculturation movement. It happened again at Xicotepec, and with the Barbados Declaration, as anthropologists forcefully brought to the attention of Mexican and Latin American bishops the damage that current missionary policies were doing to indigenous cultures.

The study has also shown the importance of worldviews in community life, that the preponderance of economic and political models can give a shortsighted view of community dynamics, especially by ignoring the agency of indigenous worldviews. The study also indicates the necessity for ethnography to widen its traditional horizons and become conversant with both the Eastern and Western philosophical-theological traditions. That the worldviews of many people and their cultures are religious is a basic fact, regardless of one's personal evaluation of this fact. The little traditions of many cultures have been influenced by the great traditions of existing or once flourishing cultures transmitted by traders, warriors, missionaries, migrants, or electronic media. Therefore, it is necessary to understand religious systems at some depth rather than using simple stereotypes or seeking to reduce religion to a series of social functions. They are not cultural epiphenomena, as the anthropologist Reichel-Dolmatoff made clear to Bishop Ruíz at Melgar. Their importance is shown in the power of the classic religious epics—in this study, the Bible and the Popol

Vuh. Religions are about meaning, values, and the moral sense of individuals and their communities. As such, they demand careful study of the way they are incarnated in human communities. This has been the effort and intent of the two volumes about the Maya and Catholicism.

Notes

Chapter 2. The Traditional Maya Worldview as Influenced by Later Evangelization

1. I have used the phrase "sustaining powers" to designate what are more commonly called "souls" in Mayan ethnographies. Mayan theology is a pantheistic system in which beings are emanations of a single being, that is, they are part of the One, not distinct from it. Pantheistic systems have concepts and logics different than those of Judeo-Christian theologies that are transcendent systems, in which beings are distinct from a creator God. It causes confusion to describe a pantheistic system using terms from the theological language of a transcendent system. It is impossible to literally translate some pantheistic concepts because they do not exist in a transcendent system, and vice versa. The root meaning of *ch'ulel* in Tzeltal and Tzotzil is "holy, sacred" (Laughlin, personal communication). This is consistent with the pantheistic concept of the emanated sustaining power of the One as an individual. In Ki'che', the Spanish loan word *anima* is used for the Judeo-Christian "soul" (Mondolch, personal communication), indicating that there is no concept in the Mayan languages for "soul" in the Judeo-Christian sense. In keeping with the pantheistic form of Maya theology, this writer prefers to use the term "sustaining power" as less confusing than "soul" when translating "ch'ulel" or its equivalents.

Chapter 7. Worldview of Tridentine Catholicism

1. "Originally an indulgence was a relaxing of the punishment (penance) imposed by the Church for transgression against some religious commandment. At first such indulgences were conferred by the popes upon those who risked their lives fighting against the infidel. Gradually, however, such active service for the common good of Christendom was extended to include financial gifts for worthy spiritual causes, even if the donor did not actively participate in them. The financing of cathedrals, monasteries and hospitals was at times designated by the papacy as warranting an indulgence. Thus good works could lead to a remission of punishment. But by now [the sixteenth century] such a remission had come to mean the relaxing of punishment not just by the Church itself, but by God, and a right to expect God's mercy" (Koenigsberger 1989: 143). The theological basis was that the Pope, as custodian of that surplus of good deeds accumulated by Christ and augmented by the saints through the ages, could grant remission of divine punishment (Denzinger 1955: 201, #550; 239, #740a).

2. There are seven sacraments in the Catholic system. They can best be viewed a passage rites to prepare and strengthen a person for moral action appropriate for the stage of the life cycle they are about to enter. Baptism is the initiation rite into the Catholic community by remitting the original sin that all men are born with. Penance, confessing one's sins, first takes place around the ages of seven to nine, when a child

is considered capable of knowing right from wrong. This is usually followed by the first reception of the Holy Eucharist. In Catholic doctrine bread and wine actually become the body and blood of Christ (transubstantiation, real presence). Penance and the Eucharist are repeated throughout the life cycle to obtain forgiveness of sins and to fortify one with food for the soul, so as to persevere in the promises of amendment made in the sacrament of penance. Confirmation is received with the dawning of adolescence, to fortify Catholics for the temptations and trials of adulthood. Matrimony declares to all the choice of a mate for life and an acknowledgement of the obligations that go with the married state. Extreme Unction, which usually includes a confession, consoles and prepares one for the final journey of the soul. For a few, Holy Orders is the sacrament for the ordination of priests, men set aside to administer the above sacraments and in this way proclaim the gospel of Jesus.

All the sacraments are rituals meant to be a means to moral action in one's life and community. Ritualism takes place when the performance of or participation in the sacramental rituals becomes the primary concern, rather than moral action.

Chapter 8. Presentation and Maya Reception of the Tridentine Worldview

1. Wallace's description of the revitalization efforts of Handsome Lake among the Iroquois during their crisis following the French-Indian War and the American Revolution has many parallels with the Maya and their revitalization movements.

"Heavenly messengers, Handsome Lake said, had told him that unless he and his fellows became new men, they were doomed to be destroyed in an apocalyptic world destruction. They must cease drinking, quarreling, and witchcraft, and henceforth lead pure and upright lives. Some of his instructions were directed toward theological and ritual matters, but the bulk of his code was directed toward the resolution of moral issues presented by the new social and economic situation of the reservation Iroquois. He told the Iroquois to adopt the white man's mode of agriculture, which included a man's working the fields (hitherto, a woman's role); he advised that some learn to read and write English; he counseled them to emphasize the integrity of the married couple and its household, rather than the old maternal lineage. In sum, his code was a blueprint of a culture that would be socially and technologically more effective in the new circumstances of reservation life than the old culture could ever have been" (Wallace 1966: 32).

2. The large diocese of Los Altos was centered in Quetzaltenango. In 1951, two new dioceses were created from part of its territories: the Diocese of San Marcos that included Huehuetenango, and the Diocese of Sololá that included El Quiché and the western part of Chimaltenango. In 1967, El Quiché and Huehuetenango were made independent dioceses.

3. The term "Catholic Action" has a restricted theological meaning and a broader one often used in Guatemala. A Maya catechist conducting a catechism session is called Catholic Action in Guatemala, but is not Catholic Action in the more-restricted sense that will be explained in chapter 9.

Chapter 13. The Bible and Its Worldview as a Cultural Document

1. For some, the anthropological treatment of the Bible in this chapter may be seen as an attempt to diminish or disrepect it in some way. The religious affiliations of two

quoted experts should be noted. James Kugel was Starr Professor of Hebrew Literature at Harvard University from 1982 to 2002 and a Professor of Bible at Bar Ilam University in Israel. He is a practicing Orthodox Jew (Kugel 2007: 45). Hans Küng is a Roman Catholic priest in good standing in a Swiss diocese, and Professor Emeritus of Ecumenical Theology at Tübingen University (Küng 2001: xvii–xviii). He is disliked by the Vatican, which has rejected some of his writings.

Chapter 17. Guatemala: The Role of the Maya in the Worldview of Marxist Insurgency

1. This photograph by Duncan Earle, was taken in San Cristóbal on January 1, 2003, as Zapatistas commemorated the Chiapas uprising of January 1, 1994 (see chapter 20). It is used here as a symbol of all Maya frustrations and protests.

Chapter 18. Guatemala: The Maya in the Military's Worldview of Counterinsurgency

1. All the figures in this paragraph are estimations based on two studies of violations of human rights during the revolutionary period in Guatemala, 1962–96. *Guatemala, Never Again!*, commonly referred to as the REMHI study (for Recovery of Historical Memory Project), is based on thousands of personal testimonies of war atrocities. It was sponsored by the Human Rights office of the Archdiocese of Guatemala and appeared in 1998. The English version is a summary of the four-volume Spanish version.

The United Nations commissioned its own study by a group known as CEH (Comisión para el Esclarecimiento Historico). Its report was entitled *Guatemala: Memoria del silencio*, although it is more frequently cited in a slightly shorter form, *Memoria del silencio*. It was based 7,388 personal testimonies as well as on collaborative information given by 20,000 others. Its findings were similar to those of the REMHI study, which were incorporated in CEH's final report along with many other research reports. The American Association of the Advancement of Science has published most of the Spanish version on its Web site, www.aaas.org. The report uses section numbers in place of page numbers. The English version is highly condensed from the twelve volumes of the Spanish CEH version and can also be found on the same web site.

The studies projected from their samples that 200,000 Guatemalans lost their lives during the Guatemalan revolutionary period, 1962–96 (Comisíon para el Esclaracimiento Historíco 1999, Mandato y Procedimiento de Trabajo, #112). As previously indicated, there were two distinct revolutions within this period: the predominantly Ladino one in the Eastern Highlands and Guatemala City from 1962 to the early 1970s, and the predominantly Maya one in the western departments from the late 1970s to 1996. A rough estimate of the casualties during the first revolution run from 6,000 to 15,000 (Centro de Investigación y Documentación Centroamericana 1980: 83). We will use 10,000 as an estimate, leaving 190,000 as the number of casualties during the uprising in the western part of the country.

How many of these were Maya? The CEH study (II, #102, graph 4) found that the Maya suffered 83.3 percent of all the violations of human rights in the sample and ladinos 16.5 percent. While violations of human rights includes more than murder, it is assumed here that the figures can also be used for one of its main categories, deaths.

Again, this figure refers to the years 1962–96. If the figures from the first revolution could be removed, it would yield a higher percentage of Maya for the period 1978–96. From graphs 1, 2, and 3, showing the individual years and those departments with the highest number of arbitrary executions (Comisíon para el Esclaracimiento Historíco 1999, II:131, 144, 153), it is clear that the highest percentage by far occurred during the second revolution. In graph 1, the scale of the number of violations during the years of the first revolution in the eastern departments only goes to 70, while the scale for the second graph, covering the height of the insurgency in the western departments, goes to 7,000. Therefore, the Maya percentage should be greater than 83.3 percent. We put the estimate at 90 percent. The figures for those responsible for the casualties come from CEH (1999, II:2, #109, graph 6). The estimate of 200,000 Maya refugees in Mexico comes from Manz (1988: 147).

Chapter 26. Liberation Catholicism: Its Relation to the Morality of Armed Rebellion

1. The material about Father Ronald Hennessey is taken from Thomas Melville's book (2005). Melville quotes conversations that Hennessey had with many people. These are reconstructed, verified quotations. Melville explains, "The use of quotations marks is not meant to claim verbatim repetition of the syntax and vocabulary employed at a given moment.... All scenes in which Hennessey appears to have dialogue ... have been reconstructed from his diaries and memory, and from the hundreds of letters he wrote to his family members.... Ron and I talked for hours, days, weeks, and years, going over this manuscript time and again as I tried to pull from a very modest and self-deprecating individual the words, reactions and emotions that he experienced in the situations described. Often he would say to me, 'What do you think I felt?' or 'What would you have said?' Whatever I answered, albeit hesitatingly, he would smile and respond, 'Right! That's it! Put it down!'" (Melville 2005: 14).

Chapter 27. Choices Faced by Catholic Maya in a Turbulent Society

1. Some clarification is helpful regarding the labels used in this and the following chapters. This writer prefers to distinguish Pentecostal Charismatics from Protestants. Others include the former as a subcategory of Protestants. Given the historical backgrounds and differences between them, this writer prefers to use "Protestant" to refer to the more traditional Protestant denominations, such as Methodists, Presbyterians and Episcopalians. But Episcopalians do not like to be designated Protestants. While rejecting the authority of the Pope, their theology retains many Catholic elements, as expressed in the alternative term for Episcopalians, Anglo-Catholics.

Bibliography

Abbott, Walter M., ed.
1966   *The Documents of Vatican II.* Translated by Joseph Gallagher. New York: The America Press.

Adams, Abigail E.
2001   Making One Our Word: Protestant Q'eqchi' Mayas in Highland Guatemala. In *Holy Saints and Fiery Preachers: The Anthropology of Protestantism in Mexico and Central America*, edited by James W. Dow and Alan R. Sandstrom, 205–34. Westport, CT: Praeger Publishers.

Ad Gentes
1966 [1965]   Decree on the Church's Missionary Activity. In The Documents of Vatican II. Abbott, Walter M., ed. Translated by Joseph Gallagher. New York: The America Press

Alejos Garcia, José
2003   The Ch'ols Reclaim Palenque, or the War of Eternal Return. In Rus, Hernández Castillo, and Mattiace 2003a, 85–102.

Angrosino, Michael
1994   The Cultural Concept and the Mission of the Catholic Church. *American Anthropologist* 96 (4): 824–32.

Annis, Sheldon
1987   *God and Production in a Guatemalan Town.* Austin: University of Texas Press.

Anonymous
1993   Experiencia teologíca de Chiapas, Mexico. In *Teología India Mayense* 1993, 231–64.

Arias, Jacinto
1991   *El mundo numinoso de los mayas: Estructura y cambios contemporáneos.* Tuxtla Guitérrez, Chiapas: Gobierno del Estado de Chiapas.

Asociación la Voz de Atitlán
1998   *La verdad está en la historia: 32 años de servicio.* Santiago Atitlán: Asociacíon la Voz de Atitlán.

Asociación para el Avance de las Ciencias Sociales en Guatemala (AVANCSO)
2002   *Se cambio el tiempo: Conflicto y poder en territorio K'iche' 1880–1996.* Serie Cuadernos de Investigación No. 17. Guatemala City: Asociación para el Avance de las Ciencias Sociales en Guatemala.

Barmeyer, Niels
2009   *Developing Zapatista Autonomy: Conflict and NGO Involvement in Rebel Chiapas.* Albuquerque: University of New Mexico Press.

Baron, Akeska
2006   "Women Don't Talk": Gender and Codemixing in an Evangelical Tzotzil Community. PhD diss., University of Washington.

Bassie-Sweet, Karen
1991   *From the Mouth of the Dark Cave: Commemorative Sculpture of the Late Classic Maya*. Norman: University of Oklahoma Press.
1996   *At the Edge of the World: Caves and the Late Classic Maya World View*. Norman: University of Oklahoma Press.

Behrens, Susan Fitzpatrick
2004   From Symbols of the Sacred to Symbols of Subversion to Simply Obscure: Maryknoll Religious Women in Guatemala, 1953–1967. *The Americas* 61 (2): 189–216.

Benedict XVI, Pope
2008   Address of His Holiness Benedict XVI to the Bishops of Guatemala on the "Ad Limina" Visit, March 6. Vatican City: Libreria Editrice Vaticana.

Benjamin, Thomas
1996   *A Rich Land, a Poor People: Politics and Society in Modern Chiapas*. Rev. ed. Albuquerque: University of New Mexico Press.

Berryman, Phillip
1984   *The Religious Roots of Rebellion*. Maryknoll, NY: Orbis Books.
1987   *Liberation Theology: Essential Facts about the Revolutionary Movement in Latin America and Beyond*. Philadelphia: Temple University Press.

Blaffer, Sarah C.
1972   *The Black-man of Zinacantán: A Central American Legend*. Austin: University of Texas Press.

Bobrow-Strain, Aaron
2001   Between a Rock and a Hard Place. In *Violent Environments*, edited by Nancy Lee Peluso and Michael Watts, 156–88. Ithaca, NY: Cornell University Press.
2007   *Intimate Enemies: Landowners, Power, and Violence in Chiapas*. Durham, NC: Duke University Press.

Boff, Leonardo, and Clovis Boff
1987   *Introducing Liberation Theology*. Translated by Paul Burns. Maryknoll, NY: Orbis Books.

Bogenschild, Thomas
1992   *Roots of Fundamentalism in Liberal Guatemala: Missionary Ideologies and Local Responses*. PhD diss., University of California at Berkeley.

Bonner, Arthur
1999   *We Will Not Be Stopped*. UPUBLISH.COM: Universal Publishers.

Bonpane, Blase
1985   *Guerrillas of Peace: Liberation Theology and the Central American Revolution*. Boston: South End Press.

Brintnall, Douglas
1979   *Revolt against the Dead*. New York: Gordon and Breach.

Brown, Raymond F., Joseph A. Fitzmyer, and Roland E. Murphy, eds.
1990   *The New Jerome Biblical Commentary*. Upper Saddle River, NJ: Prentice Hall.

Brown, Robert McAfee
1979   The Significance of Puebla for the Protestant Churches in North America. In

Puebla and Beyond: English Translation of the Official Conclusions of the Third General Council of Latin American Bishops, edited by John Eagleson and Philip Scharper, 330–46. Washington, DC: National Conference of Catholic Bishops.

Bunzel, Ruth
1959 *Chichicastenango.* Seattle: University of Washington Press.

Burguete Cal y Mayor, Ataceli
2003 The De Facto Autonomous Process: New Jurisdictions and Parallel Governments in Rebellion. In Rus, Hernández Castillo, and Mattiace 2003a, 191–218.

Calder, Bruce J.
1970 *Crecimiento y cambio de la Iglesia Católica guatemalteca, 1944–66.* Guatemala City: Seminario de Integracion Guatemalteca.
2004 Interwoven Histories: The Catholic Church and the Maya, 1940 to the Present. In *Resurgent Voices in Latin America: Indigenous People, Political Mobilization and Religious Change,* edited by Edmund L. Cleary and Timothy J. Steigenga, 93–124. New Brunswick, NJ: Rutgers University Press.

Cancian, Frank
1965 *Economics and Prestige in a Maya Community.* Stanford, CA: Stanford University Press.
1967 Political and Religious Organizations. In *Social Anthropology,* ed. Manning Nash, 283–98. Vol. 6 of the *Handbook of Middle American Indians.* Austin: University of Texas Press.
1992 *The Decline of Community in Zinacantán: Economy, Public Life, and Stratification, 1960–87.* Stanford, CA: Stanford University Press.

Carey, David
2001 *Our Elders Teach Us: Maya-Kaqchikel Historical Perspectives: xkib'ijkan qate' qatata'.* Tuscaloosa: University of Alabama Press.

Carlsen, Robert S.
1997 *The War for the Heart and Soul of a Highland Maya Town.* Austin: University of Texas Press.

Carmack, Robert M.
1988 The Story of Santa Cruz Quiché. In *The Harvest of Violence,* edited by Robert Carmack, 39–69. Norman: University of Oklahoma Press.
1995 *Rebels of Highland Guatemala: The Quiché-Mayas of Momostenango.* Norman: University of Oklahoma Press.

Cartledge, Mark
2007 *Encountering the Spirit: The Charismatic Tradition.* Maryknoll, NY: Orbis Books.

Casas, Bartolomé de Las
1958 *Apologetica Historia.* In *Obras escogidas de Fray Bartolomé de Las Casas, IV.* Biblioteca de Autores Espanoles, tomo 106. Madrid: Sucs. J. Sanchez de Ocaña y Cia.

Catechism of the Catholic Church
1994 *Catechism of the Catholic Church.* New Hope, KY: URBI et ORBI Communications.

CEH. *See* Comisión para el Esclaracimiento Histórico

Centro de Investigación y Documentación Centroamericana
1980   *Desarrollo historico de la violencia institucional en Guatemala.* Guatemala City: Editorial Universitaria de Guatemala.
Centro Nacional de Pastoral Indigena (CENAPI)
1970   *Xicotepec, indigenas en polemica sobre la iglesia.* Mexico City: Centro Nacional de Pastoral Indigena.
Chanteau, Miguel
1999   *Las andanzas de Miguel: La autobiografía del Padre expulsado de Chenaló.* San Cristóbal de Las Casas: Editorial Fray Bartolomé de Las Casas.
Chestnut, R. Andrew
2003   *Competitive Spirits: Latin America's New Religious Economy.* New York: Oxford University Press.
Chojnacki, Ruth Judith
2004   *Indigenous Apostles: Maya Catholic Catechists Working the Word in Highland Chiapas.* PhD diss., University of Chicago.
2010   *Indigenous Apostles: Maya Catholic Catechists Working the Word in Highland Chiapas.* Amsterdam: Editions Rodopi.
Christenson, Allen
2001   *Art and Society in a Highland Maya Community.* Austin: University of Texas Press.
2003   *Popol Vuh: The Sacred Book of the Maya.* Winchester, UK: O Books.
Cleary, Edward L.
1993   The Journey to Santo Domingo. In Hennelly 1993a, 3–23.
2007   The Catholic Charismatic Renewal. In *Conversion of a Continent*, edited by Timothy Steigenga and Edward Cleary, 153–71. New Brunswick, NJ: Rutgers University Press.
Cleary, Edward L., and Philip Berryman, eds. and trans.
1989   *Path from Puebla: Significant Documents from the Latin American Bishops since 1979.* Washington, DC: United States Catholic Conference.
Clinton, William J.
    1999 *Washington Post*, March 11, 1999, A1
Collier, George Allen, and Elizabeth Lowery Quaratiello
1999   *Basta!: Land the Zapatista Rebellion in Chiapas.* Rev. ed. Oakland: Food First Book, The Institute for Food and Development Policy.
Collier, Jane
1973   *Law and Social Change in Zinacantán.* Stanford, CA: Stanford University Press.
Colom, Yolanda
1998   *Mujeres en la alborada: Guerrilla y participación femenina en Guatemala, 1973–1978.* Guatemala City: Artemis and Edinter.
Colonnese, Louis M., ed.
1970   *The Church in the Present Day Transformation of Latin America in the Light of the Council: Second General Conference of Latin American Bishops.* Vol. 1, *Position Papers*; vol. 2, *Conclusions*. Washington, DC: United States Catholic Conference and the General Secretariat of CELAM.

Comaroff, Jean, and John Comaroff
1991   *Of Revelation and Revolution: Christianity, Colonialism and Consciousness in South Africa*. Vol. 1. Chicago: University of Chicago Press.
1997   *Of Revelation and Revolution: The Dialectics of Modernity on a South African Frontier*. Vol. 2. Chicago: University of Chicago Press.
Comisión Episcopal Para Indigenas
1988   *Fundamentos teológicos de la pastoral indigena en México*. Mexico City: Conferencia del Episcopado Mexicano.
Comisíon para el Esclaracimiento Historíco (CEH)
1999   *Guatemala: Memoria de silencio*. 12 vols. Guatemala City: CEH. Most of the Spanish version is available online at http://shr.aaas.org/guatemala/ceh/mds/spanish/toc; the English version, with only the Conclusion, Recommendations, and Appendices is available at http://shr.aaas.org/guatemala/ceh/report/english.
Conferencia del Episcopado Guatemalteco (CEG)
1993   500 años sembrando el evangelio. In *Teología India Mayense* 1993, 265–312.
Conferencia Episcopal de Guatemala (CEG)
2007   *Testigos fieles del evangelio*. 2nd ed. Guatemala City: Ediciones San Pablo.
Cook, Garrett
2000   *Renewing the Mayan World: Expressive Culture in a Highland Town*. Austin: University of Texas Press.
2001   The Maya Pentecost. In *Holy Saints and Fiery Preachers: The Anthropology of Protestantism in Mexico and Central America*, edited by James W. Dow and Alan R. Sandstrom, 147–68. Westport, CT: Praeger Publishers.
Corral, Enrique
1997   Introducción. In Hoyos de Asig 1997, 5–9.
Cross, Frank Moore
1973   *Canaanite Myth and Hebrew Epic: Essays in the History of Israel*. Cambridge, MA: Harvard University Press.
Davis, Shelton H.
1997   *La tierra de nuestros antepasados: Estudio de la herencia y tenencia de la tierra en el altiplano de Guatemala*. La Antigua, Guatemala: Centro de Investigaciones Regionales de Mesoamérica; South Woodstock, VT: Plumsock Mesoamerican Studies.
De Leon Ceto, Miguel
2006   *Fuentes de poder del movimiento evangélico en Nebaj, el Quiché*. Guatemala City: Universidad Rafael Ladívar, Facultad de Ciencias Políticas y Sociales.
Delgado, Arnulfo
1998   *Entre dos fuerzas*. Lima: Grafigram.
Delhaye, P.
2003   Clerical History of Celibacy. In *New Catholic Encyclopedia*, 2nd ed., 3:322–28. Detroit, MI: Gale Group.
Denzinger, Henry
1955   *The Sources of Catholic Dogma*. Translated by Roy J. Defarrari. Fitzwilliam, NH: Loreto Publications.

Diáz del Castillo, Bernal
1938   The Discovery and Conquest of Mexico, 1517–1521. Translated by A. P. Maudsley. New York: Harper and Brothers.

Diócesis del Quiché
1994   El Quiché: El pueblo y su iglesia, 1860–1980. Santa Cruz del Quiché: Diócesis del Quiché.

Douglas, Mary
2007   Thinking in Circles: An Essay on Ring Composition. New Haven, CT: Yale University Press.

Douglas, William
1968   Santiago Atitlán. In Seminario de Integración Social Guatemalteca 1968, 229–76.

Eagleson, John, and Philip Scharper
1979   Puebla and Beyond: Documentation and Commentary. Maryknoll, NY: Orbis Books.

Earle, Duncan, and Jean Simonelli
2005   Uprising of Hope: Sharing the Zapatista Journey to Alternative Development. Walnut Creek, CA: AltaMira Press.

Early, John D.
1965   The Sons of San Lorenzo in Zinacantán. PhD diss., Harvard University.
1969   Unpublished Notes.
1970a A Demographic Profile of a Maya Community—The Atitecos of Santiago Atitlan. Milbank Quarterly 48 (3): 167–78.
1970b The Structure and Change of Mortality in a Maya Community. Milbank Quarterly 48 (3): 179–201.
1973   Education via Radio among Guatemalan Highland Maya. Human Organization 32 (4): 221–29.
1982   The Demographic Structure and Evolution of a Peasant System: The Guatemalan Population. Boca Raton: University Presses of Florida.
1983   Some Ethnographic Implications of an Ethnohistorical Perspective of the Civil-Religious Hierarchy among the Highland Maya. Ethnohistory 30 (4): 185–202.
2000   La estructura y evolución demográfica de un sistema campesino: La población de Guatemala. Translated by Anne M. Luna and Eddy H. Gaytán; with a Preface by Ricardo Falla. South Woodstock, VT: Plumsock Mesoamerican Studies/CIRMA. [1982 Spanish translation]
2006   The Maya and Catholicism: An Encounter of Worldviews. Gainesville: University Press of Florida.
2011 [2006] Los mayas y el catolicismo. Serie Autores Invitados No. 21. Guatemala: Asociación para el Avance de las Ciencias Sociales en Guatemala (AVANCSO).
2017 [2012] Las cultures maya y catholica en crisis. Serie Autores Invitados. Guatemala: Asociación para el Avance de las Ciencias Sociales en Guatemala (AVANCSO).

Early, John D., and Thomas Headland
1998   Population Dynamics of a Philippine Rain Forest People: The San Ildefonso Agta. Gainesville: University Press of Florida.

Early, John D., and John F. Peters
2000   *The Xilixana Yanomami of the Amazon: History, Structure, and Population Dynamics*. Gainesville: University Press of Florida.

Eaton, Joseph W., and Albert J. Mayer
1954   *Man's Capacity to Reproduce: The Demography of a Unique Population*. Glencoe, IL: The Free Press.

Eber, Christine
2000   *Women and Alcohol in a Highland Maya Town: Water of Hope, Water of Sorrow*. 2nd ed. Austin: University of Texas Press.
2003   Living Their Faith in Troubled Times. In *Women of Chiapas: Making History in Times of Struggle and Hope*, 113–29. New York: Routledge.
2004   Buscando Una Nueva Vida: Liberation through Autonomy in San Pedro Chenalhó, 1970–1998. In Rus, Hernández Castillo, and Mattiace 2003a, 135–60.
2011   *The Journey of a Tzotzil-Maya Woman of Chiapas, Mexico: Pass Well over the Earth*. Austin: University of Texas Press.

Eber, Christine, Megan Snedden, and Megham Dallin
2006   Are We Standing on Rock or Sand. *Practicing Anthropology* 28 (3): 34–38.

Edmunson, Monro S., ed. and trans.
1982   *The Ancient Future of the Itza: The Book of Chilam Balam of Tizimin*. Austin: University of Texas Press.
1986   *Heaven Born and Its Destiny: The Book of Chilam Balam of Chumayel*. Austin: University of Texas Press.

Ehlers, Tracy Bachrach
2000   *Silent Looms: Women and Production in a Guatemalan Town*. Rev. ed. Austin: University of Texas Press.

Elbow, Gary Stewart
1972   *Cultural Factors in the Spatial Organization of Three Highland Guatemalan Towns*. Ph.D. diss., University of Pittsburgh.

Eliade, Mircea
1958   *Patterns in Comparative Religion*. New York: Sheed and Ward.
1959   *Cosmos and History: The Myth of the Eternal Return*. New York: Harper and Brothers.

Evans, Timothy Edward
1990   *Religious Conversion in Quzaltenango, Guatemala*. PhD diss., University of Pittsburgh.

Falla, Ricardo
1983   *Voices of the Survivors: The Massacre at Finca San Francisco, Guatemala*. Boston: Cultural Survivor and Anthropology Resource Center.
1985   Dinamicas religiosas. In *Revolución en la selva del Ixcán*. Unpublished manuscript.
1988   Struggle for Survival in the Mountains: Hunger and Other Privations Inflicted on Internal Refugees from the Central Highlands. In *Harvest of Violence: The Maya Indians and the Guatemalan Crisis*, edited by Robert Carmack, 235–55. Norman: University of Oklahoma Press.
1994   *Massacres in the Jungle: Ixcán, Guatemala, 1975–1982*. Boulder, CO: Westview Press.

1998   *The Story of a Great Love: Life with the "Communities of Population in Resistance": A Spiritual Journey*. Washington, DC: Ecumenical Program on Central America and the Caribbean.

2001 [1978]   *Quiché Rebelde: Religious Conversion, Politics, and Ethnic Identity in Guatemala*. Translated by Phillip Berryman. Austin: University of Texas Press.

2008   Fernando Hoyos, amor, revolución y eso que dicen dios. In Hoyos de Asig, Blanco Carballo, and Corral Alonso 2008, 223–30.

Farriss, Nancy M.
1984   *Maya Society under Colonial Rule*. Princeton, NJ: Princeton University Press.

Fazio, Carlos
1994   *Samuel Ruiz, el caminante*. Mexico City: Espasa Calpe Mexicana.

Fernández Fernández, José Manuel
1988   *El comité de unidad campesina: Origen y desarrollo*. Guatemala City: Centro de Estudios Rurales Centroamericanos.

Fischer, Edward F.
2001   *Cultural Logics and Global Economics: Maya Identity in Thought and Practice*. Austin: University of Texas Press.

Fischer, Edward F., and R. McKenna Brown, eds.
1996   *Maya Cultural Activism in Guatemala*. Austin: University Of Texas Press.

Flores Reyes, Gerardo
1996   Algunos aspectos importantes de la teología india. In *Teología India Mayense* 1996a, 253–64.

Floyd, J. Charlene
1997   The Government Shall Be upon Their Shoulders: The Catholic Church and Democratization in Chiapas, Mexico. PhD diss., The City University of New York.

Foster, George M.
1965   Peasant Society and the Image of the Limited Good. *American Anthropologist* 67 (2): 293—315.

Freidl, David, Linda Schele, and Joy Parker
1993   *Three Thousand Years on the Shaman's Path*. New York: William Morrow.

Freire, Pablo
2001 [1970]   *Pedagogy of the Oppressed*. New York: Continuum.

Freyermuth Enciso, Graciela
2003   Juana's Story. In *Women of Chiapas*, edited by Christine Eber and Christine Kovic, 31–36. New York: Routledge.

Frison, Bruno R.
1992   *Los Maya-Quiches: Expresiones de su religiosidad*. Quito: Ediciones Abya-Yala.

García, Tomás
1996   Reflexión. In *Teología India Mayense* 1996a, 90. Guatemala City: n.p.

Garrard-Burnett, Virginia
1998   *Protestantism in Guatemala: Living in the New Jerusalem*. Austin: University of Texas Press.

Gaudium et Spes
1966 [1965] Pastoral Constitution on the Church in the Modern World. In The Docu-

ments of Vatican II. Abbott, Walter M., ed. Translated by Joseph Gallagher. New York: The America Press.

Geaney, D. J.
2003   Catholic Action. In *New Catholic Encyclopedia*, 2nd ed., 3:275–78. Detroit, MI: Gale Group.

Gil Tebar, Pilar
2003   Irene, A Catholic Woman in Oxchuc. In *Women of Chiapas*, edited by Christine Eber and Christine Kovic, 150. New York: Routledge.

Goldin, Lilian, and Brent Metz
1991   An Expression of Cultural Change: Invisible Converts to Protestantism. *Ethnology* 30 (4): 325–337.

González, Matilde
2002   *Se cambió el tiempo: Conflicto y poder en el territorio K'iche', 1880–1996*. Serie de Cuadernos de Investigación, no. 17. Guatemala City: Asociación para el Avance de Ciencias Sociales en Guatemala (AVANCSO). [The author's name is not stated on the cover, title page, or publication page of the book, but mentioned in the Presentación. Most library catalogs and bibliographic databases list it with no author. It can only be found under the title of the book.]

Gossen, Gary H.
1986   Mesoamerican Ideas as Foundation for Regional Synthesis. In *Symbol and Meaning beyond the Closed Community: Essays in Mesoamerican Ideas*, edited by Gary H. Gossen, 1–8. Albany: Institute for Mesoamerican Studies, State University of New York.
2002   *Four Creations: An Epic Story of the Chiapas Mayas*. Norman: University of Oklahoma Press.

Green, Linda
1999   *Fear as a Way of Life: Mayan Widows in Rural Guatemala*. New York: Columbia University Press.

Guiteras-Holmes, Calixta
1961   *Perils of the Soul: The World View of a Tzotzil Indian*. New York: Free Press of Glencoe.

Gutierrez, Gustavo
1968   Toward a Theology of Liberation. In Hennelly 1990, 62–76.
1973   *A Theology of Liberation: History, Politics, Salvation*. Maryknoll, NY: Orbis Books.

Harvey, Neil
1994   *Rebellion in Chiapas: Rural Reforms, Campesino Radicalism, and the Limits to Salinismo*. University of California, San Diego: Center for U.S-Mexican Studies.
1998   *The Chiapas Rebellion: The Struggle for Land and Democracy*. Durham, NC: Duke University Press.

Henck, Nick
2007   *Subcommander Marcos: The Man and the Mask*. Durham, NC: Duke University Press.

Hennelly, Alfred T., ed.
1990   *Liberation Theology: A Documentary History*. Maryknoll, NY: Orbis Books.

1993a   *Santo Domingo and Beyond.* Maryknoll, NY: Orbis Books.
1993b   A Report from the Conference. In Hennelly 1993a, 24–36.
1993c   Message of the Fourth General Council to the Peoples of Latin America and the Caribbean, 63–155. In Hennelly 1993a, 63–155.

Hernández Ixcoy, Domingo
1984   Las raices de una realidad practica. Entrevista realizada por François Lartigue. *Civilización* 2 (2): 291–345.

Hernández Pico, Juan
2008   Desde dónde y como vi a Fernando Hoyos. In Hoyos de Asig, Blanco Carballo, and Corral Alonso 2008, 177–84.

Hill, Robert M., II
1992   *Colonial Cakchiquels: Highland Maya Adaptations to Spanish Rule, 1600–1700.* Fort Worth, TX: Harcourt, Brace, Jovanovich.

Hill, Robert M., II, and John Monaghan
1987   *Continuities in Maya Social Organization: Ethnohistory in Sacapulas, Guatemala.* Philadelphia: University of Pennsylvania Press.

Hinshaw, Robert E.
1975   *Panajachel: A Guatemalan Town in Thirty-Year Perspective.* Pittsburgh: University of Pittsburgh Press.

Holland, Clifton L.
2008   *Expanded Status of Christianity, Country Profile: Guatemala, 1980, revised 2008.* San Pedro, Costa Rica: Latin America Socio-Religious Studies Program. Available at the Religion-In-The Americas (RITA) Database Web site, www.prolades.com.

Horst, Oscar
1956   *An Analysis of Land Use in the Rio Samala Region of Guatemala.* Ph. diss., Ohio State University.

Hoyos, Fernando
1973   *Violence in Guatemala: The Position of the Church Concerning the Use of Violence by Oppressors and Its Use by Those Seeking Liberation From It.* Unpublished manuscript.
n.d.   *Utopía Tomás Moro.* Unpublished manuscript.

Hoyos de Asig, María Pilar
1997   *¿Donde Estás? Fernando Hoyos.* Guatemala City: Fondo de Cultural Editorial.
2008   En el corazon del pueblo. In Hoyos de Asig, Blanco Carballo, and Corral Alonso 2008, 79–142.

Hoyos de Asig, María Pilar, Antonio Blanco Carballo, and Enrique Corral Alonso, eds.
2008   *En la memoria del pueblo: Homenaje a Fernando Hoyos.* Santiago de Compostela: Fundación 10 de Marzo.

Hunter, Neale, and Deirde Hunter
1966   Christian China. In *Soundings: Occasional Notes on Christianity and International Service.* Washington, DC: Council of International Lay Associations.

Hurtado Paz y Paz, Margarita
2009   *Organización y lucha rural, campesina y indigena: Huehuetenango, Guatemala,*

1981. Rio de Janeiro: Congreso 2009 de la Asociación de Estudios Latinoamericanos.

Iglesia Guatemalteca en el Exilio
1992   *Nosotros concemos nuestra historia: 500 años de resistencia indigena, negra y popular*. 2nd ed. Mexico City: Iglesia Gutemaltenca en en Exilio.

Ignatius, of Loyola, Saint
1990   *The Spiritual Exercises of St. Ignatius of Loyola*. Translated by Elder Mullan. Boulder, CO: Net Library.

Iribarren Pascal, Pablo
1985   *Proceso de la diócesis de San Cristobal de Las Casas, Chiapas, Mexico, 1960–85*. Ocosingo, Chiapas: n.p.

Irwin, K. W.
2003   Sacramental Theology. In *New Catholic Encyclopedia*, 2nd ed., 12:465–79. Detroit, MI: Gale Group.

Jedin, H.
2003   Council of Trent. In *New Catholic Encyclopedia*, 2nd ed., 14:168–76. Detroit, MI: Gale Group.

Jegen, M. E.
2003   Catechesis, III (Reformation). In *New Catholic Encyclopedia*, 2nd ed., 3:232–36. Detroit, MI: Gale Group.

John Paul II, Pope
1990   *Redemptoris Missio*. Available online at The Vatican Web site, http://www.vatican.va/holy_father/john_paul_ii/encyclicals/documents/hf_jp-ii_enc_07121990_redemptoris-missio_sp.html.

Judd, Stephen P.
1993   From Lamentation to Project: The Emergence of an Indigenous Theological Movement in Latin America. In Hennelly 1993a, 226–35.

Kelly, Dave C.
1970   *Maryknoll History, Part I, 1943–1969, Guatemala El Salvador Region*. Guatemala City: Regional Council Secretariat, unpublished notes.
1979   *Maryknoll History, Part II, 1969–1978, Guatemala El Salvador Region*. Guatemala City: Regional Council Secretariat, unpublished notes.

Kilmartin, E. J.
2003   Ex Opere Operato. In *New Catholic Encyclopedia*, 2nd ed., 5:501–2. Detroit, MI: Gale Group.

Kobrak, Paul
2003   *Huehuetenango: Historia de una guerra*. Huehuetenango: Centro de Estudios y Documentación de la Frontera Occidental de Guatemala.

Koenigsberger, H. G., George L. Mossee, and G. Q. Bowler
1989   *Europe in the Sixteenth Century*. 2nd ed. London: Longman.

Koizumi, Junji
1981   *Symbol and Context: A Study of Self and Action in a Guatemalan Culture*. PhD diss., Stanford University.

Konefal, Betsy
2010   For Every Indio Who Falls: A History of Maya Activism in Guatemala, 1960–1990. Albuquerque: University of New Mexico Press.
Kovic, Christine
2003   Demanding Their Dignity as Daughters of God. In *Women of Chiapas*, edited by Christine Eber and Christine Kovic, 131–46. New York: Routledge.
2004   Mayan Catholics in Chiapas Mexico: Practicing Faith on Their Own Terms. In *Resurgent Voices in Latin America: Indigenous Peoples, Political Mobilization, and Religious Change*, edited by Edward L. Cleary and Timothy J. Steigenga, 187–209. New Brunswick, NJ: Rutgers University Press.
2005   *Maya Voices for Human Rights: Displaced Catholics in Highland Chiapas*. Austin, TX: University of Texas Press.
Kovic, Christine, and Francisco Arguelles Paz y Puente
2003   Today, the Women. In Kovic 2005, 147.
Kugel, James L.
1997   *The Bible As It Was*. Cambridge, MA: The Belknap Press of Harvard University Press.
2007   *How to Read the Bible: A Guide to Scripture, Then and Now*. New York: Free Press.
Küng, Hans
1995   *Christianity: Essence, History, and Future*. Translated by John Bowden. New York: Continuum Publishing.
2001   *The Catholic Church: A Short History*. Translated by John Bowden. New York: Modern Library.
LaFarge, Oliver
1947   *Santa Eulalia*. Chicago: University of Chicago Press.
Landa, Diego de
1941   *Landa's relación de las cosas de Yucatan: A Translation*. Translated and edited by Alfred M. Tozzer. Cambridge, MA: Peabody Museum of American Archaeology and Ethnology, Harvard University.
Laughlin, Robert M.
1977   *Of Cabbages and Kings*. Washington, DC: Smithsonian Institution Press.
Le Bot, Yvon
1995   *La guerra en tierras Mayas: Comunidad, violencia, modernidad en Guatemala (1970–1992)*. Mexico City: Fondo de Cultura Economica.
1997   *El sueño zapatista: Entrevistas con el subcomandante Marcos, el mayor Moisés y el comandante Tacho, del Ejército Zapatista de Liberación Nacional*. Mexico City: Plaza y Janés.
Legorreta Díaz, María del Carmen
1998   *Religión, política y guerrilla en las Cañadas de la Selva Lacandona*. Mexico City: Cal y Arena.
León-Portilla, Miguel
1973   *Time and Reality in the Thought of the Maya*. Boston: Beacon Press.
Leyva Solano, Xóchitl
1998   Catequistas, misioneros, y tradiciones en Las Cañadas. In *Chiapas: Los rumbos de otra historia*, edited by Juan Pedro Viquerira and Mario Humberto Ruz,

375–406. Mexico City: Centro de Estudios Mayas del Instituto de Investigaciones Filológias y Coordinación de Humanidades (UNAM).
2003   Regional, Communal, and Organizational Transformation in Las Cañadas. In In Rus, Hernández Castillo, and Mattiace 2003, 161–83.

Leyva Solano, Xóchitl, and Gabriel Ascencio Franco
1996   *Lacandonia al filo del agua*. Mexico City: Fondo de Cultura Economica.

López Hernández, Eleazar
1993   Teología de hoy. In *Teología India Mayense* 1993, 137–42.
1996   La biblia y teología india. In *Teología India Mayense* 1996a, 111–23. Guatemala City: n.p.
2000   *Teología india: Antología*. Buenos Aires, Argentina: Editorial Guadalupe.
2006   "Datos sobre el caso del P. Eleazar López Hernandez," June 7. Conselho Indigenista Missionário Web site, http://www.cimi.org.br/site/pt-br/index.php?system=news&action=read&id=1691.

Los Tzeltales
    1972   Estamos Buscando La Libertad: Los Tzeltales de La Selva Anuncian La Buena Nueva. San Cristobal de Las Casas, Chiapas: Diocesis de San Cristobal.

Lovell, W. George
1982   Collapse and Recovery: A Demographic Profile of the Cuchumatan Highlands of Guatemala (1520–1821). In *The Historical Demography of Highland Guatemala*, edited by Robert Carmack, John Early, and Christopher Lutz, 103–20. Albany: Institute for Mesoamerican Studies, State University of New York at Albany.
2000   *A Beauty that Hurts*. 2nd ed. Austin: University of Texas Press.
2005   *Conquest and Survival in Guatemala*. 3rd ed. Montreal: McGill-Queens University Press.

Lovell, W. George, and Christopher H. Lutz
1994   Conquest and Population. *Latin American Research Review* 29 (2): 133–40.

Lumen Gentium
1966 [1965]   Dogmatic Constitution on the Church. In The Documents of Vatican II. Abbott, Walter M., ed. Translated by Joseph Gallagher. New York: The America Press.

MacLeod, Murdo
1973   *Spanish Central America: A Socioeconomic History, 1520–1720*. Berkeley, CA: University of California Press.

Manz, Beatriz
1988   *Refugees of a Hidden War: The Aftermath of Counterinsurgency in Guatemala*. Albany: State University of New York Press.
2004   *Paradise in Ashes: A Guatemalan Journey of Courage, Terror, and Hope*. Berkeley, CA: University of California Press.

Marcos, Subcomandante, and Yvon Le Bot
1997   *El sueño zapatista*. Barcelona: Plaza y Janés.

Mattiace, Shannan L.
2003   Regional Renegotiations of Space: Tojolabal Ethnic Identity in Las Margaritas, Chiapas. In Rus, Hernández Castillo, and Mattiace 2003a, 109–34.

Maurer, Eugenio
1984   *Los Tseltals ¿Paganos o cristianos? Su religion ¿Sincretismo o sintesis?* Mexico City: Centro de Estudios Educativos.
1996   Tseltal Christianity. In *The Indian Face of God in Latin America*, by Manuel M. Marzal et al.; translated by Penelope R. Hall, 23–66. Maryknoll, NY: Orbis Books.

McBrien, Richard P.
2008   *The Church: The Evolution of Catholicism*. New York: Harper One of Harper Collins.

McBryde, F. Webster
1933   *Sololá: A Guatemala Town and Cakchiquel Market-Center*. New Orleans: Tulane University.

McCormick, R. A., and D. Christenson
2003   Morality of War. In *New Catholic Encyclopedia*, 2nd ed., 14:635-44. Detroit, MI: Gale Group.

McCreery, David
1994   *Rural Guatemala, 1760–1940*. Stanford, CA: Stanford University Press.

McGovern, Arthur F.
1989   *Liberation Theology and Its Critics: Toward an Assessment*. Maryknoll, NY: Orbis Books.

Melotto, Angelico
1984   Homily at Memorial Mass for Father Stanley Rother. In Rother 1984, 101–4.

Melville, Thomas
2005   *Through a Glass Darkly: The U.S. Holocaust in Central America*. www.Xlibris.com: Xlibris Corporation.

Melville, Thomas, and Marjorie Melville
1971   *Whose Heaven, Whose Earth?* New York: Alfred A. Knopf.

Menchú, Rigoberta
1984   *I, Rigoberta Menchú: An Indian Woman in Guatemala*. Edited and translated by Elizabeth Burgos-Debray. Translated by Ann Wright. New York: Verso.

Mendelson, E. Michael (Nathaniel Tarn)
1965   *Los escandolos de Maximon*. Seminario de Integración Social, no. 19. Guatemala City: Tipografía Nacional.

Metz, Brent E.
2006   *Ch'orti-Maya Survival in Eastern Guatemala: Indigeneity in Transition*. Albuquerque: University of New Mexico Press.

Meyer, Jean
2000   *Samuel Ruiz en San Cristóbal*. 2nd rev. ed. Mexico City: Tusquets Editores Mexico.

Miller, Donald
2009   *Global Pentecostalism: The New Face of Christian Social Engagement*. Berkeley, CA: University of California Press.

Moksnes, Heidi
1995–96   Unpublished Field Notes.
2003   *Mayan Suffering, Mayan Rights: Faith and Citizenship among Catholic Tzotziles in Highland Chiapas, Mexico*. Uppsala, Sweden: Göteborg University.
Molina, Alonso de
1941   Doctrina Cristiana. In *Codice Franciscano, Informe de la Provincia del Santo Evangelio al visitador Lic. Juan de Ovando*, 30–54. Mexico City: Editorial Salvador Chavez Hayhoe.
Monahan, David
1984   Foreword. In Rother 1984, 1–10.
Mondloch, James L.
1980   K'e?s: Quiché Naming. *Journal of Mayan Linguistics* 1 (2): 9–25.
Montagu, Roberta
1970   Autoridad, control y sanción en la zona cental de Chiapas. In *Ensayos de antropología en la zona central de Chiapas*, edited by Norman McQuown y Julian Pitt-Rivers, 345–69. Mexico City: Instituto Nacional Indigenista.
Montejo, Victor
1987   *Testimony: Death of a Guatemalan Village*. Translated by Victor Perera. Willimantic, CT: Curbstone Press.
1999   *Voices from Exile: Violence and Survival in Modern Maya History*. Norman: University of Oklahoma Press.
2001   *El Q'anil: Man of Lightning*. Translated by Wallace Kaufman and Susan G. Rascón. Tuscon: University of Arizona Press.
2002   The Multiplicity of Mayan Voices: Mayan Leadership and the Politics of Self Representation. In Warren and Jackson 2002, 123–39.
2005   *Maya Intellectual Renaissance: Identity, Representation, and Leadership*. Austin, TX: University of Texas Press.
Montejo, Victor, and Q'anil Akab'
1992   *Brevissima relación testimonial de la continua destrucción del Mayab'*. Providence, RI: Maya Scholars Network.
Morales, Mardonio
1994   Relato del jesuita Mardonio Morales, fragmentos de la entrevista. In *¿Porque Chiapas?*, by Luis Panzos, 138–45. Mexico City: Editorial Diana.
Morales Bermudez, Juan
1995   El congreso indigena de Chiapas: Un testimonio. *America Indigena* 55 (1/2): 305–40.
Morales Jiménez, Carlos
1985   Testimonio de una comunidad de la selva lacandon. *Textual: Analisis del Medio Rural* 19 (March): 179–85.
Morgan, Jesse
2005   Standing at the Crossroads: Cultural Change in Nahualá as Seen through the Eyes of Javier, a Maya Elder. In *Roads to Change in Maya Guatemala: A Field School Approach to Understanding the K'iche'*, edited by John P. Hawkins and Walter Randolph Adams, 61–98. Norman: University of Oklahoma Press.

Morrissey, James Arthur
1978   *A Missionary Directed Resettlement Project among the Highland Maya of Western Guatemala*. PhD diss., Stanford University.
Nash, June
1970   *In the Eyes of the Ancestors: Belief and Behavior in a Mayan Community*. Prospect Heights, IL: Waveland Press.
Neal, Sister Marie Augusta
1965   *Values and Interest in Social Change*. Englewood Cliffs, NJ: Prentice-Hall.
Nick, Elizabeth
1975   *Some Factors Related to and Knowledge about Fertility Control in a Traditional Mayan Village in Guatemala*. Master's thesis, Florida Atlantic University.
n.d.   Unpublished Diary.
Norget, Kristin
2004   "Knowing Where We Enter": Indigenous Theology and the Popular Church in Oaxaca, Mexico. In *Resurgent Voices in Latin America*, edited by Edward L. Leary and Timothy J. Steigenga, 154–86. New Brunswick, NJ: Rutgers University Press.
Noth, Martin
1981 [1967]   *The Deuteronomistic History*. Sheffield: Journal for the Study of the Old Testament, Supplement Series 15.
Nouwen, Henri J. M.
1985   *Love in Fearful Land: A Guatemalan Story*. Notre Dame, IN: Ave Maria Press.
Oakes, Maude
1951   *The Two Crosses of Todos Santos*. Bollingen Series, 27. New York: Pantheon Books.
O'Malley, John W.
2008   *What Happened at Vatican II*. Cambridge, MA: The Belknap Press of Harvard University Press.
Payeras, Mario
1983   *Days of the Jungle: The Testimony of a Guatemalan Guerrillero, 1972–76*. New York: Monthly Review Press.
1987   *El trueño en la ciudad: Episodios de la lucha armada urbana de 1981 en Guatemala*. Mexico City: Juan Pablos Editor.
1991   *Los fusiles de Octobre: Ensayos y articulos militares sobre la revolución guatemalteca, 1985–1988*. Mexico City: Juan Pablos Editor.
1997 [1984]   *Los pueblos indigenas y la revolución guatemalteca: Ensayos étnicos, 1982–1992*. Guatemala City: Luna y Sol.
Pazos, Luis
1994   *¿Porque Chiapas?* Mexico City: Editorial Diana.
Perera, Victor
1993   *Unfinished Conquest: The Guatemalan Tragedy*. Berkeley, CA: University of California Press.
Pieper, Jim
2002   *Guatemala's Folk Saints: Maximon/San Simon, Rey Pascual, Judas, Lucifer, and Others*. Albuquerque: University of New Mexico Press.

Pitarch, Pedro
2010 [1996]   *The Jaguar and the Priest: An Ethnography of Tzeltal Souls.* Austin: University of Texas Press.
Pop Caal, Antonio (author later dropped his first name)
1992   *Li Juliisil Kirisyaanil ut li Minok ob'* (Judeo-cristianismo y colonizacíon). Guatemala City: Semenario Permanente de Estudios Maya, Cuaderno No. 2.
Porras Castejón, Gustavo
2009   *Las Huellas de Guatemala.* 2nd ed. Guatemala City: F&G Editores.
Prechtel, Martín
1998   *Secrets of the Talking Jaguar: Memoirs from the Living Heart of a Mayan Village.* New York: Penguin Putnam.
1999   *Long Life, Honey in the Heart: A Story of Initiation and Eloquence from the Shores of a Mayan Lake.* New York: Penguin Putnam.
Redfield, Robert
1941   *The Folkculture of Yucatan.* Chicago, IL: University of Chicago Press.
1946   *Notes on San Antonio Palopó.* Chicago, IL: University of Chicago Microfilm Collection of Manuscripts on Cultural Anthropology.
Reeves, René
2006   *Ladinos with Ladinos, Indians with Indians: Land, Labor, and Ethnic Conflict in the Making of Guatemala.* Stanford, CA: Stanford University Press.
REMHI, Recovery of Historical Memory Project; the Official Report of the Human Rights Office, Archdiocese of Guatemala
1999   *Guatemala, Never Again!* Maryknoll, NY: Orbis Books.
Remijnse, Simone
2002   *Memories of Violence: Civil Patrols and the Legacy of Conflict in Joyabaj, Guatemala.* Amsterdam: Rosenberg Publishers.
Restall, Matthew
2003   *Seven Myths of the Spanish Conquest.* Oxford: Oxford University Press.
Ricard, Robert
1966 [1933]   *The Spiritual Conquest of Mexico.* Translated by Lesley Bryd Simpson. Berkely: University of California Press.
Richard, Pablo
1993   La teología india y la biblia cristiana. In *Teología India Mayense* 1993, 385–402.
Richard, R. L., and W. J. Hill
2003   Trinity, Holy. In *New Catholic Encyclopedia*, 2nd ed., 14:189–201. Detroit, MI: Gale Group.
Ríos Figueroa, Julio
2002   *Siglo XX: Muerte y resurrección de la Iglesia Católico en Chiapas.* San Cristóbal de Las Casas: Programa de Investigaciones Multidisciplinarias sobre Mesoamérica y el Sureste, UNAM.
Rodriguez, Jeanette, and Ted Fortier
2007   *Cultural Memory: Resistance, Faith, and Identity.* Austin: University of Texas Press.

Rojas Lima, Flavio
1968 Los otros pueblos del lago. In Seminario de Integración Social Guatemalteca 1968, 277–340.
Rother, Stanley
1984 *The Shepherd Cannot Run: Letters of Stanley Rother, Missionary and Martyr*. Edited by David Monahan. Oklahoma City, OK: Archdiocese of Oklahoma City.
Rother, Stanley, and Juan Mendoza Lacan
1991 *Traduccion de Nueva Testamento, Tzutuhil: Hecho por el equipo parrochial para la querida gente Tzutujhil*. Santiago: La Parroquia.
Rovira, Guiomar
1994 *¡Zapata Vive! La rebelión indigena de Chiapas contada por sus protagonistas*. Barcelona: Virus Editorial.
Ruíz García, Samuel
1968 Evangelization in Latin America. In *The Church in the Present Day Transformation of Latin America in the Light of the Council*. Vol. I, *Position Papers*, 153–77. English edition edited by Louis M. Colonnese. Washington, DC: Latin American Bureau of the United States Catholic Conference and the General Secretariat of CELAM.
1973 Los cristianos y la justicia en America Latina. *Cristus* 36 (Octubre): 32–37.
1975a *Teología bíblica de la liberación*. Mexico City: Libreria Parroquial.
1975b Salvacíon y liberacíon. *Estudios Indigenas* 5 (1): 3–10.
1994 En esta hora de gracia (In This Hour of Grace). English translation in *Origins* 23 (34), February 10: 591–602.
1999a *La búsqueda de la libertad*. San Cristóbal: Editorial Fray Bartolomé de Las Casas.
1999b *Indigenous Cultures: A New Springtime for the Church*. Plenary Address of the CTA (Call to Action) National Conference, November 7, 1999, Milwaukee. Available at the Call to Action Web site, http://www.cta-usa.org/reprint11-99/garcia.html.
2000 "Seeds of the Word" in Chiapas. *National Catholic Reporter*, February 18, 2000; and online in the *Reporter* archives, natcath.com/archives.
2003 *Como me convirtieron los indigenas*. 2nd ed. Santander: Editorial Sal Terrae.
2006 *An Interview with Charles Hardy by James W. Russell*. http:///www.curbstone.org/ainterview.cfm?AuthID=166.
Rus, Jan, and George Collier
2003 A Generation of Crisis in the Central Highlands of Chiapas: The Cases of Chamula and Zinacantán, 1974–2000. In Rus, Hernández Castillo, and Mattiace 2003b, 33–61.
Rus, Jan, Rosalva Aída Hernández Castillo, and Shannan Mattiace
2003a *Mayan Lives, Mayan Utopias: The Indigenous People of Chiapas and the Zapatista Rebellion*. Lanham, MD: Rowman and Littlefield.
2003b Introduction. In Rus, Hernández Castillo, and Mattiace 2003a, 1–26.
Sable, Thomas F.
2005 Uniate Churches. In *Encyclopedia of Religion*, 2nd ed., 14:9462–66. Detroit: Macmillan Reference USA.

Sacerdotes Mayas
1993   La religión Maya. In *Teologia India Mayense* 1993, 143–62. Quito: Ediciones Abya-Yala.
Salomon, J. B., L. F. Mata, and J. Gordon
1968   Malnutrition and Communicable Diseases of Childhood in Rural Guatemala. *American Journal of Public Health* 58 (3): 505–16.
Samandú, Luis, Hans Siebers, and Oscar Sierra
1990   *Guatemala: Los retos de la Iglesia Católica en una sociedad en crisis*. San Jose, Costa Rica: Departamento Ecuménico de Investigaciones.
Samson, C. Matthews
2007   Re-Enchanting the World. Tuscaloossa AL: University of Alabama Press
Santa Cruz Mendoza, Santiago
2004   *Insurgentes: Guatemala, la arrancada*. Santiago de Chile: LOM Ediciones.
Santos, Carlos
2007   *Guatemala, el silencio del gallo: Un misionero español en la guerra mas cruenta de América*. Barcelona: Random House Mandadori.
Schirmer, Jennifer
1998   *The Guatemalan Military Project: A Violence Called Democracy*. Philadelphia: University of Pennsylvania Press.
Scotchmer, David G.
1991   *Symbols of Salvation: Interpreting Highland Maya Protestantism in Context*. Albany: PhD diss., State University of New York at Albany.
2001   Pastors, Preachers, or Prophets? Cultural Conflict and Continuity in Maya Protestant Leadership. In *Holy Saints and Fiery Preachers: The Anthropology of Protestantism in Mexico and Central America*, edited by James W. Dow and Alan R. Sandstrom, 235–62. Westport, CT: Praeger Publishers.
Segundo, Juan Luis
1976   *Liberation of Theology*. Maryknoll, NY: Orbis Books.
1990 [1983]   Two Theologies of Liberation. In Hennelly 1990, 353–66. Also published in *The Month*, 17 October 1984, 321–27.
1992   *The Liberation of Dogma: Faith, Revelation, and Dogmatic Teaching Authority*. Maryknoll, NY: Orbis Books.
Seminario de Integración Social Guatemalteca
1968   *Los Pueblos del Lago de Atitlán*. Seminario de Integración Social Guatemalteca, no. 23. Guatemala City: Tipografía Nacional.
Sexton, James
1992   *Ignacio: The Diary of a Maya Indian of Guatemala*. Philadelphia: University of Pennsylvania Press.
Shapiro, Judith
1981   Ideologies of Catholic Missionary Practice. *Comparative Studies in Society and History* 23 (1): 130–49.
Sharer, Robert J.
1994   *The Ancient Maya*. 5th ed. Stanford, CA: Stanford University Press.

Siller, Clodomiro L.
1993 Simbolos y mito en Teología India. In *Teología India Mayense* 1993, 365—84.
1996 El mito en la Teología India. In *Teología India Mayense* 1996a, 267–87.n.p.

Simon, Jean-Marie
1987 *Guatemala: Eternal Spring, Eternal Tyranny*. New York: W. W. Norton.

Sobrino, Jon
1993 The Winds in Santo Domingo and the Evangelization of Culture. In Hennelly 1993a, 167–83.

Stoll, David
1993 *Between Two Armies: In the Ixil Towns of Guatemala*. New York: Columbia University Press.
1999 *Rigoberta Menchú and the Story of All Poor Guatemalans*. Boulder, CO: Westview Press.
2008 Human Rights, Land Conflicts, and Memory of the Violence in the Ixil Country of Northern Quiché. In *Human Rights in the Maya Region: Global Politics, Cultural Conventions, and Moral Engagements*, edited by Pedro Pitarch, Shannon Speed, and Xochitl Leyva Solano. Durham, NC: Duke University Press.

Sullivan, Kathleen
1998 *Religious Change and the Recreation of Community in an Urban Setting among the Tzotzil Maya of Highland Chiapas, Mexico*. PhD diss., City University of New York.

Tarn, Michael, with Martín Prechtel
1997 *Scandals in the House of Birds: Shamans and Priests on Lake Atitlán*. New York: Marsilio Publishers.

Tavanti, Marco
2003 *Las Abejas: Pacifist Resistance and Syncretic Identities in a Globalizing Chiapas*. New York: Routledge.

Tax, Sol
1946 *The Towns of Lake Atitlán*. Chicago: University of Chicago Microfilm Collection of Manuscripts on Cultural Anthropology.

Tedlock, Dennis, trans.
1996 *The Popol Vuh: The Mayan Book of the Dawn of Life*. New York: Simon and Schuster.

Tello Díaz, Carlos
2001 *La rebelión de las cañadas: Origen y ascenso del EZLN*. 2nd rev. and enl. ed. Mexico City: Cal y Arena.

Teología India Mayense
1993 *Teología India Mayense*. (Primer y Secundo Encuentro de Teología India de Región Mayense). Quito: Ediciones Abya-Yala.
1996a *Teología India Mayense*. (Memoria del Tercer Encuentro de Teología India Región Mayense). Guatemala City: n.p. Available at the Proyecto Filosofía en Español Web site, http://www.filosofia.org/ave/001/a230.htm.
1996b Mito o palabra antigua. In *Teología India Mayense* 1996a, 41–146.

Thompson, Sir Robert
1966   *Defeating Communist Insurgency: The Lessons of Malaya and Vietnam*. New York: Frederick A. Praeger.
Toj Medrano, Emeterio
2008   Las luchas campesinas, el Comité de Unidad Campesinoa (CUC). In Hoyos de Asig, Blanco Carballo, and Corral Alonso 2008, 144–46.
Torres Restrepo, Camilo
1965   Mensaje a los cristianos. *Frente Unido* (August 25): 3.
Tzeltales, Los
1972–74   *Estamos buscando la libertad: Los tzeltales de la selva annuncian la buena nueva*. Ocosingo, Chiapas: Misión de Ocosingo-Altamirano.
Voces Episcopales de Mexico
1993   Mensaje de Cuaresma 1993. In *Teología india Mayense* 1993, 313–38.
Veblen, Thomas T.
1982   Native Population Decline in Totonicapan, Guatemala. In *The Historical Demography of Highland Guatemala*, edited by Robert Carmack, John Early, and Christopher Lutz, 81–102. Albany: Institute for Mesoamerican Studies, State University of New York at Albany.
Vogt, Evon Z.
1976   *Tortillas for the Gods: A Symbolic Analysis of Zinacanteco Rituals*. Cambridge, MA: Harvard University Press.
von Rad, Gerhard
1962   *Old Testament Theology*. New York: Harper.
Vos, Jan de
1980   *La paz de Dios y del Rey. La conquista de la Selva Lacandon por los españoles, 1525–1821*. Chiapas: Fonapas; Gobierno del Estado de Chiapas. (A 2nd edition was published in 1988 by the Fondo de Cultura Economica in Mexico City.)
1988   *Oro verde: La conquista de la Selva Lacandon por los maderos Tabasqueños, 1822–1949*. Mexico City: Fondo de Cultura Economica.
2002   *Una tierra para sembrar sueños: Historia reciente de la Selva Lacandona*. Mexico City: Fondo de Cultura Economica.
Wagley, Charles
1969 [1941]   *Economics of a Guatemalan Village*. New York: Kraus Reprint Co.
1949   *The Social and Religious Life of a Guatemalan Village*. Memoir Series of the American Anthropological Association, no. 71. [Menasha, Wis.]: American Anthropological Association.
Wallace, Anthony F. C.
1966   *Religion: An Anthropological View*. New York: Random House.
Walzer, Michael
1985   *Exodus and Revolution*. New York: Basic Books.
Warren, Kay B.
1978   *The Symbols of Subordination: Indian Identity in a Guatemalan Town*. Austin, TX: University of Texas Press.

1985 Creation Narratives and the Moral Order: Implications of Multiple Models in Highland Guatemala. In *Cosmogony and Ethical Order: New Studies in Comparative Ethics*, edited by Robin W. Lovin and Frank E. Reynolds, 251–76. Chicago: University of Chicago Press.
1998 *Indigenous Movements and Their Critics: Pan-Maya Activism in Guatemala*. Princeton, NJ: Princeton University Press.
2002 Voting against Indigenous Rights in Guatemala: Lessons from the 1999 Referendum. In Warren and Jackson 2002, 149–89.

Warren, Kay, and Jean Jackson, eds.
2002 *Indigenous Movements, Self-Representation, and the State in Latin America*. Austin: University of Texas Press.

Watanabe, John
1992 *Maya Saints and Souls in a Changing World*. Austin, TX: University of Texas Press.

Weber, Max
1949 *On the Methodology of the Social Sciences*. Translated and edited by Edward A. Shils and Henry A. Finch. Glencoe, IL: The Free Press of Glencoe, Illinois.

Wilson, Richard
1995 *Maya Resurgence in Guatemala: Q'eqchi' Experiences*. Norman: University of Oklahoma Press.

Womack, John
1999 *Rebellion in Chiapas*. New York: The New Press; distributed by W. W. Norton.

Woods, Clyde
1968 San Lucas Tolimán. In Seminario de Integración Social Guatemalteca 1968, 201–28.

Ximénez, Fray Francisco
1929 *História de la Provincia de San Vicente de Chiapa y Guatemala, Orden de Predicatores*, Tomo I, Libro I. Biblioteca "Goathemala," Sociedad de Geografía y História de Guatemala, 1+2. Guatemala.
1965 *História de la Provincia de San Vicente de Chiapa y Guatemala*. 4 vols. with continuous numeration of pages. Guatemala City: Ministerio de Educación.

Zatyrka, Alexander P., SJ
2003 *The Inculturation of Christian Ministries in the Indian Cultures of America. A Case Study: The Bachajón Mission and the Diocese of San Cristóbal de Las Casas, Mexico*. Innsbruck, Austria: PhD diss., Leopold Franzens Universität, Innsbruck.

Index

Page numbers in *italics* refer to illustrations.

Abejas, 150, 305–8, 388; biblical foundation, 250, 308, 440. *See also* Acteal Massacre
Abel, 183
Abraham, 183
Acteal Massacre, 306–7. *See also* Abejas
Action Catholicism, 85, 132, 147, 159, 279; as evolving phase, 151, 171, 237, 351, 357, 361, 367, 395, 402; decline of, 271, 312; dissenters, 156, 264, 398, 401; as ladinization, 173, 396; and pan-Maya, 457; and personal renewal, 397; in Santiago Atitlán, 333, 339, 346; as revitalization, 398, 406, 407, 409. *See also* Catholic Action; Liberation Movement; Theology
Acts of the Apostles, 205, 268, 444. *See also* Apostles
African bishops, 229, 435, 446
Aguacatán, Huehuetenango, 61, 101, 112
Alcohol, 51, 67, 68, 110, 112, 403; abuse of, 37, 51, 66, 70, 79, 119, 169; condemnation of, 202, 282; legitimation of abuse, 68–69; ritual use of, 114, 212
Alcoholics Anonymous, 403
Alianza Nacional Campesina Independiente Emiliano Zapata (ANCIEZ), 276, 284
Alonso, Juan, 332
Altamirano, Chiapas, *100*, 146, 217
Alta Verapaz, *5*, 81, 112, 238, 257, 374
Ancestors, 15, 45, 212, 226–27, 306, 324; and Catholicism, 113, 117; cultural founders, 33, 81, 405, 441, 444; and Jesus, 209; and land, 20, 49, 79, 80; and morality, 16, 19, 36, 44, 47, 49, 79; and rituals, 20, 33, 114, 119; and saints, 32; and shamans, 19, 30, 39. *See also* Wisdom of the ancestors
Animism, 12. *See also* Pantheism
Anthropologists, 32, 417–20, 425–28; critique of missionary activity by, 229, 418, 459; and Bishop Ruíz, 412, 414, 420. *See also* Anthropology; Barbados, Declaration of; Xicotepec conference
Anthropology, 329, 354, 401, 425, 427–28; and religious paradigms, 458–59; and Bishop Ruíz, 413–14, 420. *See also* Anthropologists; Barbados, Declaration of; Xicotepec conference
Apocalyptic literature, 184, 462n1
Apostles, 94, 117, 205, 274, 440. *See also* Acts of the Apostles
Arbenz, Jacobo (president), 109, 115, 235, 385, 387, 403
Arévalo, Juan José (president), 59, 115, 235, 385
Arias, Jacinto, 12, 18–20
Armas, Castillo (president-general), 235
Arvelo de Jiménez, Nelly, 429
Assassination, 237, 278, 368. *See also* Murder
Assyrians, *180*, 183, 186

Babylonia, 183
Bachajón, Chiapas, 100, 195, 356, 410, 423; Jesuits in, 101–2, 146, 170, 194, 355, 410, 424, 444. *See also* Chilón-Sitalá, Chiapas
Banaszek, Stanley, 399
Baptism, 228, 324–25, 340, 446–47, 461n2; Maya interpretation of, 31, 104, 322–23; Charismatic, 156, 402
Barbados, Declaration of, 418–20, 425–29, 459

Barillas, Huehuetenango, 351, 399
Bible, 115, 169–70, 245, 274, 410; and catechisms, 97, 126, 190, 229–31; and literacy, 120, 175; selection of texts from, 196, 215, 285, 296; translations of, 223–25, 343, 440, 444
—compilation of, 178, 181–86; and Hebraic culture, 88, 177–79; Deuteronomistic themes of moral instruction, 178–86, *180*, 188, 189, 436
—empowerment from, 186, 294, 298, 301, 308, 321, 371, 377, 459; for Catholic Action, 126, 363, 371; for CUC, 245–46, 377; for CPRs, 268–70; for liberation movement, 152, 229–31; to reject violence, 274, 287; for women, 152, 213; for Zapatismo, 308
—interpreted by: Action Catholics, 126, 139, 146, 151, 153, 161, 175, 187, 189–90, 229–31, 455; Bishop Ruíz's method, 191–93, 221, 287, 423; catechists, 117, 146, 150, 153, 196–97, 202–5, 371, 445; fundamentalists, 190, 436; Luther, 90; Marcos, 288; Teología India, 440, 455; Tijwane questions, 194–96; Tridentine Catholics, 91–93, 97, 121, 150, 349, 436; women, 210, 213; Zapatistas, 285–86, 289
Births, 15, 23, 104; birthday, 23, 123–24, 309, 338; birth intervals, 54; birth rates, 54; of infants, 15, 16; phase in Maya cycles, 13, 14, 19, 31, 117, 154, 453–54; virgin birth, 215. *See also* Cycles; Rebirth
Bonilla, Héctor Daniel, 429
Bonpane, Blase, 317, 330
Breaking points (cultural transitions), 90, 139, 152, 215, 308, 313, 396, 426; Tridentine Catholic from Traditional Maya, 107–9, 113–15, 160, 200, 395
Brown, Robert McAfee, 128
Buddhism, 33

Cabricán, 324–26, 328, 383
Cain, 183
Cal, Pop (formerly Antonio), 432
Calvin, John (bishop), 90
Cámara, Helder (bishop), 315
Canaan, 183

Cañadas, Lacandón, 274, 275, 292, 294, 297; Catholics and Marxists in, 278–89; migration to, *217*, 217–18
Canonization, 339, 342. *See also* Rother, Stanley
Cargo/cofradía system, 38, 63, 74, 80, 104, 114, 166, 199, 335; and alcohol, 66, 67; critique and reform of, 111, 113, 117–18, 119, 200, 201, 205–6; focus of conflict, 107–8, 115; and ritual, 80, 110, 200, 205; time and cost of, 64, 66, 71, 73, 81, 112, 212; and women, 66, 69. *See also* Civil-religious system
Cargueros. *See* Cargo/cofradía system
Carlin, Ramon, 162, *163*, 197, 335, 336, 343, 344
Casariego, Mario (cardinal), 157, 244, 353
Castellanos, Absalon (general), 277
Castro, Fidel (president), 310
Catechesis, 47, 87, 138, 162; Catholic Action, 197, 204–7, 215; Liberation Movement, 139, 209, 226
—Tridentine, 101–3, 105, 215, 331, 395; and Bible, 162, 190, 210; critiques of, 220–21, 226, 415, 420–21, 431; as focus of conflict, 109, 113, 115; as continuation of traditional Maya covenant, 99, 106, 112, 161; Maya crisis not considered by, 121, 123, 130, 154; as revitalized new cycle, 112, 116, 118, 119, 161
Catechisms, 96–98, 104, 120, 159, 229; and Bible, 126, 190
Catechists: and charismatics, 399–400; Chiapas before Bishop Ruíz, 100–101; communicating Catholic worldview, 152; compose catechism, 219–27; critique of Catholic Tridentine effort, 121; critique Maya traditionalists, 117–18; during counterinsurgencies, 257, 259, 385; early effort in Santiago Atitlán, 103–10; importance of outside influences on, 111–14, 351; join EPG, 372, 374, 392; join EZLN, 274, 283, 285–87, 374, 392; Las Casas celebration, 123–24; Marxian influence on, 242–46, 279–87; other religious duties, 150; promote civil rights groups, 205, 302, 303; promote cooperatives, 205–7; school for, 102–3; and Teología India,

441, 444; training, 149–50; transition to Action Catholicism, 147–49, 194–95; translators, 97; under militarization, 268, 271. *See also* Cristina; Rudolfo Ruíz Santis

Catholic Action, 85, 131–32, 146; alternative names for, 127, 136; and charismatics, 399–400, 407; Chiapas origin, 411, 424; and counterinsurgency, 251, 268; and the CUC, 245–47; defections from, 173; discontent within, 241, 243, 370; and EPG, 239, 241; European origin, 126–27; and exhumation, 267; and EZLN, 287; Guatemalan decline, 271–72; Guatemalan origin, 132, 145; as ladinization, 173, 396; and liberation movement, 135–36; observe-judge-act method, 146, 161, 279, 294; and pan-Maya movement, 402–6; and REMHI, 267; as revitalization, 173–74; and Vatican Council II, 129–31; and Zapatismo, 287. *See also* San Andrés Semetabaj, Sololá; Santiago Atitlán, Sololá; Theology

Catholic Church, 127–29, 184, 266, 342. *See also* Catechesis; Catechists; Catholic Action; Nuns; Pastoral workers; Priests, Catholic; Theology

Catonsville, Maryland, 329

Cave, 20–23, 28, 29, 31, 355

CEH. *See* Commission for Historical Clarification

CELAM. *See* Latin American Episcopal Conferences (CELAM)

Cerro Verde, Lacandón, 298–301

Chajul, 241, 247, 351, 399, 400, 402

Chamá Mountains, Alta Verapaz, 257

Chamula, Chiapas, 59, 102, 207, 407, 424, 447, 482

Chanteau, Miguel, 209, 305, 307–8, 356

Charismaticism, 156–57, 159, 397; and Catholic orthodoxy, 401–2; during wartime, 397, 399; history in Guatemala, 157, 398–400; as Maya revitalization, 397, 453; observations of, 158; within charismatic groups, 398

Charles V (emperor; also King Charles I of Spain), 91

Chase Sardi, Miguel, 429

Chichicastenango, El Quiché, 58, 255, 266

Chilam Balam, books of, 17, 24

Chilón-Sitalá, Chiapas, 146, 170, 172, 355, 444

Chimaltenango, Chimaltenango, 5, 238, 255, 373, 374

China, 172, 321, 386, 458

Chol region, Chiapas, 100, 124, 218, 274, 280, 281, 410

Chortí region, Guatemala, 357

Christian Democratic Party, 150, 239, 240, 250–51, 281, 391

Christianity, 125–27, 130, 321. *See also* Theology

Christian Marxists, 151, 279, 284, 363, 365

CIA (Central Intelligence Agency), 235

Civil patrols, 263–66, 400

Civil-religious system, 36, 81. *See also* Cargo/cofradía system

Clinton, William (president), 261

Cobán, Alta Verapaz, 257

Cofrades. *See* Cargo/cofradía system

Colom, Yolanda, 317, 379, 380

Colonial compact, 41, 50, 51

Colonialism, 134–35, 348

Colonization: European, 421, 425; Ixcán, 172, 325, 351, 399; Petén, 326, 329, 330, 383, 385; Santa María Tzejá, 240, 242, 348

Comillas University, 365

Comité Clandestino Revolucionario Indígena (CCRI), 283, 290–91, 295–96

Commission for Historical Clarification (CEH), 257, 259, 260, 265, 463n1

Committee for Campesino Unity (CUC), 243–47, 368, 388, 392. *See also* Cristina; Hernández Ixcoy, Domingo; Hoyos, Fernando

Communism, 115–16, 236, 317–18, 353, 396

Communities of Populations in Resistance (CPRs), 268–70

Community development, 54, 153, 171, 174, 445; original focus of Catholic Action, 131, 162, 169, 372

Concuá, Baja Verapaz, 236

Conflict resolution, 38, 306

Conversion experience, 136, 138, 311, 321, 348, 414, 424
Conversos, 91
Cooperatives, 174, 239, 255, 314, 356, 372, 452
Corral, Enrique, 263, 363, 365–66, 373
Cotzal, El Quiché, 241, 247, 271
Council of Trent, 89–96, 98, 125, 135, 435, 452. *See also* Theology
Counterinsurgency
—Chiapas, 297, 300, 308. *See also* Acteal massacre
—Guatemala, 246, 251; death toll, 260, 463n1; Maya soldiers, 252–53; phase one (100% solution), 250, 251–53; phase two (ashes), 254–57; savagery of, 253, 258; United States involvement in, 236, 261. *See also* Guatemalan military
Covenant: similarity of Maya and biblical, 186
—biblical: expressed as historical phases, 161–85; literal vs. symbolic interpretations, 162; theme in Deuteronomistic pattern, 179, *180*, 186
—traditional Maya, 19–20; and conflict mediation, 38; incorporated Spanish gods, 31–32; non-fulfillment of, 26; plurality of, 27; questioning of, 71–72, 79, 84; rituals of, 20–26, 38
—revitalized Maya: by pan-Maya, 7; by Tridentine Catholicism at Conquest, 32, 119; by Tridentine Catholicism in twentieth century, 83, 108, 112, 118
CPRs. *See* Communities of Populations in Resistance
Cráter, 319, 328–30
Credit Union National Association, 349
Crisis: Maya, 74–83; Catholic, 125
Cristina, 370–72
CUC. *See* Committee for Campesino Unity
Cuchumatán (Kuchumatán) mountains, 50, 240, 256, 351, 383
Cultural shock, 348
Curandero. *See* Shamans
Cursillos de Capacitación Social, 317–19, 321
Cycles, 33, 378, 453; similarity of Maya and biblical, 117, 209, 406

—biblical, 154, 179, 308. *See also* Bible: compilation of
—Maya, 16–19, 27, 30, 119, 393, 407, 453–54. *See also under* Covenant

David (king), 183, 308
Death, 13, 16, 453; Black Death, 83; and Council of Trent, 92, 95; fear of, 20; of gods, 18; and land disposition, 37, 51, 63, 305; and Maya cycle, 13–14, 16, 18–19, 47, 83, 453; prediction of, 110, 324; rates of, 52, 54; and rituals, 81; threats of, 241–42; and witchcraft, 65. *See also* Covenant; Crisis: Maya; Cycles; Malnutrition
Delgado, Arnulfo, 400
Deposit of faith, 93, 436–37
Deuteronomist history, 179. *See also* Bible: compilation of; Cycles: Biblical
Diaconate, 201, 204, 274, 296, 340, 445–46
Diocesan assembly (San Cristóbal), 146, 304
Divination, 29, 115. *See also* Shamans
Dominican nuns, 217, 248
Dominican priests, 123, 146, 207, 217
Don Quixote, 310
Dreams, 18, 158, 271, 366, 396, 457; from ancestors, 30, 39, 42
Drunkenness. *See* Alcohol

Earthquakes, 7, 213, 241, 244, 398, 413–14
Earth tremors, 26
Ecclesiastical documents, 229, 435
Economic depression, 207, 240
EGP. *See* Guerrilla Army of the Poor (EGP)
Egypt, 183
Ejidos, 90, 155, 174, 218, 277, 282; and NAFTA, 219, 278
Election, biblical, 179–81. *See also* Bible: compilation of
Elections, political, 84, 304, 311, 378; nullified, 240, 249, 368, 373, 384, 403
Ellacuría, Ignacio, 365
El Salvador, 134, 155, 248, 252, 277, 382
Emanation, 12, 14, 17–18, 23, 26, 29–30;

absorbing interpretative principle, 119, 228, 454; biblical interpretation by, 45–46, 48; enduring concept, 393; land as, 79; and rituals, 33, 36. *See also* Cycles; Pantheism

Encyclicals, 127–28, 487

Enlightenment, the, 50, 93, 125

Epidemics, 22, 32, 52. *See also* Malnutrition

Epistemology, 83, 435

Ethnocentrism, 333, 434, 458; Marxian, 381–82; Tridentine theological, 94, 325, 422; western theological, 313, 325

Ethnography, 382, 489

Evangelicals, 115, 239, 300, 301. *See also* Pentecostals

Exodus (biblical), 302, 307, 444; cited by Maya, 210, 224, 226, 230; to understand current reality, 244, 245, 355, 368; used in biblical pattern, 180, 183, 185–87; and violence, 285, 286. *See also* Bible

Exodus (migration), 50

EZLN. *See* Zapatista Army of National Liberation (EZLN)

Falla, Ricardo, 269–70, 353–55, 367

FAR. *See* Fuerzas Armadas Rebeldes (FAR)

Faudree, Marcella, 343

Fertility, 52, 54; of land, 17, 36, 50, 79, 80, 222, 230. *See also* Births

Fertilizer, 55, 206, 240, 276, 370, 372; loans for, 168, 170, 174

Fiscales, 422, 447, 456

FLN. *See* Forces of National Liberation

Forces of National Liberation (FLN), 276, 281–82, 284, 288–90, 310–11. *See also* Marcos (subcomandante); Zapatista Army of National Liberation (EZLN)

Franciscans, 100, 410

Francisco Marroquín Language Institute, 344

Freire, Paulo, 86, 134, 164, 170, 186, 243, 366. *See also* Literacy

Frustration: within Catholic Action, 376, 378; to communicate catechesis, 373; from government corruption and failure of reform, 241, 278, 281, 292, 312, 322, 374; and non-violence, 391; from poverty, 82; with using peaceful means, 383

Fuerzas Armadas Rebeldes (FAR), 276, 281–82, 284, 288–90, 310–11

Fundamentalists, 17, 83, 93, 190, 435

Future life, 95, 228

García, Tomás, 431

Genesis, Book of, 181–82, 205, 222–24, 226–27, 230

Gerardi, Juan (bishop), 245, 332, 352–53, 414

Gerontocracy, 63, 71

Global society, 434, 458

Goliath, 308

González Estrada, Rafael (bishop), 101, 112–15

Good Samaritan story, 154, 192, 327

Grace, state of, 116–18, 130, 288, 364, 397, 402, 435

Gran, Jose María, 351, 370

Grande, Rutilio, 368

Grandparents, 13, 44–45, 197, 440

Gregorian University, 410

Grünberg, Georg, 429

Guadalupe, Our Lady of, 303, 388–90, 439

Guatemalan military: intelligence unit, 248, 381; officer corps, 249, 251–52; soldiers, 252–54, 256, 263. *See also* Kaibiles

Guerrilla Army of the Poor (EGP), 236–37, 239, 242–43, 247, 254; in Alta Verapaz, 257; in Guatemala City, 247–48; in Huehuetenango, 240, 256, 260, 268; in Ixcán, 240, 250; in Ixil area, 241; under army control, 262–63. *See also* Colom, Yolanda; Fuerzas Armadas Rebeldes (FAR); Hoyos, Fernando; Payeras, Mario; Porras, Gustavo

Guevara, Ernesto (Che), 297, 310, 361

Gurriarán, Javier, 349

Gurriarán, Luis, 270–71, 348–50, 353, 370, 388; and Ixcán colonization (Santa Maria Tzejá), 240, 242, 264, 350; and

Gurriarán, Luis—*continued*
  Melville group, 328–30, 350; organizes cooperatives and study centers, 169, 349
Gutiérrez, Gustavo, 415

Hanukkah, 184
Harmonious Balance: in Catholic catechesis, 116–17, 228; in traditional Maya worldview, 13–17, 19, 26, 32, 36, 116
Hebrew: Bible, 88; culture, 92, 178; language, 88, 190
Hennessey, Ronald, 383–87, 464n1
Hermosillo, Lacandón, 150, 207–8
Hernández Ixcoy, Domingo, 245
Hinduism, 33
Homicide, 65. *See also* Murder
Hoyos, Fernando, 243–45, 354–55, 363–73, 392
Hutterites, 54

Ideal type, 32, 119
Ignacio (pseud.), 157–58
Inculturation, 409, 434, 446, 459; of contemporary Christianity in Maya culture, 143–44, 415, 424, 439, 443, 448; of early Christianity in Greek intellectual culture, 89; Latin American bishops on, 438, 448, 453; of Maya in ladino culture, 396; resistance to, 420, 424, 455, 458; Vatican II and encyclical on, 435–37. *See also* Deposit of Faith; Theology: Teología India
Indian Congress (1974), 123, 147, 153, 170, 278–79, 447
Indirect rule, 40, 50
Indonesia, 236, 250, 332
Indulgences, 90, 461n1
Infectious disease, 52, 76, 78, 174. *See also* Malnutrition
Inquisition, 91
Insurgents. *See* EPG; EZLN
Interpreters of Bible. *See* Bible: compilation of
Irapuato, Guanajuato, 409–10
Iroquois, 112, 462n1
Islam, 91
Israel, *180*, 182–83, 186, 305, 381

Jacaltenango, Huehuetenango, 152, 256, 316, 400
Jesuits, 19, 340, 458; in Bachajón, 101–2, 146, 170–72, 194, 356, 423–24, 444–47. *See also* Falla, Ricardo; Hoyos, Fernando; Marcos (subcomandante)
Jesus, 88, 89, 92, 94–95, 348, 362–63
—biblical presentation: defined by Jewish theological terms, 178, *180*, 184, 191; emphasis on golden rule, 192; in Jewish culture, 88, 414. *See also* Bible: compilation of
—Maya view of: invoked in traditional prayers, 3, 200; in biblical reflections by, 201–5, 209, 214–15, 223–28, 230, 285; in cycles, 45–47, 117, 154, 173; expressing Maya moral sense, 187, 197, 227; inculturation by Maya Catholic priesthood, 440; as liberator, 136, 192, 268; as Maya ancestor, 209; as rejecting violence, 274
Jews, 90–91, 183–86
John Paul II (pope), 288, 342, 386, 437
John XXIII (pope), 127, 129, 162, 318, 386
Jordan River, 180, 182–83
Joyabaj, El Quiché, 260, 265–67, 332
Judah, 183
Judaism, 88, 178, 184
Judges, Book of, *180*, 183
Judith, Book of, 185–86, 270
Just war criteria, 377–78. *See also* Morality

Kaibiles, 253
King, Martin Luther, Jr., 188
Koine Greek, 88, 190
Korea, 164
Kugel, James L., 178
Kwashiorkor, 75, 76. *See also* Malnutrition

Labor, forced, 30, 41–42, 51–52, 60
Lacandón, Chiapas, 146, 240, 277, 290, 298, 312, 376; EZLN strength in, 274, 282; migration to, 207, 216–18
Ladinization, 263, 431, 453; and Catholic Action, 173, 226, 396
La Estancia, El Quiché, 370–72
LaFarge, Oliver, 99

Land: central role in Maya life, 35, 36, 79–80; conflicts about, 80, 218–19, 277, 305, 390; disposition at death, 37, 63; eviction from, 377; illegal seizures of, 43, 130, 170; invasions of, 155, 171; loss of, 30, 50–59, 134; reform, 171, 235, 275, 284; results of loss, 72, 74, 80, 187, 243; as sacred, 36, 46, 50, 79, 80, 222; traditional usage of, 49–50; types of land rights, 36–43, 50, 205, 208, 218, 224, 230. *See also* Ejidos
Lansing, Donald, 344
Lanz, Andrés, 373
Last Judgment, 90, 186, 225, 321
Latin American Episcopal Conferences (CELAM), 141, 417; Medellín, 133, 139–41, 144–45, 169, 310, 362, 413, 437; Puebla, 140–41, 437, 442; Santo Domingo, 96, 143–44, 438, 442, 455
Lay apostolate, 127, 130–31. *See also* Catholic Action
Legitimate, 134–35, 275, 290, 408; by bible, 92; liberation movement, 141, 169; injustice, 136, 139, 249; observance of traditional rituals, 107
Lias, Padre, 104, 107
Liberal governments, 3, 51
Liberation movement, 6, 131–33, 135–37; and biblical reflection, 189–90; ecclesiastical recognition of, 139–43; as evolving Catholicism, 395; and inculturation, 144; and option for the poor, 151; origin of, 137–39, 144; and use of violence, 251. *See also* Action Catholicism; Bible; Catholic Action
Lichtenberg, Joan, 343
Life cycle, 13, 16; and Catholic sacraments, 461n1; in Maya worldview, 17, 19, 29, 45, 116, 405
Life expectancy, 74–75, 164, 452
Life table, 74–75, 79
Línea Proletaria (LP), 275, 280–81
Literacy, 102, 153, 161, 172, 221, 313, 347; for defense, 161; in El Quiché, 169; and Hoyos, 243–44, 365–66; in Jesuit area, 170; and ladinization, 396; in Santiago Atitlán, 164–68, 333–34, 347. *See also* Freire, Paulo

Literary forms, 177–78, 181, 187, 343
Liturgy, 96, 335–36, 355, 365, 431, 434, 448; in indigenous form, 197, 453–54. *See also* Ritual
Lockett, James (Diego), 340–41
López Hernández, Eleazar, 433, 439–40, 456
Los Altos, diocese of, 144–45, 462n2
LP. *See* Línea Proletaria (LP)
Madres. *See* Nuns
Magdalenas, Chiapas, 72, 150, 199–200, 204–7, 230, 274. *See also* Rudolfo Ruíz Santis
Malaysia, 226, 250
Malnutrition, 76–78, 164, 263. *See also* Infectious disease; Kwashiorkor; Marasmus
Manuel stories, 400. *See also* Theology: Teología India
Marasmus, 75, 76. *See also* Malnutrition
Marcos (subcomandante), 281–84, 287–91, 293–302, 308–12. *See also* EZLN
Margaritas, Chiapas, 207–8, 216–17, 298
Marist Brothers, 102, 217
Markets, 245, 277, 339; diminish Maya isolation, 406, 453; injustices against Maya in, 124, 219, 225; and ladinization, 173. *See also* Merchants
Marriage, 227, 283, 296, 396; and inculturated church, 448, 457; traditional petition for, 63, 69, 150
Marx, Karl, 125, 357
Marxism, 247, 309–11, 318; Christian form of, 279, 284, 363–65
Marxists, 138, 281–82, 355, 373, 376, 380–82
Mary, Virgin, 202, 213–14
Maryknoll, 101, 152, 157, 172, 255, 400, 433. *See also* Banaszek, Stanley; Bonpane, Blase; Colonization; Hennessey, Ronald; Lansing, Donald; Melville, Marjorie; Melville, Thomas; Woods, William
Maximon, 109–11, 162
Mayalán, El Quiché, 158, 352
McKenna, Donald, 372
McSherry, Thomas, 339
Medellín, Colombia. *See* Latin American Episcopal Conferences (CELAM)

Melotto, Angelico (bishop), 343
Melville, Arthur, 328
Melville, Marjorie, 316–18, 322, 326–30, 361, 366
Melville, Thomas, 320–22, *326*, 329–30, 361, 366, 383
Memory of Silence (CEH), 257–61, 265. *See also* Commission for Historical Clarification
Menchú, Rigoberta, 146
Mendoza Lacán, Juan, 343
Merchants, 60–61, 70, 74, 81, 228, 336, 349; among first catechists, 111–12, 257
Metaphysics, 33, 188, 202, 215; Marxian, 8, 151, 365; Maya, 151, 174, 184, 186, 365, 405–7, 453–54
Mexican bishops, 437–38, 456–57
Mexican revolutionary tradition, 283
Momostenango, Totonicapán, 114
Montana dioceses, 357
Montejo, Victor, 253, 256. *See also* Jacaltenango, Huehuetenango
Montes, César, 237, 315, 320, 326, 350–51
Montessori school, 164–68, 333–34, *457*
Moral code, 19, 80, 187, 227, 287
Morales, Ignacio, 355
Morales, Mardonio, 170–72, 287, 355–56, 376
Moral instruction, 178, 181–82, 187
Morality, 93, 96, 118, 126, 136, 310–11, 412; crisis of, 74, 82; of just war, 286, 377–79, 383–84, 392, 452; as purpose of bible, 181, 231, 229
Moral justification, 287, 378, 283
Moral order, 43–48, 398, 427
Moral sense, 168, 173, 197, 395, 409, 453, 460
Morán, Rolando, 238, 382
Morbidity, 52, 419
Mortality, 52, 54, 72, 74, 84, 130, 164, 419; in sixteenth century, 94; and theological paradigm, 91. *See also* Death
Mosonyi, Esteban Emilio, 429
Mount Sumal, Guatemala, 268
Murder, 14, 17, 37, 182, 278, 292, 305, 308, 339, 400, 452; of priests, 260, 265–66, 331–33, 342, 345–46, 353, 365, 368, 385, 399; as tactic of counterinsurgency, 236, 250, 254–56, 337, 350, 353, 372, 380, 384, 386; United States involvement in, 261. *See also* Alonso, Juan; Assassination; Ellacuría, Ignacio; Gerardi, Juan (bishop); Gran, Jose María; Grande, Rutilio; Homicide; Romero, Oscar (archbishop); Rother, Stanley; Villanueva, Faustino; Woods, William
Music, 18, 64, 147, 156, 397, 399
Muslims, 91
Myth, 33, 142, 342–43, 442. *See also* Symbols

NAFTA (North American Free Trade Agreement), 219, 278, 291
Nahualá, Sololá, 158, 344, 357, 402
Nebaj, El Quiché, 169, 241, 268, 400, 402, 407
New Testament, 88, 315, 317, 444
New Ulm, Minnesota, 357
Nicaragua, 252, 277, 318, 332, 350, 368, 391; influence on EPG, 382, 385; influence on military, 252, 391; influence on pastoral workers, 155, 286
Nick, Elizabeth, *167*, 344
Noah, *180*, 182, 269–70
Nonviolence, 246, 391–92
Non-western cultures, 95, 125, 143–44, 313, 408, 458. *See also* Inculturation
Nuns, 137, 150–53, 191, 217, 333–34, 341, 348; formation as ladinization, 433; health programs by, 49, 152, 298, 356; legal assistance by, 210; literacy programs by, 167, 344; organize cooperative, 171; oppose Zapatismo, 302; teach bible, 213; teach catechists, 102, 146, 341; teach sewing, 213; teach women's liberation, 213–14

Oath, 92–94
Ocosingo, Chiapas, 117, 217, 302; and EZLN, 291, 373; Marxist advisors in, 278, 280, 284, 285, 355; the parish, 101, 146, 217, 229, 278, 424. *See also* Dominican priests; Vargas, Javier
Oklahoma diocese, 162, 196, 333; mission in Santiago Atitlán, 162, 168, 196, 334,

339. *See also* Carlin, Ramon; Faudree, Marcella; Lichtenberg, Joan; Nick, Elizabeth; Rother, Stanley; Stafford, Thomas; Westerman, Robert

Old Testament, 214, 305, 306, 349, 440. *See also* Bible: compilation of

Option for the poor, 154, 156, 362, 454; CELAM on, 140–41, 143; Bishop Ruíz and, 146, 151, 424

Organización del Pueblo en Armas (ORPA), 237–38, 253–54, 334, 336

ORPA. *See* Organización del Pueblo en Armas (ORPA)

Pakay, Pascual, 109
Palenque, Chiapas, 146, 216, 302–3, 373
Panajachel, Sololá, 61, 64, 335
Pan-Maya movement, 7, 402; and defections from Catholic Action, 173, 402, 404–7, 457; a Maya cycle, 173; as reaction to Catholic seminaries, 433; as revitalization, 453
Pantheism, 12, 45, 406, 461n1; difference from transcendent god, 105; in Maya theology, 119, 228. *See also* Emanation; Theology: Maya
Papal volunteers, 164
Papantla, Veracruz, 454
Paradigms, theological, 120, 139, 393, 424, 438, 458–59
Partido Guatemalteco de Trabajo (PGT), 236, 238
Partido Revolucionario Institucional (PRI), 123, 134, 275, 278, 300, 305, 307, 424
Pastoral workers, 126, 131, 145, 150–55, 186, 209–10
Patzún, Chimaltenango, 58
Pauline churches, 127
Paul (saint), 88, 184, 186, 237, 362
Payeras, Mario, 240, 247, 379, 382–83
Pentecostals, 306, 397–98, 407
Peter, Marian. *See* Melville, Marjorie
PGT. *See* Partido Guatemalteco de Trabajo (PGT)
Phoenix program, 236
Pius IX (pope), 93. *See also* Syllabus of Modern Errors

Pius X (pope), 92; "Oath Against the Errors of Modernism," 93
Pius XI (pope), 127. *See also* Encyclicals
Plantations, 43, 59–61, 68, 113, 224, 277, 327
Population increase/decrease, 40, 52, 54, 72, 74, 79
Population pressure, 50, 55, 58
Porras, Gustavo, 319, 379
Prayer formulas, 23, 96, 97, 335, 339; exact recitation of, 24, 31, 104
Prechtel, Martín, 343
PRI. *See* Partido Revolucionario Institucional (PRI)
Priests, Catholic, 85, 119, 154, 260, 373; incorporated into traditional Maya religion, 3, 31–32, 104, 339; Maya Catholic priests, 424, 430–31, 439, 441, 446, 448, 455; typology of, 357–60
Priests, Maya, 22, 26, 28–30. *See also* Shamans
Processions, 22, 79, 118, 388, 404, 447
Prolactin, 54
Prophets, 89, 137, 179, *180*, 307–8, 358; injustice denounced by biblical, 183, 184, 192, 210, 211, 230, 452; Jesus as, 126, 136; John the Baptist as, 203
Puebla, Mexico. *See* Latin American Episcopal Conferences (CELAM)
Pueblo Creyente, 303–5, 388, 391
Punishment, *180*, 186, 229, 324, 400. *See also* Bible: compilation of

Quetzalcoatl, 439
Quetzaltenango, 58, 70, 238, 352, 403, 431
Quiptic, 207, 279–81, 294

Rabinal, Baja Verapaz, 237, 260
Ramírez, Ricardo. *See* Morán, Rolando
Rebirth, 13–14, 19, 31, 83, 154, 186, 417. *See also* Cycles
Recinos, Alfredo, 107–10
Reciprocity. *See* Covenant
Recovery of Historical Memory (REMHI), 260, 265–67, 353, 463

Reform, 142, 229, 459
—religious, 118–19, 125–26, 458; of Judaism, 88, 178, 184; personal vs. social, 152, 157, 313, 397, 399, 401; by the Reformation, 90–93, 125; of traditional Maya theology, 83, 106, 111–13, 119, 160, 205, 403, 405; of Tridentine Catholic theology, 125, 452;
—social: frustration of political, 236, 248, 275, 278, 374, 377, 387, 392; land, 59, 115, 171, 235, 275, 280, 281, 284, 318, 385, 403; legal, 52, 171; political, 239, 251, 282, 317, 370, 373, 391, 452
Refugees, 264; Acteal, 307; CPRs, 269; in Guatemalan army centers, 257, 262; Guatemalan in Mexico, 152, 155, 256, 264, 271–72, 288. See also Communities of Populations in Resistance (CPRs)
Regeneration, 13, 16, 33, 117, 186, 209, 453–54. See also Cycles; Rebirth; Revitalization
Regional Seminary of the Southeast, 454
Reichel-Dolmatoff, Gerardo, 413–14, 416, 459
REMHI. See Recovery of Historical Memory (REMHI)
Revitalization: of Catholicism, 125, 174; and charismatic renewal, 396, 398; among Iroquois, 112, 462n1; phase in Maya cycles, 186, 230, 453. See also Covenant: Revitalized Maya; Cycles; Rebirth
Ribeiro, Darcy, 429
Ritual: defense from cultural threats, 42; and Maya covenant, 16, 20–24, 81–83; prayer formulas in, 31, 104; as sustaining emanation, 18, 26; and witchcraft, 65; worldview expressed in, 11, 105, 187. See also Alcohol; Cargo/cofradía system; Covenant; Priests, Catholic; Priests, Maya; Sacraments
Ritualism, 96, 359, 462n2
Romero, Oscar (archbishop), 368
Rossell y Arellano, Mariano (archbishop), 115–16, 132
Rother, Stanley, 254, 332–39, 342–46
Rudi, Padre, 266–67
Rudolfo Ruíz Santis, 72–74, 199–201, 205–6, 274, 340, 446

Ruíz García, Samuel (bishop): accusations about insurgency, 274; attacks Tridentine theology, 420–21; becomes bishop, 408; book attribution, 190, 229; criticized as naive, 355, 392; death, 408; diocesan assembly, 146; dissuasion of insurgency, 274, 288; education of, 410; first Maya contact, 102, 411; and Guatemalan refugees, 271; inculturated Maya Catholic church, 454; Indian congress (1974), 124; influence of anthropology on, 459; invites leftists, 280; Marcos's view of, 312; and Maya seminarians, 432; Medellín conference, 416–17; Melgar conversion, 414–16; ordains Maya deacons, 446; parents and early life, 409; as priest, 410; in protest marches, 303; request for married Maya Catholic priests, 447–48; Vatican Council II, 229, 411; Vatican opposition, 455; waverings about violence, 286; Xicotepec meeting, 417–18; Zapatista view of, 286
Ruíz Santis, Rudolfo. See Rudolfo Ruíz Santis

Sabanilla, Chiapas, 281
Sacraments, 461n2; and charismatics, 401–2; Council of Trent on, 92, 95–96; loss of function, 96; and Maya deacons, 446; objects of medieval piety, 89. See also Priests, Catholic; Ritualism
Sacred Heart Congregation, 69, 266, 348–49. See also Alonso, Juan; Gran, Jose María; Gurriarán, Javier; Gurriarán, Luis; Lanz, Andrés; Villanueva, Faustino
Saints: Catholic calendar of, 23; covenant gods, 39; each town with patron, 47; as emanated tribal gods, 23, 83; fused with ancestors, 32; objects of medieval piety, 89. See also Cargo/cofradía system
Saliège (cardinal), 131
Salvation, 89–90, 173, 305, 415, 433; catechists view of, 117, 121, 203, 226–29; Council of Trent on, 92–95, 139, 156; theme in pattern, 179, 182. See also Covenant: Biblical: theme in Deuteronomistic pattern

San Andrés Larraínzar, Chiapas, 199, 290, 340, 446. *See also* Lockett, James (Diego); Rudolfo Ruíz Santis
San Andrés Semetabaj, Sololá, 46, 101, 112–13, 116, 160, 169, 402
San Antonio Ilotenango, El Quiché, 61, 101, 112–13, 115, 243, 354
San Bartolomé Jocotenango, El Quiché, 101, 112–13, 265
San Cristóbal Diocese. *See* Ruíz García, Samuel (bishop)
San Cristóbal Las Casas, Chiapas, 102, 124, 152, 273, 356; protest marches in, 284, 301, 305
San Emiliano, Lacandón, 294–97, 299
San Francisco (pseud.), Huehuetenango, 113, 335
San Juan Ixcoy, Huehuetenango, 323–24, 326
San Juan La Laguna, Sololá, 58, 157–58, 400, 402
San Lucas Tolimán, Sololá, 58, 357
San Mateo Ixtatán, Huehuetenango, 256, 283
San Miguel Acatán, Huehuetenango, 322, 326
San Pedro Chenalhó, Chiapas, 30, 64, 147, 150, 218, 274, 301, 307; biblical reflection in, 209, 230, 305, 441. *See also* Chanteau, Miguel; Yibeljoj, Chiapas
Santa Catarina Palopó, Sololá, 58
Santa Cruz, El Quiché, 159, 243, 245, 332, 349–50, 388
Santa Eulalia, Huehuetenango, 56, 433
Santa Maria Tzejá, El Quiché, 240, 242, 255, 264, 348, 350, 376. *See also* Colonization; Gurriarán, Luis
Santa Maria Visitación, Sololá, 357
Santiago Atitlán, Sololá, 24, 30, 56, 104–5; insurgency in, 253, 337; Maximon saga, 106–9, 112; Oklahoma project, 162, 169, 333; physical survival in, 52, 54, 58, 74–76. *See also* Carlin, Ramon; Faudree, Marcella; Lichtenberg, Joan; Nick, Elizabeth; Rother, Stanley; Stafford, Thomas; Westerman, Robert
Santiago Chimaltenango, Huehuetenango, 24, 64
Santo Domingo, Dominican Republic. *See* Latin American Episcopal Conferences (CELAM)
Scientific revolution, 93, 125
Seeds of the Word, 191, 415, 420, 435. *See also* Theology: Teología India
Seleucids, 183
Self-identity, 45, 69, 74, 80, 82–83, 187
Seminaries, 365, 368, 410, 432, 454, 455; attended, 340, 352, 356, 410; critiques of, 348, 365, 421, 431, 432, 433, 454
Semitic languages, 178
Sexual abuse, 264–65
Shamans, 39, 104, 205, 339, 404, 441; as ancestors, 16, 17; and Catholic priests, 113, 339; and charismatic experience, 396–97, 453, 457; as curers, 26, 39, 64; as diviners, 17, 23, 26, 27, 29; and modern medicine, 84; use of church, 114, 323; and witchcraft, 26, 64
Simojovel, Chiapas, 303, 373
Sisay, Juan, 108–9
Sisters. *See* Nuns
Sisters of the Divine Shepherd, 102, 171, 424
Sitalá, Chiapas. *See* Chilón-Sitalá, Chiapas
Slohp, 287
Smelser, Neil, 357
Soccer, 104, 301
Social consciousness, 6, 239, 320, 366; and biblical reflection, 6, 239, 281, 292, 312; Catholic efforts to raise, 6, 169, 237, 284, 317, 349, 370–71; target of insurgency recruitment, 171, 237, 248, 292
Social justice: in biblical phases, *180*, 184, 192; and Bishop Ruíz, 424–25; Catholic Action, 126; Catholic church, 125–26; and charismatics, 157; Council of Trent, 452; human rights organizations, 304–5, 307–8; Indian Congress (1974), 143; insurgencies seeking, 236–37, 239, 243–44, 246, 281–82, 285; and just war criteria, 377–78, 383, 386–87; Las Casas celebration, 124; and Marcos, 302, 310–11; option for the poor, 146; pastoral workers, 313–14, 328, 331–73; procession for, 388, 391; United States paranoia about, 235; Vatican Council II, 130, 412; without violence, 287–88, 295

Soil erosion, 55
Somoza family, 134, 248
Songs, 20, 101, 147–48, 221, 318, 341, 419; charismatic, 156, 399; women's group, 214
Soul, 461n1. *See also* Sustaining power
Spanish embassy, 246, 368
Spiritual assistants, 445
Spiritual Exercises of St. Ignatius, 353, 355, 363–64. *See also* Jesuits
Spirituality, 34, 95, 126, 156–57, 363–64; Maya, 102, 404, 409
Spokane Diocese, 357
Stafford, Thomas, *136*, 162–63, 334–36, 342–44, 346
Stetter, Carlos, 352, 399
Strikes, labor, 241, 244, 388
Sumpango, Sacatepéquez, 58
Sunday service, 147, 150, 152, 196–97
Sustaining power, 12–13, 15–18, 21, 24, 26, 30–31, 33; translation of, 461n1
Syllabus of Modern Errors, 93
Symbols, 12, 21, 33, 45, 89, 410; in bible, 178, 181–82, 184, 187; catechist use of, 209; differing interpretations of, 105, 113; and inculturation, 423, 434, 439, 441–42; during insurgency, 241, 259, 267; Maya in Catholic liturgy, 197, 336; and myth making, 342; in protest processions, 391–92; and ritualism, 359
Syphilis, 89

Theology, 4, 178, 436; Charismatic, 156–59, 271, 396–402, 453, 457
—Action Catholic (Catholic Action, Liberation Catholic), 135–39, 143, 146–47, 149, 430; biblical basis, 228, 356; demands equality, 210; includes Maya Christian theology, 454; moral center of community, 264, 301; need for, 424; negates passive pietism, 192; observe-judge-act formula, 126, 139, 146, 279, 287, 297; not utopian, 192; on violence, 376. *See also* Catechesis; Catechists
—biblical, 182, 190, 322; basis of Catholic Action and liberation movement, 6; concern for present time, 185; demand for equality, 210; Deuteronomist pattern in, 179; empower Maya, 349; sustain community, 308; not utopian, 192. *See also* Bible
—Christian: fusion from different cultures, 92; historical development of, 87, 89; passive pietism in, 7, 125
—Maya, 416, 461n1; not unified, 120. *See also* Covenant; Cycles; Emanation
—Tridentine Catholic: lack of relevance, 349, 435, 452; rejection by Bishop Ruíz, 420–21; sacramental emphasis, 96, 98; used as ideology, 136, 138. *See also* Catechesis; Catechists
—Teología India (Maya Christian), 7, 143–44, 313, 336, 408, 439–43, 448; ambivalent reception by Maya Catholics, 443–44; based on wisdom of ancestors, 439; dangers of, 443; draws from bible, 305; draws from Maya culture, 305, 453; Maya tradition as an Old Testament, 305–6, 439; need for, 430–32; stresses action, 202; summary of, 442; on value of Maya culture, 226
Todos Santos, Huehuetenango, 58
Tojolabal region, Chiapas, 124, 216, 218, 298
Torreblanca, Lucio (bishop), 100
Torres Restrepo, Camilo, 142, 315, 361–62, 366
Totonicapán, 5, 52, 101, 402
Transcendent god, 12, 105, 443, 461. *See also* Emanation
Tridentine Catholicism. *See* Catechesis; Theology
Trinity, 89, 92, 159, 215
Tula, Hidalgo, 454
Turcios Lima, Luis, 236–37, 315, 319–20, 382
Tuxtla Gutiérrez, Chiapas, 145, 291, 298, 303–4, 341, 410
Tzeltal region, Chiapas, 15, 100, 121, 124, 129, 447; Jesuit parishes, 171–72; migration, 216. *See also* Catechisms
Tzotzil region, Chiapas, 15, 124, 148, 274, 280, 281; language, 12, 38, 340

Unidos en la Esperanza, 244
Unión del Pueblo (UP), 275, 278–81, 285
United Nations. *See* Memory of Silence (CEH)

United Nations Educational, Scientific, and Cultural Organization (UNESCO), 219
Universidad Centroamericana José Simeón Cañas, 365
Universidad Iberoamericana, 420, 445
Universidad Rafael Landívar, 366
University of Texas, 354
University of Wisconsin, 349
UP. *See* Unión del Pueblo (UP)
Uspantán, El Quiché, 246

Valle, Nestor, 326
Varese, Stefano, 429
Vargas, Javier, 217, 279, 286, 298
Vatican Council I, 93
Vatican Council II, 6, 129, 132, 145, 448, 452, 456. *See also* Ruíz García, Samuel (bishop)
Vázquez Leal, Juan Antonio, 265–66
Vesey, John, 339
Vietnam, 236, 250, 329, 380
Villanueva, Faustino, 265–66, 332

Warren, Kay, 45, 116
Weber, Max, 8, 11, 357
Westerman, Robert, 162, 349
Wisdom of the ancestors, 45, 82, 195, 423; basis of Maya Christian theology, 408, 416, 422, 439, 440, 453; as moral sense and code, 117, 173, 187, 409. *See also* Ancestors
Witchcraft, 37, 65–66, 462n1
Women, Maya, 22, 54, 69; as insurgents, 247, 282–83, 296, 299; issues, 66, 69, 70, 209, 264, 265, 305; work, 35–36, 43, 66, 69, 206, 296
—in Action Catholicism: in biblical reflection groups, 147–48, 210; as catechists, 102, 150; cooperatives, 171; diocesan organization for, 212–15; education of, 166, 168; as pastoral workers, 151–52. *See also* Cristina
Woods, William, 269, 251
Words, power of, 24
Worldview: analytical importance of, 7, 8; cohesive force, 71; of counterinsurgency, 250–53, 262–63, 293, 300–301; and fundamentalism, 83; source of self-identity, 45, 74, 80, 187; theological foundations of, 4
—Action Catholic, 137–39, 199–231. *See also* Catholic Action; Liberation Movement; Theology
—biblical, 177–88. *See also* Theology
—insurgents, 236–239, 280–87. *See also* Guerrilla Army of the Poor (EGP); Zapatista Army of National Liberation (EZLN)
—Maya, 3, 4, 11–34. *See also* Ancestors; Covenant; Cycles; Emanation; Land; Pantheism; Priests, Maya; Ritual; Shamans; Theology
—Teología India (Maya Christian), 442–43. *See also* Theology
—Tridentine Catholic, 87–98. *See also* Theology

Xalbal, Ixcán, 241
Xicotepec conference, 103, 417–22, 429, 433, 459
Xi' Nich', 302, 388, 391
Xólotl story, 440. *See also* Theology: Teología India

Yajalón, Chiapas, 373
Yibeljoj, Chiapas, 113, 147, 149, 187, 210, 356, 444; reflection on woman's status, 196, 215
Yon Sosa, Marco Antonio, 236–37, 315

Zapatista Army of National Liberation (EZLN), 273–74; and Bishop Ruíz, 286; pastoral workers and catechists involvement, 285–86; recruitment, 282–84; rejection of, 287–88; training, 282–83; uprising, 273–74, 291–92, 379; war decision, 290–91; war pressures on, 289
Zapatistas (also Zapatismo), 292, 294–95, 299–302, 305–7
Zinacantán, Chiapas, 35–36, 59, 64, 66
Zwingli, Ulrich, 90

John D. Early, professor emeritus of anthropology at Florida Atlantic University in Boca Raton, is the author of *The Maya and Catholicism: An Encounter of Worldviews* and the coauthor of several books about population dynamics, including *The Xilixana Yanomami of the Amazon* and *The Dynamics of a Philippine Rain Forest People,* which received a Choice Outstanding Academic Book Award.

www.ingramcontent.com/pod-product-compliance
Lightning Source LLC
Chambersburg PA
CBHW021414300426
44114CB00010B/488